AF718275

WITHDRAWN

SOMATIC CELL GENETICS AND MOLECULAR GENETICS OF TREES

FORESTRY SCIENCES

Volume 49

The titles published in this series are listed at the end of this volume.

Somatic Cell Genetics and Molecular Genetics of Trees

Edited by

M. RAJ AHUJA

Institute of Forest Genetics,
Federal Research Centre for Forestry and Forest Products,
Grosshansdorf, Germany

WOUT BOERJAN

Laboratorium voor Genetica,
Flanders Interuniversity Institute for Biotechnology,
Gent, Belgium

and

DAVID B. NEALE

Institute of Forest Genetics,
USDA Forest Service,
Berkeley, CA, U.S.A.

KLUWER ACADEMIC PUBLISHERS
DORDRECHT / BOSTON / LONDON

A C.I.P. Catalogue record for this book is available from the Library of Congress

ISBN 0-7923-4179-1

Published by Kluwer Academic Publishers,
P.O. Box 17, 3300 AA Dordrecht, The Netherlands.

SD
403.5
.S64
1996

Kluwer Academic Publishers incorporates
the publishing programmes of
D. Reidel, Martinus Nijhoff, Dr W. Junk and MTP Press.

Sold and distributed in the U.S.A. and Canada
by Kluwer Academic Publishers,
101 Philip Drive, Norwell, MA 02061, U.S.A.

In all other countries, sold and distributed
by Kluwer Academic Publishers Group,
P.O. Box 322, 3300 AH Dordrecht, The Netherlands.

Printed on acid-free paper

All Rights Reserved
© 1996 Kluwer Academic Publishers
No part of the material protected by this copyright notice may be reproduced or
utilized in any form or by any means, electronic or mechanical,
including photocopying, recording or by any information storage and
retrieval system, without written permission from the copyright owner.

Printed in the Netherlands

Contents

Section II. Transformation and gene expression

Preface

This proceedings is based on a joint meeting of the two IUFRO (International Union of Forestry Research Organizations) Working Parties, Somatic Cell Genetics (S2.04-07) and Molecular Genetics (S2.04-06) held in Gent, Belgium, 26-30 September, 1995. Although a joint meeting of the two Working Parties had been discussed in the past, this was the first such meeting that became a successful reality. In fact this meeting provided an excellent forum for discussions and interactions in forest biotechnology that encouraged the participants to vote for a next joint meeting.

In the past decade rapid progress has been made in the somatic cell genetics and molecular genetics of forest trees. In order to cover recent developments in the broad area of biotechnology, the scientific program of the meeting was divided into several sessions. These included somatic embryogenesis, regeneration, transformation, gene expression, molecular markers, genome mapping, and biotic and abiotic stresses.

The regeneration of plants, produced by organogenesis or somatic embryogenesis, is necessary not only for mass cloning of forest trees, but also for its application in genetic transformation and molecular biology. Although micropropagation has been achieved from juvenile tissues in a number of forest tree species, *in vitro* regeneration from mature trees remains a challenging problem in most hardwoods and conifers. The mechanisms involved in the transition from juvenile to mature phase in woody plants are poorly understood. This transition can now be investigated at the molecular level.

By employing *Agrobacterium*-mediated gene transfer and particle gun DNA delivery systems, it has been possible to genetically transform tissues from forest tree species, and to regenerate transgenic plants. In addition, the molecular characterization of transgenic tissues as well as plants has been accomplished in several hardwood and conifer species. Gene expression studies indicate that transgenes affecting different traits, for example, insect resistance, lignin biosynthesis, or leaf phenotype, are expressed at the post transcriptional level. However, transgenes stability remains a major concern in transgenic research whether they are annuals or woody plants.

Gene expression studies are contributing to the understanding of gene regulation in development and differentiation of tissues and wood in the forest trees. Molecular markers have opened up a Pandora's box for studies in population genetics, genetic variation, genetic charactzerizaion, and genome mapping in trees. A large set of molecular markers has been developed in the last 5 years. These include, restriction fragment length polymorphisms (RFLPs), random amplified polymorphic DNAs (RAPDs), amplified fragment length polymorphisms (AFLPs), restriction landmark genomic scanning (RLGS), simple sequence repeats (SSRs), single-strand conformation polymorphisms (SSCPs), and PCR-codominant markers. Some of these markers have been effectively utilyzed for monitoring genetic diversity in tree populations, mapping of genes for simply inherited Mendelian traits as well as quantitative trait loci (QTL) for complex traits, and understanding of genome organization and evolution in the forest tree species. Isolation and characterizations of genes is another area of interest in the molecular biology of trees. In addition, stress-related gene expression in trees is providing useful information on the mechanisms involved in host-parasite interactions.

These and other related areas in genetics and molecular biology of forest trees were covered at this joint meeting. We hope this proceeding will provide a broad view of the latest developments in the biotechnology of woody plants.

M. Raj Ahuja
Wout Boerjan
David B. Neale

Section I.

Somatic embryogenesis and regeneration

CLONAL PROPAGATION OF CONIFERS VIA SOMATIC EMBRYOGENESIS

P.K. GUPTA, R. TIMMIS, K.A. TIMMIS, J.A. GROB, W.C. CARLSON AND D.E. WELTY
Weyerhaeuser Company
Weyerhaeuser Technology Center, Tacoma, WA 98477, USA

1. Abstract

The technology to initiate, maintain, and develop somatic embryos and emblings of conifer species via somatic embryogenesis (SE) is now well established. ABA and osmolality of the medium are important factors controlling embryo development. Good quality embryos have been produced on fibrous pads imbibed with liquid medium containing a combination of polyethylene glycol (MW 4,000-8,000), ABA and activated charcoal. High osmolality of the medium in the pad has been maintained by adding a second pad (soaked with higher osmolality medium) beneath the first. This method yields 10-60 cotyledonary embryos/ml settled embryonal suspensor masses (ESM). We have grown emblings from several genotypes of Norway spruce (*Picea abies*) and Douglas-fir (*Pseudotsuga menziesii*) in soil for field testing, and established over 1100 emblings of Norway spruce (*P. Abies*) and 600 emblings of Douglas-fir from several genotypes on typical Weyerhaeuser forest regeneration sites. The emblings exhibit uniform growth and phenology within a clone, compared with seedlings. The morphology of all trees appears normal.

2. Introduction

The diversity and the extent of the world's forests are declining and yet the demand for wood worldwide is expected to double by the 21st century. To accommodate this demand, the forest productivity of our remaining lands will have to be increased, while other areas are set aside untouched. Advances in biotechnology will accelerate tree improvement. In particular, somatic embryogenesis offers new ways for faster multiplication of high-value clones for reforestation, which will help in the race to increase forest productivity. It also lends itself to automation, artificial seed production, long term storage and genetic manipulation. In forestry, the production of artificial seed throughout the year could provide a complementary technology to reduce risk where seed production is limited and uncertain.

Since 1985 a great deal of progress has been made in the development of somatic embryogenesis technology for conifers. Papers have been published on somatic embryo development, germination and conversion to plantlets with different conifer species (Gupta and Durzan, 1995, Attree and Folk, 1993, Tautorus, 1991). In this chapter we will describe plantlet regeneration of Norway spruce (*Picea abies*), Douglas-fir (*Pseudotsuga menziesii*) and loblolly pine (*Pinus taeda*) via somatic embryogenesis, their establishment in the soil, and cryopreservation.

3. Norway spruce (*Picea abies*)

Mature embryos were used for induction of embryogenic cultures of Norway spruce. Seeds were surface sterilized by methods described earlier by Gupta et al, (1995). Embryos were excised aseptically, and cultured on semi-solid medium (Verhagen and Wann, 1989). Embryonal suspensor masses (ESMs) initiated from hypocotyl regions of mature embryos. ESM cultures were maintained on semi-solid

M. R. Ahuja et al. (eds.), Somatic Cell Genetics and Molecular Genetics of Trees, 3–9.
© 1996 *Kluwer Academic Publishers. Printed in the Netherlands.*

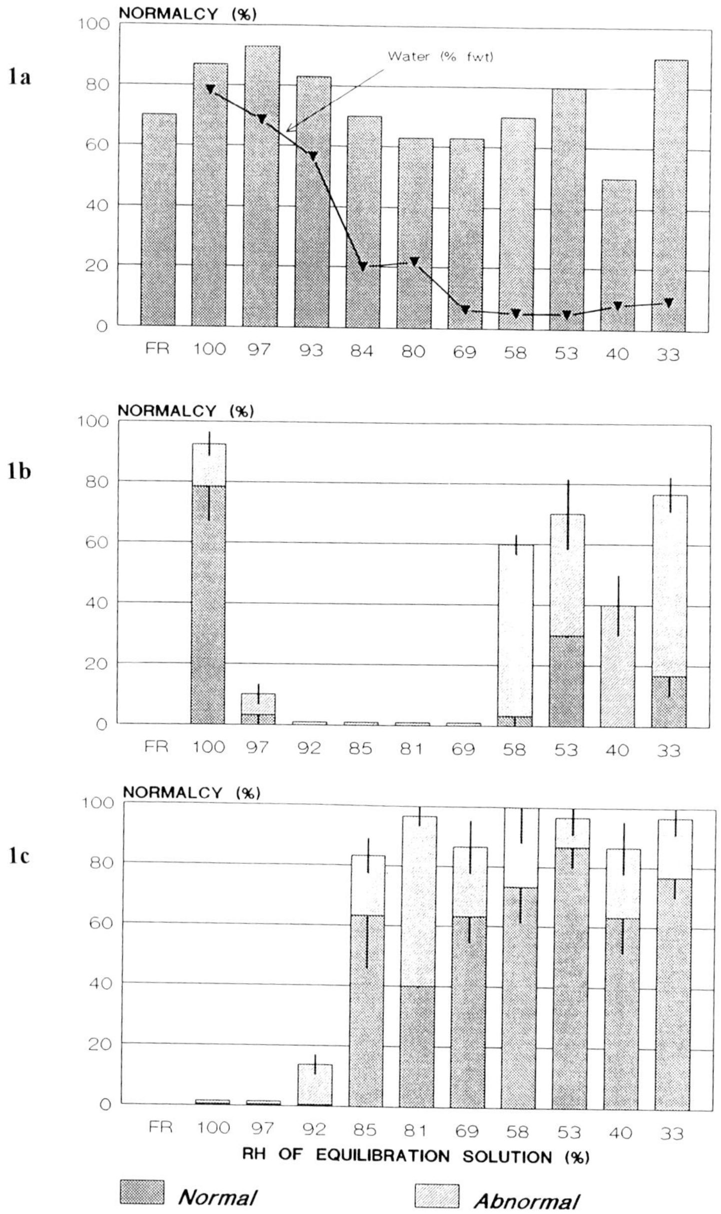

Figure 1 Germination of desiccated Norway Spruce somatic embryos before and after 3 months storage, (1a) immediately after desiccation to constant weight (FR = Fresh embryo), (1b) after 3 months storage at room temperature over the desiccation solutions, and (1c) after 3 months storage in a freezer at ~ -18°C following desiccation to constant weight. Standard error bars are based on 3 plates (of 10 embryos) per treatment.

medium (developed by Gupta and Durzan 1986) by subculture at two-week intervals. ESM cultures were also grown in suspension culture in 1L Erlenmeyer flasks and maintained by subculture every week.

During seed development many conifers show multiple zygotic embryos within the seed (Singh, 1978). Multiple embryos may arise through simple polyembryony when multiple fertilizations occur, or by cleavage polyembryony when embryos cleave to form several identical embryos. Cleavage polyembryony is not found in Norway spruce seeds. However, *in vitro* ESM cultures of Norway spruce multiplied through the pine type (Singh, 1978) of cleavage polyembryony in the presence of auxins (2,4-D, NAA) and cytokinins (KN, BAP).

Embryo development occurs following removal of auxins and cytokinins and the addition of Abscisic acid (ABA). Abscisic acid inhibits cleavage polyembryony, and allows singulation and further development and maturation of embryos. We have produced 60-100 cotyledonary embryos per mL settled ESMs by combining ABA (50 mg/L) with activated charcoal (1.25 mg/L; Pullman and Gupta 1991). Combining ABA and charcoal improved embryo quality and reduced the labor required for several transfers.

Norway spruce embryos germinated best after 7 days in the dark followed by continuous light. We have achieved 80-90% germination of selected embryos from several genotypes on semi-solid medium (Gupta et al, 1991).

Experiments conducted in 1989, and not previously reported, demonstrated desiccation tolerance (to 6-10% water content) of mature somatic embryos of Norway spruce. Selected embryos were placed in empty petri-plates above saturated solutions of known relative humidity in sealed vessels until they attained constant weight. This usually took about 4 days. Embryos were then placed directly in germination medium. Embryos germinated normally after desiccation, with a yield equaling but not significantly exceeding that of fresh controls (Fig. 1a). However, embryos could not survive in this unprotected dry condition indefinitely at room temperature: after 3 months only those stored over pure water retained full capacity for germination (Fig. 1b). The advantage of desiccation to water contents at or below about 20% of fresh weight was that it permitted long term storage of naked somatic embryos in a regular freezer at -20°C (Fig. 1c), and therefore probably in liquid nitrogen.

In 1991 we established over 4,000 somatic seedlings in soil from 17 genotypes (Gupta et al, 1991). A subset of these comprising 1,100 somatic seedlings of Norway spruce are now growing on a typical Weyerhaeuser regeneration site. The somatic seedlings exhibit uniform growth and phenology within a clone, compared with seedlings. The morphology of all trees appears normal.

4. Douglas-fir (*Pseudotsuga menziesii*)

Immature embryos at pre-cotyledonary stage were used for the initiation of embryogenic cultures. Seeds were collected immediately from the fresh cones after collection, and surface sterilized as described earlier (Gupta et al, 1995). The early-stage embryos were excised, with suspensors still attached to the female gametophyte, and placed onto the induction medium (Gupta et al, 1995). ESM cultures (white, mucilaginous, translucent growth) were initiated from embryonal heads of the immature embryos. After 5-6 weeks, they were removed from the initiation medium and transferred to maintenance medium. ESMs were maintained by subculture every 12-15 days on the this medium and incubated in the dark at 23°C.

As with *Picea* species, and unlike *Pinus*, cleavage polyembryogenesis has not been observed in zygotic embryos of Douglas-fir. However, in culture, ESMs multiplied by cleavage polyembryony typical of other conifers (Gupta et al, 1995). Embryonal suspensor masses were also established in liquid in 1L Erlenmeyer flasks and maintained by regular weekly subculture (Gupta et al, 1994). An increased osmolality of the maintenance medium is important for the growth of large-sized embryonal heads of early stage embryos. These in turn permit development of good quality cotyledonary embryos. Without increasing osmolality many genotypes of Douglas-fir produced smaller-sized embryonal heads which did

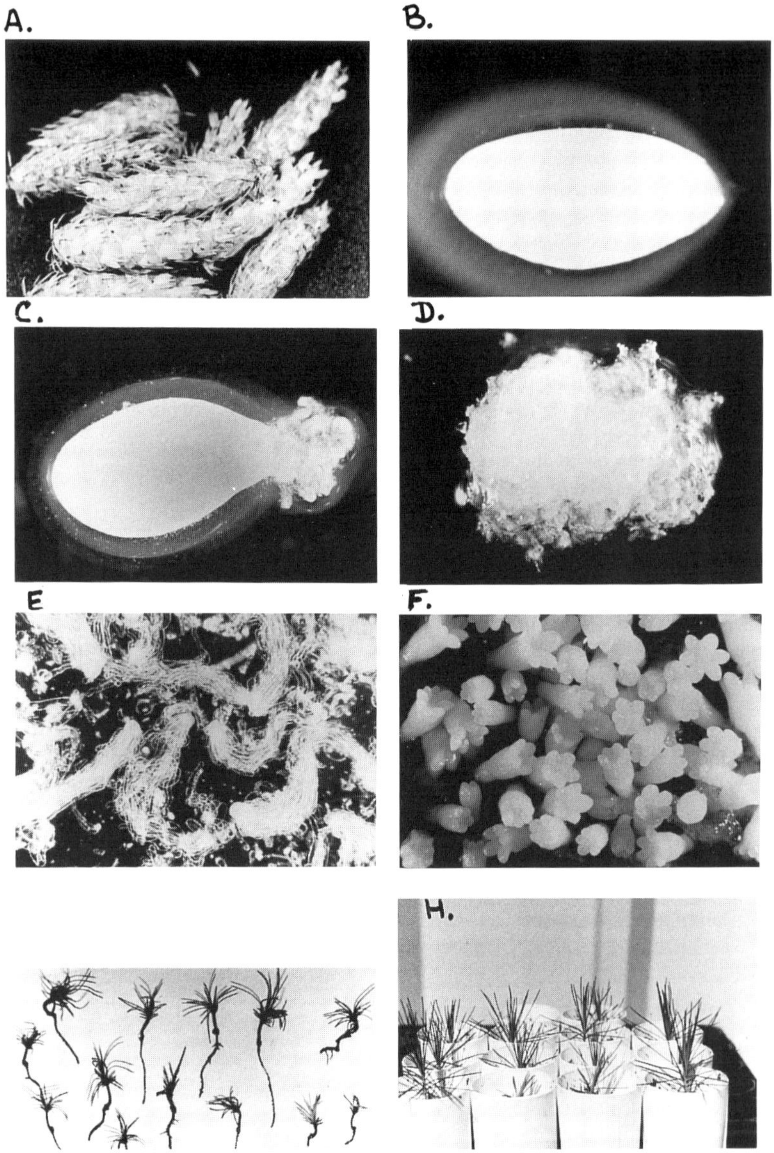

Figures 2.(Loblolly pine:A - H) : Fig.A. Immature cones. Fig.B. Excised femal gametophyte. Fig.C. Extrusion of embryogenic tissue from micropyler end of the female gametophyte. Fig.D. Growth of embryonal suspensor mass (ESM) on semi-solid medium. Fig.E. Early stage of embryos in suspension culture. Fig.F. Cotyledonary embryo development on medium with PEG, ABA and activated charcoal. Fig.G. Germinated embryos after 8 weeks on semi-solid germination medium. Fig.H. Somatic seedlings of loblolly pine growing in a greenhouse.

not develop good quality cotyledonary embryos. Osmolality of the maintenance medium was increased (to 190-220 mM/kg) by adding myo-inositol (Gupta & Pullman, 1990).

Embryonal suspensor mass cultures from the maintenance medium were transferred to liquid medium amended with ABA (5 mg/l). Two weekly subcultures were made with ABA in the liquid medium. After two weeks of ABA treatment, ESM cultures were allowed to settle in a measuring cylinder for 30 min. The supernatant was discarded, and 1.0-1.5 mls of settled ESM cultures were transferred to the polyester pads, wetted with 9-10 ml liquid medium, for the development of cotyledonary embryos.

Embryo development was improved with the combination of activated charcoal (1.25 mg/l) and ABA (30 mg/l) (Gupta and Pullman, 1990). Abscisic acid alone did not inhibit the precocious germination of cotyledonary embryos. An increased osmolality of the medium was found necessary for late-stage embryo development and maturation. The osmolality of the late embryo development and maturation medium was increased to 550-650 mM/kg by adding polyethylene glycol (PEG) MW 4,000-8,000, which was found to be the best osmoticum for good quality cotyledonary embryo development (Gupta and Pullman, 1991). Several workers have reported improved conifer embryo development with PEG (Attree et al, 1992).

The quality of cotyledonary embryos was further improved by a combination of an increased osmolality with ABA (5-50 mg/l), $GA_{4/7}$ (5-50 mg/L) and activated charcoal (1.25 g/l) (Pullman and Gupta, 1994). High osmolality of the medium on the pad has been maintained by adding a second pad (soaked with higher osmolality medium) beneath the first. Twenty to fifty cotyledonary embryos were produced from 1 ml settled ESM suspension culture on development medium after 5-6 weeks incubation in the dark. The variation in both quality and number of cotyledonary embryos produced from 1 ml settled ESM cultures, was due to genotypic differences.

In 1992, we established over 2,000 somatic seedlings from 10 genotypes in soil in a greenhouse. Strikingly uniform growth was observed among somatic seedlings within a clone, compared with seedlings from germinated zygotic embryos or seed. The morphology appeared normal, and over 600 somatic seedlings have been grown on a typical Weyerhaeuser forest regeneration site.

Cryogenic storage is a key step in implementing somatic embryogenesis technology because it allows preservation of ESM cultures for a long period of time while clones are field tested. It has been shown that ESM cultures are amenable to cryopreservation. We have cryostored ESM of 500 genotypes of Douglas-fir from superior full-sib families, and regenerated somatic seedlings from samples of several genotypes. Cryogenic storage significantly improved embryo yield, germination percent and conversion to soil. However, once established in soil, both cryogenically stored and non-stored somatic seedlings have remained similar in both morphology and growth.

5. Loblolly pine (*Pinus taeda*)

Immature embryos at precotyledonary stage were used for the initiation of embryogenic cultures. Cones were collected 4-6 weeks after fertilization and opened in the laboratory Fig. 2A). After surface sterilization, seed coats were removed, and female gametophytes (Fig. 2B) with embryos still attached were cultured as described earlier (Gupta & Pullman, 1990). After 4-5 weeks, white glossy and mucilaginous extrusions (ESM) were observed from the female gametophytes at the micropylar end (Fig. 2C). Extrusions were relatively infrequent and occurred with an overall frequency of 1-3%. These ESM originated from immature (precotyledonary) embryos inside the female gametophyte. A similar type of initiation from immature embryos has also been reported in several other *Pinus* species (Jain et al, 1989). Embryonal suspensor masses contained only embryos with early stages and suspensors. Each embryo had elongated cells at one end (suspensors) and smaller cells with large nuclei and dense cytoplasm at the embryonal end.

Embryonal suspensor masses were separated from female gametophyte and transferred to a medium amended with lower concentration of 2,4-D, and KN, and BAP (Gupta and Durzan, 1987). They were maintained by subculture every 12 - 15 days on the same medium, and incubated in the dark at 23°C (Fig. 2D). Natural cleavage polyembryony has been observed in *Pinus* species. Early-stage embryos *in vitro* also multiplied by true-to-pine-type cleavage polyembryony (Gupta & Durzan, 1987). Somatic polyembryogenesis occurred continuously as long as the ESM cultures were maintained on the 2,4-D medium. Embryonal suspensor masses were also grown in suspension culture in a 1L Erlenmeyer flasks and maintained by weekly subculture (Fig. 2E). Keeping the osmolality higher (180-250 mM/kg) by applying osmoticants (myo-inositol, sorbitol, sugars etc.) in maintenance culture increased the embryonal head size. This higher osmolality treatment (in maintenance culture) eventually lead to good quality cotyledonary embryo development of several loblolly pine genotypes (Gupta and Pullman, 1990).

One mL aliquots of settled ESMs cultures were plated directly on semi-solid development medium. Abscisic acid alone or with activated charcoal was not effective in producing good quality cotyledonary embryos. A combination of increased osmolality (using PEG 4,000-8,000) with ABA and activated charcoal produced good quality cotyledonary embryos (Gupta et al, 1993). Thirty to hundreds somatic embryos per ml settled ESMs were produced from several genotypes on medium with PEG 8000 (100-130 mg/L), ABA (20-25 mg/L), activated charcoal (1.0-1.25 mg/L), and maltose (20 g/L) (Fig. 2F). Earlier, maltose had been shown to be a better sugar then sucrose for embryo maturation (Uddin, 1993).

Embryos were kept for germination on semi-solid medium (Gupta and Pullman, 1990). After 8 weeks of germination (Fig. 2G) in light, germinants (2-10 mm epicotyl and 1-2 cm root length) were transferred into soil. Somatic seedlings were successfully grown in a greenhouse (Fig. 2H). There are few other reports on the growth of somatic seedlings of loblolly pine in soil (Becwar and Pullman, 1995). Embryonal suspensor masses have been successfully cryostored in liquid nitrogen as described for Douglas-fir (Gupta et al, 1995).

6. Conclusion

It is evident from this review and other published reports (Gupta and Durzan, 1995, Attree and Folk, 1993, Tautorus, 1991) of somatic embryogenesis of Norway spruce, Douglas-fir and loblolly pine, that the technology has a high potential for commercialization. However, despite these achievements, significant losses occur through various steps from culture initiation to somatic seedling establishment. Understanding the cause of these problems and working with a large number of genotypes which reliably produce uniform, zygotic-like somatic embryos at low cost, are major goals. Progress thus far suggests that somatic embryogenesis is a promising technology for clonal propagation of selected genotypes.

References

Attree, S.M. and Fowke, L.C (1993) Embryogeny of gymnosperms: advances in synthetic seed technology of conifers. *Plant Cell Tiss. Org. Cult.* **35**, 1-35.

Attree, S.M., Moore, D., Sawhney, V.K. and Fowke, L.C. (1991b) Enhanced maturation and desiccation tolerance of white spruce (*Picea glauca*) somatic embryos: Effect of non-plasmolyzing water stress and abscisic acid, *Annals. Bot.* **68**, 519-525.

Becwar, M.R. and Pullman, G.S. (1995) Somatic embryogenesis in loblolly pine, in Douglas-fir, in S.M. Jain, P.K. Gupta and R.J. Newton (eds.), *Somatic Embryogenesis in Woody Plant Vol. 3*, Kluwer Academic Publishers, Dordrecht, pp. 287-302.

Gupta, P.K. and Durzan, D.J. (1986a) Plantlet regeneration via somatic embryogenesis from subcultured callus of mature embryos of *Picea abies* (Norway spruce), *In Vitro Cell and Devel. Biol.* **22**, 685-688.

Gupta, P.K. and Durzan, D.J. (1987) Biotechnology of somatic polyembryogenesis and plantlet regeneration in loblolly pine, *Bio/Technology* **5**, 1477-151.

Gupta, P.K. and Pullman, G.S. (1991a) Method for reproducing coniferous plants by somatic embryogenesis using abscisic acid and osmotic potential variation, U.S. Patent No. 5,036,007.

Gupta, P.K. and Pullman, G.S. (1990) Method for reproducing coniferous plants by somatic embryogenesis, U.S. Patent No. 4, 957, 866.

Gupta, P.K. and Grob, J.A. (1995) Somatic embryogenesis in conifers, in S.M. Jain, P.K. Gupta and R.J. Newton (eds.), *Somatic Embryogenesis in Woody Plant Vol. 1*, Kluwer Academic Publishers, Dordrecht, pp. 81-98.

Gupta, P.K., Timmis, R., Timmis K., Carlson, W.C., and Welty, D.E. (1995) Somatic embryogenesis in Douglas-fir, in S.M. Jain, P.K. Gupta and R.J. Newton (eds.) *Somatic Embryogenesis in Woody Plant Vol. 3*, Kluwer Academic Publishers, Dordrecht, pp. 303-314.

Gupta, P.K., Timmis, R., Pullman, G., Yancey, M., Kreitinger, M., Carlson, W.C. and Carpenter, C. (1991) Development of an embryogenesis system for automated propagation of forest trees, in I.K. Vasil (ed.), *Scale-up and Automation on Plant Propagation*, pp. 76-90, Academic Press, New York.

Gupta, P.K., Timmis, R., Timmis, K., Carlson, W.C., Grob, J.A. and Welty, D.E. (1994) Plantlets regeneration via somatic embryogenesis in Douglas-fir, *TAPPI Biological Science Symposium*, pp. 35-40.

Gupta, P.K., Pullman, G., Timmis, R., Kreitinger, M., Carlson, W.C., Grob, J. and Welty, D.E. (1993) Forestry in the 21st Century: The biotechnology of somatic embryogenesis, *Bio/Technology* **II,** 454-459.

Jain, S.M., Dong, N. and Newton, R.J. (1989) Somatic embryogenesis in slash pine (*Pinus elliottii*) from immature embryos cultured in vitro, *Plant Sci.* **65**, 233-241.

Pullman, G.S. and Gupta, P.K. (1991) Method for reproducing coniferous plants by somatic embryogenesis using absorbent materials in the development stage media, U.S. Patent No. 5,034,326.

Pullman, G.S. and Gupta, P.K. (1994) Method for reproducing conifers by somatic embryogenesis using mixed growth hormones for embryo culture, U.S. Patent No. 5,294,549.

Tautorus, T.E., Fowke, L.C. and Dunstan, D.I. (1991) Somatic embryogenesis in conifers, *Can. J. For. Res.* **69**, 1873-1899.

Uddin, M.R. (1993) Somatic embryogenesis in gymnosperms, U.S. Patent No. 5,187,092.

Verhagen, S.A. and Wann, S.R. (1989) Norway spruce somatic embryogenesis, High frequency initiation from light-cultured mature embryos, *Plant Cell Tiss. Org. Cult.* **16**, 103-111.

EFFECTS OF GLUTAMINE AND OSMOTICUM ON SOMATIC EMBRYO MATURATION IN NORWAY SPRUCE (*Picea abies*) (L.) Karst.

TODD G. GUEVIN[1] & EDWARD G. KIRBY
Department of Biological Sciences
Rutgers University
University Heights,
Newark, NJ 07102, USA.
[1]Current address: *Union Camp Corp., Princeton, New Jersey 08543 USA*

Abstract. The effect of glutamine, nitrate, and osmotic treatments on embryo maturation frequency was examined in cultured embryo suspensor masses of Norway spruce. The frequency of maturation was increased when embryo suspensor masses were cultured on media containing glutamine (10-60 mM), as compared to culture on equal molar levels on nitrate. Glutamine permitted somatic embryos to mature at control levels, even when embryos were exposed to osmotic treatments that reduced maturation frequency of somatic embryos cultured on medium containing nitrate.

1. Introduction

Earlier we reported that osmotic stress as caused by exposure to polyethylene glycol (PEG) inhibited growth of Douglas-fir cell cultures and that glutamine supplements to the medium facilitated adjustment to osmotic stress (Leustek and Kirby 1988). Furthermore, cells grown on media supplemented with glutamine had decreased cell size and lower osmotic potentials than control cells, indicating active osmoregulation (Leustek and Kirby 1990).

During maturation treatments somatic embryos of conifers are often submitted to increased medium osmolarities, similar to osmolarities that inhibit growth of Douglas-fir cells. Therefore, we postulate that culture of somatic embryos on maturation medium containing PEG and supplemented with glutamine will enhance embryo maturation frequencies. In this study, we have examined the relationship between glutamine and osmolarity on the development and maturation of somatic embryos of Norway spruce.

2. Materials and Methods

Embryo suspensor masses (ESMs) of Norway spruce (*Picea abies* [L.] Karst) were induced as described by von Arnold (1987) and proliferated on medium (Verhagen and Wann 1989) containing 4.5 μM BAP and 10 μM NAA, 60 mM sucrose, 5 mM glutamine, 0.67 mM asparagine, and 0.25% (w/v) Gel-Rite gellan gum (Kelco) at pH 5.8.

M. R. Ahuja et al. (eds.), Somatic Cell Genetics and Molecular Genetics of Trees, 11–16.
© 1996 *Kluwer Academic Publishers. Printed in the Netherlands.*

The effect nitrogen form on embryo maturation was examined by culture of ESMs on medium supplemented with glutamine or potassium nitrate. The levels of potassium in the glutamine-supplemented medium were increased to the levels in potassium nitrate-supplemented medium by addition of KCl. Asparagine was omitted from experimental media (Table 1). All maturation media contained 40 μM abscisic acid (ABA) (Sigma Chemical Co., St. Louis MO, Cat. No. A1049, dissolved in 0.1N NaOH). Experiments consisted of 5 replicate samples containing three ESMs of approximately 75 mg each.

TABLE 1. Milimolar levels of glutamine, potassium nitrate, and potassium chloride in maturation media. Culture medium was supplemented with glutamine or potassium nitrate. Levels of KCl in the glutamine-supplemented medium were increased so that the total level of potassium was equal in all media

Glutamine mM	KNO_3 mM	KCl mM
10	0.5	60.5
20	0.5	60.5
40	0.5	60.5
60	0.5	60.5
0	10.5	50.5
0	20.5	40.5
0	40.5	20.5
0	60.5	0

The effect of the addition of polyethylene glycol (PEG 3350)(Sigma Chemical Co., St. Louis MO) to medium containing 20 mM glutamine (with potassium levels adjusted as described in Table 1), or 20 mM potassium nitrate was examined. Experiments consisted of four replicate samples containing three ESMs, each of approximately 75 mg. All experiments were repeated three times.

Cultures were maintained at 24° C and supplemented with low light (30 mM m^{-2} sec^{-1}; 16-h photoperiod; General Electric 20W Cool-White fluorescent bulbs). The number of cotyledonary-stage somatic embryos was determined after 8 weeks.

Osmolarity measurements: Medium osmolarity was measured by freezing point depression using an Osmette S (Precision Instruments, Sudbury MA). Values were converted from osmolarity to MPa using the van't Hoff equation (Borowitzka 1981). Values represent means of three independent determinations.

3. Results

After 8 weeks in culture the number of cotyledonary stage embryos per gram initial fresh weight was higher on media supplemented with glutamine, as compared to media supplemented with nitrate at equal molar levels (Figure 1).

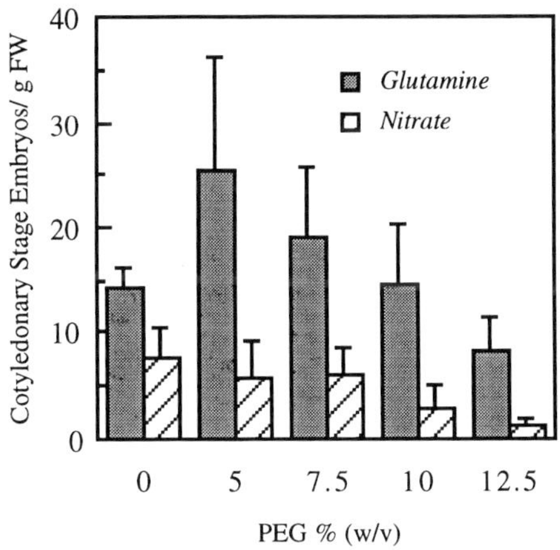

Figure 1. The effect of glutamine (solid bars) and potassium nitrate (hatched bars) on embryo maturation. The number of cotyledonary stage embryos per gram initial fresh weight was measured. Data are the mean of three experiments. Error bars represent the SEs of the means for each treatment.

The effect of increased osmolarity on somatic embryo maturation by culture of ESMs on media containing glutamine or nitrate was also examined. The concentration of KNO_3 giving maximum maturation frequency (20 mM) was compared to an equal level of glutamine. To compare accurately the effect of PEG on medium osmolarity, the osmotic potentials of media were measured by freezing point depression (Figure 2).

ESMs cultured on glutamine-supplemented media developed greater numbers of mature somatic embryos than ESMs cultured on nitrate-supplemented media at all PEG concentrations tested (Figure 3). In addition, the combination of PEG and glutamine did not significantly alter the numbers of cotyledonary-stage embryos up to 10% PEG (w/v) (-0.98 MPa). However, the combination of PEG and KNO_3 reduced embryo maturation frequencies at PEG levels of 10% PEG (w/v) (-0.87 MPa) or greater (Figure 3).

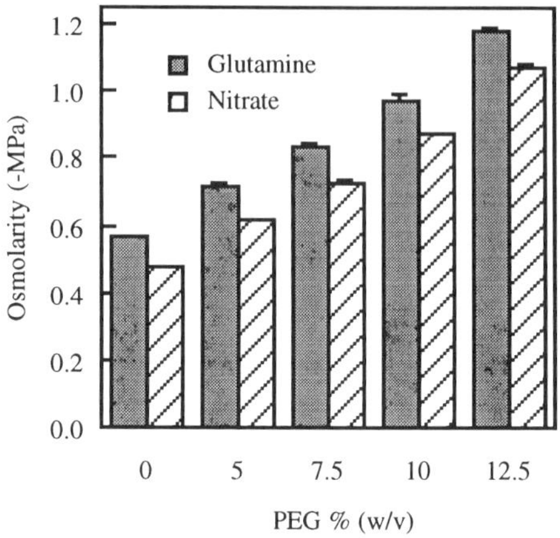

Figure 2. Osmolarity of maturation media. Media consisted of one-half strength BLG, as described, supplemented with 20 mM glutamine or 20 mM potassium nitrate. Osmolarities of media were calculated from freezing point depression data using the van't Hoff equation (Borowitzka 1981).

4. Discussion

We observed that glutamine promoted the frequency of maturation of somatic embryos of Norway spruce. The beneficial effect of glutamine on embryo maturation could be accounted for by the ability of glutamine to supply amino nitrogen for synthesis of nitrogen-containing molecules necessary for maturation of embryos, including seed storage proteins. Presumably, when glutamine is present in the maturation medium, the need for reduced nitrogen is met and the energetically expensive process of nitrate reduction is avoided. We speculate that synthesis of substances necessary for maturation is facilitated because of the reduced consumption of high energy molecules, such as NAD(P)H and ATP.

We postulated that ESMs of Norway spruce cultured on media of increased osmolarity have enhanced maturation frequency if the medium was supplemented with glutamine. ESMs developed approximately equal numbers of cotyledonary-stage embryos at 10% PEG and in the absence of PEG, provided that glutamine was present in the maturation medium. On medium containing nitrate alone, however, addition of 10% PEG caused a 59% reduction in the number of mature embryos formed, as

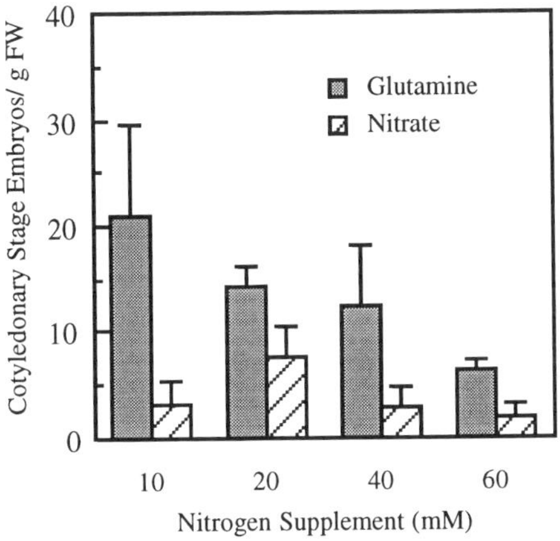

Figure. 3. The effect of polyethylene glycol (PEG 3350) on maturation of somatic embryos of Norway spruce. PEG was added to maturation media containing glutamine (solid bars) and potassium nitrate (hatched bars). Data are the mean numbers of cotyledonary stage embryos produced per gram initial fresh weight from three experiments. Error bars represent the SEs of the means for each treatment.

compared to ESMs grown on medium lacking PEG. These results demonstrate that glutamine sustained embryo maturation frequency during PEG induced water stress.

The mechanism whereby glutamine stimulates maturation of somatic embryos is not known. We have proposed that glutamine provides developing embryos an energetically inexpensive form of amino nitrogen. Evidence for the involvement of glutamine in active osmoregulation has been presented for cell suspensions of Douglas-fir (Leustek and Kirby 1990). Furthermore, Lai et al. (1992) have suggested that glutamine may have a role in the transcriptional or translational regulation of seed storage protein genes in alfalfa. In white spruce seed storage protein genes are transcriptionally regulated by ABA, but proteins are not translated at high levels in the absence of PEG (Misra et al. 1993). Future work in our laboratory will examine the mechanism whereby seed storage proteins accumulate in response to PEG.

5. Acknowledgment: We thank Professor Sara von Arnold for kindly providing us with the Norway spruce cell line used in these experiments.

6. References

Borowitzka, L.J. (1981) Solute accumulation and regulation of cell water activity, in L.G. Paleg and D. Aspinall (eds.), *The Physiology and Biochemistry of Drought Resistance in Plants*, Academic Press, New York, pp. 97-130.

Lai, F., T. Senaratna, and B.D. McKersie (1992) Glutamine enhances storage protein synthesis in *Medicago sativa* L. somatic embryos, *Plant Sci.* **87**, 69-77.

Leustek, T., and E.G. Kirby (1988) The influence of glutamine on growth and viability of cell suspension cultures of Douglas-fir after exposure to polyethylene glycol, *Tree Physiol.* **4**, 371-380.

Leustek, T., and E.G. Kirby (1990) Selection and physiology of cell cultures of Douglas-fir grown under conditions of water stress, *Tree Physiol.* **6**, 317-327.

Misra, S., S.M. Attree, I Leal, and L.C. Fowke (1993) Effect of abscisic acid, osmoticum, and desiccation on synthesis of storage proteins during the development of white spruce somatic embryos, *Ann. Bot.* **71**, 11-22.

Verhagen, S.A., and S.R. Wann (1989) Norway spruce somatic embryogenesis: high-frequency initiation from light-cultured mature embryos, *Plant Cell Tiss. Org. Cult.* **16**, 103-111.

von Arnold, S., and I. Hakman (1987) Regulation of somatic embryo development in Picea abies by abscisic acid (ABA), *J. Plant Physiol.* **132**, 164-169.

EARLY EVENTS IN SOMATIC EMBRYOGENESIS INDUCTION

M.C. PEDROSO[1], R. TAVARES[2], T. LINO-NETO[2], K.A. ROUBELAKIS-ANGELAKIS[3], M.S. PAIS[1]

[1] Centro de Biotecnologia Vegetal, Fundação da Faculdade de Ciências de Lisboa, Bloco *C2, Piso 1, Campo Grande, P-1700 Lisboa, Portugal*

[2] Departamento de Biologia, Universidade do Minho, 4719 Braga Codex, Portugal

[3] Department of Biology, University of Crete, P.O. Box 1470, 71110 Heraklion, Greece

1.Abstract

In Plant Kingdom, one of the strategies for plants to survive is the spontaneous production of direct somatic embryos [1]. Examples can be found in angiosperms, both in dicots and monocots. In artificial conditions, somatic embryogenesis is achieved using different types of initial explants and following different types of stress [for a review see, 1,8,9,11,12]. Although both direct and indirect somatic embryogenesis have been successfully achieved for many species [9], the knowledge on the early events occurring during the induction of embryogenesis is still scarce. Here, we describe some of the molecular and elemental changes occurring during the early stages of direct somatic embryogenesis using, as model plant, the woody species *Camellia japonica*. Our previous results shown that direct embryogenesis from camellia leaves is region specific [5]. This specificity was suggested to be related to pre-established elemental gradients, namely, of calcium, sodium and sulphur [4,6]. After the induction treatment, two types of response were detected: a "fast" and a slower one [6]. The first one, common to all parenchyma cells, was suggested to be a stress response, due to leaf excision and to the induction treatment. The following morphogenic or non-morphogenic responses, that were cell or explant-region specific, were suggested to be dependent on the cell-state of each cell at the time of leaf isolation [6]. Some data on the possible effect of stress on embryogenic induction is presented. The role of Ca^{2+}, Mg^{2+} and competing polyamines (free and conjugated) on cell division and morphogenesis is also discussed.

M. R. Ahuja et al. (eds.), Somatic Cell Genetics and Molecular Genetics of Trees, 17–22.
© 1996 *Kluwer Academic Publishers. Printed in the Netherlands.*

2. Stress and Somatic Embryogenesis

Microscopy and histological results suggest that plant defense mechanims are activated following the treatment to induce direct somatic embryogenesis in *Camellia japonica* [3]. Studies on the specific activities of the oxidative stress enzymes, superoxide dismutase (SOD), ascorbate peroxidase (Apx), catalase (Cat) and glutatione reductase (GR), seem to corroborate this hypothesis.

3. Ion Regulation

Morphogenesis can be defined as development of shape, but is known to be a change of metabolic pattern too [2]. In embryogenic development a whole series of switches of homeostasis may occur, each following a period of homeostatic growth, i.e. growth without change of shape. Embryogenic development is a gene-switching process. Morphogenesis may be related to a running change of gene products leading to a new stable homeostatic set. By inspecting inorganic ions in embryogenic induction we may envisage a very tight control of mineral elements in this phase. Relative ion flow changes may be related to morphogenesis. Electron probe X-ray microanalysis showed that the levels of Ca^{2+}, Mg^{2+}, Na^+, K^+, Fe^{2+}, S^{2-} (Fig. 1) and Zn^{2+} (Fig. 2), change during the induction of direct embryogenesis from camellia leaves. The knowledge on the role of inorganic elements in living cells [2], enabled us to suggest possible roles for the above referred elements in camellia embryogenesis [4,6] namely: in controlling cell shape and mechanical properties (Ca2+ and Mg2+); in cell/cell communication and osmotic control (Na+ and K+); in controlling redox activity (Fe2+); in the stability of extracellular proteins and in homeostatic control (S2-); and in cell division and extracellular activities linked to growth (Zn2+). The increase in zinc levels detected during camellia embryo development (see Fig. 2), suggests that zinc may play a role in plant embryogenesis as it does in animal organized development.

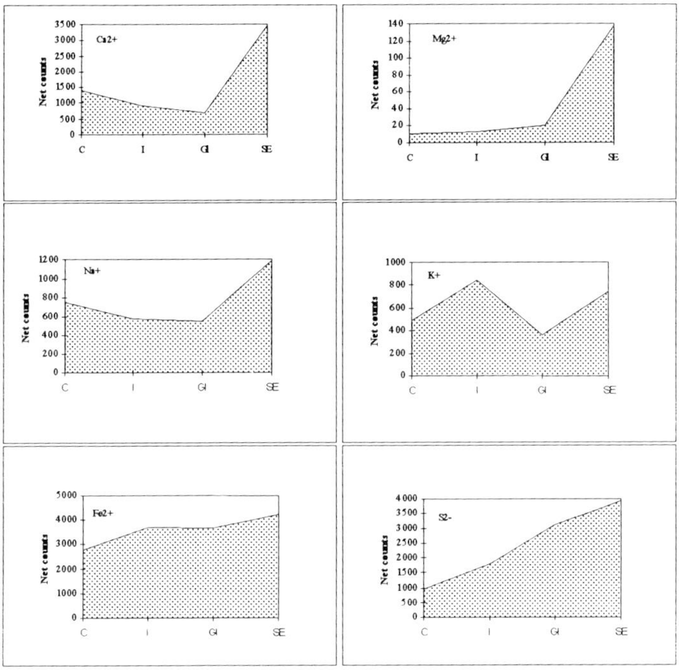

Figure 1. Calcium (Ca2+), magnesium (Mg2+), sodium (Na+), potassium (K+), I ron (Fe2+) and sulphur (S2-) changes (expressed in net counts), detected by electron probe X-ray microanalysis, during the induction of direct somatic embryogenesis in *Camellia japonica* leaves. C, control; I, induced; Gl, globular somatic embryo; SE, cotyledonary-stage somatic embryo.

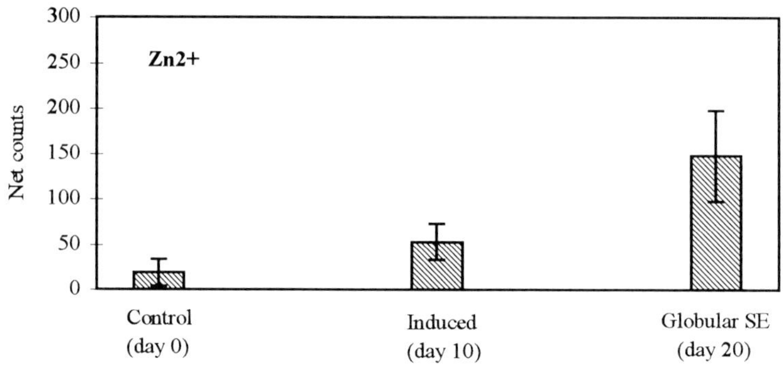

Figure 2. Zinc changes, detected by electron probe X-ray microanalysis, during the induction of direct globular embryo formation in *Camellia japonica*. X-ray spectra were collected: before the induction treatment, Control (day 0); 10 days after the induction treatment, Induced (day 10); and 20 days after the induction treatment during globular somatic embryo (SE) formation, Globular SE (day 20).

4. Polyamines in Somatic Embryogenesis

Polyamines have been reported as playing a role in cell division and morphogenesis [7]. The variations in Mg2+ levels, detected after the induction of somatic embryogenesis (in Fig. 1), were suggested to be associated with changes in the levels of competing polyamine cations [6] which vary during the cell cycle [7]. The results on the study of free and conjugated (soluble and insoluble) putrescine, spermine and spermidine contents during the induction of somatic embryogenesis, in embryogenic and non-embryogenic leaf regions of camellia explants (Fig. 3), show that these polyamines are not determining/conditioning leaf region-specific embryogenic response and that conjugated putrescine and soluble conjugated spermidine are directly or indirectly involved in somatic embryo development.

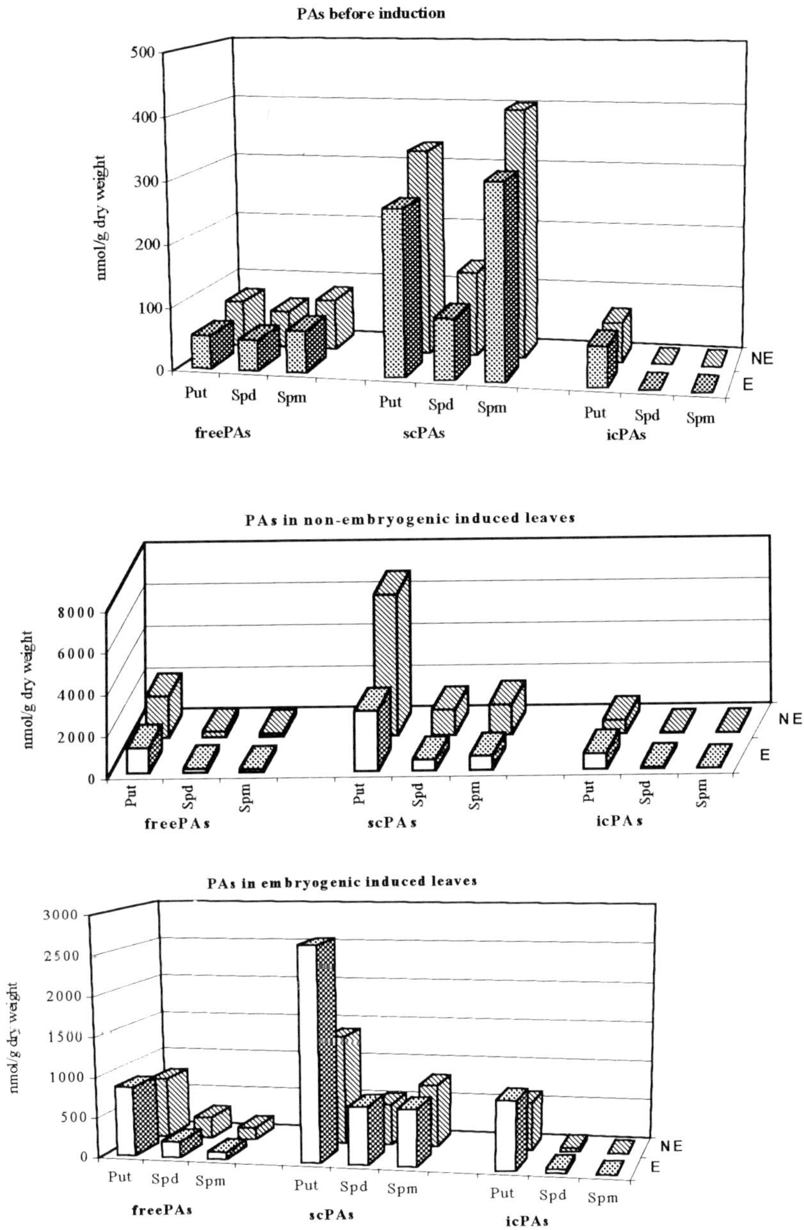

Figure 3. Free, soluble and insoluble conjugated putrescine (Put), spermidine (Spd) and spermine (Spm) in embryogenic (E) and non-embryogenic (NE) leaf regions of *Camellia japonica*. A, before induction; B and C, after induction. B, in non-embryogenic induced leaves; C, in embryogenic induced leaves. PAs, polyamines; freePAs, free polyamines; scPAs, soluble conjugated polyamines; icPAs, insoluble conjugated polyamines.

5. References

1. Emons, A.M.C.: Somatic embryogenesis: cell biological aspects. *Acta Botanica Neertherlandica* 43 (1994), 1-14.
2. Fraústo da Silva, J.J.R. and Williams, R.J.P.: *The Biological Chemistry of The Elements: The Inorganic Chemistry of Life*. Clarendon Press, Oxford, 1991.
3. Pedroso-Ubach, M.C.: *Somatic Embryogenesis in Camellia Japonica L.: A Search for Markers*. PhD Thesis. Faculdade de Ciências da Universidade de Lisboa, Lisboa, 1994.
4. Pedroso, M.C. and Pais, M.S.: A scanning electron microscope and X-ray microanalysis study during the induction of morphogenesis in *Camellia japonica* L. *Plant Science* 87 (1992), 99-108.
5. Pedroso, M.C. and Pais, M. S.: Direct embryo formation in leaves of *Camellia japonica* L. *Plant Cell Reports* 12 (1993), 639-643.
6. Pedroso, M.C. and Pais, M.S.: Early detection of embryogenic competence and of polarity in *Camellia japonica* L. by electron probe X-ray microanalysis. *Plant Science* 96 (1994), 189-201.
7. Slocum, R.D. and Flores, H.E.: *Biochemistry and Physiology of Polyamines in Plants*. CRC Press, Inc., Florida, pp. 143-148; 163-168; 176-182, 1991.
8. Thorpe, T.A.: In Vitro Organogenesis and somatic Embryogenesis: physiological and biochemical aspects. In: K.A. Roubelakis-Angelakis, K. Tran Than Van, eds., *Morphogenesis in Plants: Molecular Approaches*, NATO ASI Series, Vol. 253, Plenum Press, New York, 1993; 19-38.
9. Thorpe, T.A.: Morphogenesis and regeneration. In: I.K. Vasil, T.A. Thorpe, eds., Plant Cell and Tissue Culture, Kluwer Academic Publishers, Dordrecht, 1994; 17-36.
10. Wardlaw, C.W.: *Organization and Evolution in Plants*, 1st ed., Longmans, Green and Coltd., London, 1965; 132-275.
11. West, M.A.L. and Harada, J.J.: Embryogenesis in higher plants: an overview. *Plant Cell* 5 (1993), 1361-1369.
12. Zimmerman, J.L.: Somatic embryogenesis: a model for early development in higher plants. *Plant Cell* 5 (1993), 1411-1423.

STABILITY IN PLOIDY LEVEL DURING SOMATIC EMBRYOGENESIS IN *QUERCUS CANARIENSIS*

M. A. BUENO[1], A. GÓMEZ[1], O. VICENTE[3] AND J. A. MANZANERA[2]
[1] *INIA-CIFOR, Ap 8111, 28080 Madrid, Spain;* [2] *Department of Silvopascicultura, E.T.S.I. Montes, Universidad Politecnica de Madrid, Ciudad Universitaria s.n., 28040 Madrid, Spain;* [3] *Institute of Microbiology and Genetics, University of Vienna, Vienna Biocenter, Dr. Bohrgasse 9, A-1030 Vienna, Austria*

1. Summary

For the first time, somatic embryos have been obtained in *Quercus canariensis.* Immature embryos were isolated from acorns collected during July 1994, and cultured in an established solid medium. Initiation of the embryogenic process was studied monitoring the inductive effect of 2,4 dichlorophenoxyacetic acid (2,4-D) and the effect of organic N (amino acid). Development of somatic embryos was observed 3 weeks after initiation of the culture both in the presence or in the absence of auxin and glutamine. Although the frequency of embryogenic induction varied among the different treatments, the main factor influencing somatic embryo formation was the state of development of the initial explants, those collected at earlier stages being the most embryogenic.

Ploidy level and DNA content of somatic embryos were analyzed by flow cytometry. Embryos were purely diploid after one year in culture, in contrast to observations in other species.

Key words: somatic embryogenesis, *Quercus canariensis*, ploidy level.

2. Introduction

Quercus canariensis is a tall tree that can reach 30 m high in good stands, with a strong, regular and straight stem, clean of branches below the crown. Endemic of the Western Mediterranean, thrives in South Portugal, South Western Spain, Toledo Mountains, North Eastern Spain and in Northern Africa (Ruiz de la Torre, 1979). Very often, this oak appears in mixed stands, together with other oaks (*Q. suber*, *Q. pyrenaica*, *Q. faginea* and *Q. ilex*). It is an important tree for soil improvement and preservation, protective and ornamental and landscape purposes.

M. R. Ahuja et al. (eds.), Somatic Cell Genetics and Molecular Genetics of Trees, 23–28.
© 1996 *Kluwer Academic Publishers. Printed in the Netherlands.*

Flowering takes place in April-May and acorn maturation in October-November. Hybridization easily takes place with *Q. suber*, *Q. petraea*, and most often with *Q. pubescens*.

Like many other forest species, this oak is difficult to regenerate by conventional means of vegetative propagation, which has led us to look for *in vitro* methods.

As far as we know, this is the first report on attempting tissue culture of this oak. For the first time, somatic embryos have been obtained in *Q. canariensis*.

3. Material and methods

3.1. IN VITRO CULTURE

Immature acorns from Las Majadas forest (Cortes de la Frontera, Málaga, Spain) were collected at two stages of development, during the growing season, on 15 and 30 July 1994. Acorns were sterilized in 70% ethanol for 30 s followed by 0.8% NaOCl with a few drops of Tween 20 for 20 minutes, and washed three times in distilled sterile water for 5 minutes each.

Immature zygotic embryos were aseptically isolated and cultured in Petri dishes in Woody Plant Medium (WPM) macronutrients formula (Lloyd & McCown, 1981) with MS micronutrients (Murashige & Skoog, 1962) and 88 mM sucrose, solidified with 8 g/l agar. The medium was autoclaved at 0.7 atmospheres (115°C) for 20 minutes. Abscisic acid (ABA) and glutamine (Gln) were filter sterilized (0.22 µ) and added after autoclaving.

Cultures were incubated at 25°C ± 2 day temperature and 20°C ± 2 at night, with a 16 hour photoperiod and 80 µmol/m^2.s irradiance, provided by Osram cool white 18 w fluorescent lamps.

Somatic embryogenesis induction treatments were applied during 3 weeks, in which 2,4-D (absent or at 4.5 µM concentration) and glutamine, either absent or present at 3.4 mM, were tested in a combined design.

For somatic embryo formation and development, basal medium with 3.4 mM Gln was used. After 4 months, maturation was assayed, comparing basal medium as a control with WPM containing 3.8 µM ABA. Somatic embryos were treated for two weeks at 4°C in darkness and thereafter transferred to germination medium (basal medium containing 0.89 µM 6-benzylaminopurine, BA) at the normal 16 h photoperiod at 25°C by day and 20°C by night.

3.2. FLOW CYTOMETRY

Somatic embryos of *Q. canariensis* were chopped with a razor blade in the presence of staining solution, containing 0.2 M Tris-HCl, 4 mM $MgCl_2$, 0.5% Triton X-100 and 4 µg/ml diamidino-2-phenylindole (DAPI), and the suspension was filtered through a 40µm nylon mesh. DNA content measurements of the isolated nuclei were performed with a PAS 2 arc lamp-based flow cytometer (Partec, Germany), equipped with a filter combination optimal for excitation and detection of the DNA-DAPI complexes (excitation filter UGI, dichroic beam splitters TK420 and TK590, and longpass filter GG435). Nuclei isolated from roots of *Cerastium eriophorum*, which has a diploid DNA amount 2C = 2.6 pg/nucleus (Monica Boscaiu, Institute of Botany, University of Vienna, personal communication) were used as a reference of the nuclear DNA content. At least 10^5 nuclei were measured for each sample.

3.3. STATISTICAL ANALYSIS

For statistical calculations of percentages, a log-linear model was fitted using a Chi-square test. Asypmtotic standard errors of the parameter estimates were computed by the delta method (Lee, 1977). The ratio of the log-linear parameter estimate to its standard error was used to obtain the frequency significance level.

4. Results and discussion

4.1. INDUCTION

Immature zygotic embryos are by far the most used initial explants for induction of somatic embryogenesis in recalcitrant plants, in contrast to easy plants, in which somatic embryogenesis may be induced in other tissues, like for instance on root phloem in carrot. We used developing zygotic embryos (2 to 3 months after anthesis) for our experiments.

Globular structures appeared 3 weeks after induction onset. These structures developed later to heart- and torpedo-shaped somatic embryos. After 2 months, frequency of somatic embryo formation on the initial explants was scored and is shown in figure 1. Collection date was the single factor showing statistical significance at the 0.01 level, the earlier one being the best. A similar behaviour has been observed in related species, such as American chestnut (Merkle, 1991), in which the stage of zygotic embryo development is also important, and the later the embryos are isolated, the lower the frequency of regeneration. Similarly, in pine (Beckwar *et al*, 1990), developmental stage of the explant has proven to be a critical factor for establishing regenerating cell cultures.

Auxin-like growth regulators are commonly used as a requirement for induction of somatic embryos (Sung *et al.*, 1988). Nevertheless, in our case 2,4-D was not necessary (fig. 1). The same can be said about glutamine, probably because the explants had sufficient endogenous levels of these promoters of embryogenesis.

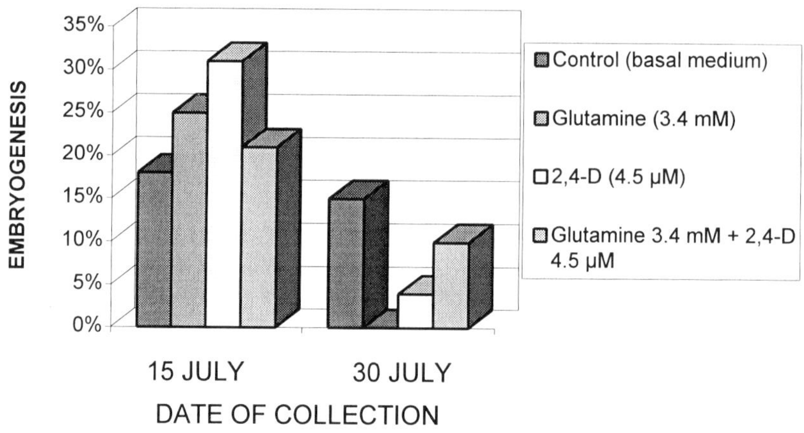

Figure 1. Effect of collection date, 2,4-D and glutamine on the frequency of somatic embryo formation. WPM was used as basal medium in all treatments.

4.2. DEVELOPMENT AND GERMINATION

Somatic embryos grew and developed in basal medium supplemented with glutamine. ABA did not improve embryo maturation at the concentration tested, as it has been the case in cork oak (Bueno *et al.*, 1992). Six months later, the embryos were subjected to a 2-week chilling period at 4°C in darkness before they were transferred to germination medium. After 2 weeks root growth was observed in the embryos, although germination rate was low (5.7%).

4.3. PLOIDY LEVEL AND DNA CONTENT OF SOMATIC EMBRYOS

To determine the ploidy level in somatic embryos clonally propagated for more than one year in culture, samples of embryo nuclei were subjected to flow cytometry analysis.

The expected pattern showing a differentiated G1 peak and a smaller G2 peak was obtained (fig. 2a). No polyploidy was detected. This method is very sensitive and allows detection of mixed ploidy levels. For instance, in samples of a *Q. suber* somatic embryo culture previously established (Bueno *et al.*, 1992; Manzanera *et al.*, 1993), three peaks are visible (fig. 2b). The first one corresponds to the diploid G1 phase of the cell cycle (2C). The second one, corresponding to a double DNA amount (4C), is unusually high for G2, indicating the presence of tetraploid nuclei in G1, which is confirmed by a third peak in 8C, corresponding to tetraploid nuclei in G2 phase.

Comparing figures 2a and 2b, it is obvious that *Q. canariensis* has a higher amount of DNA than *Q. suber*. We have quantified their DNA content using *Cerastium eriophorum* as an internal standard (2C = 2.6 pg). *Q. canariensis* DNA content at G_1 (2C) was calculated to be 2.2 ± 0.1 pg per nucleus, while in *Q. suber* is around 1.9 pg, and in *Q. petraea* is about 1.8 pg (Greilhubert, 1988).

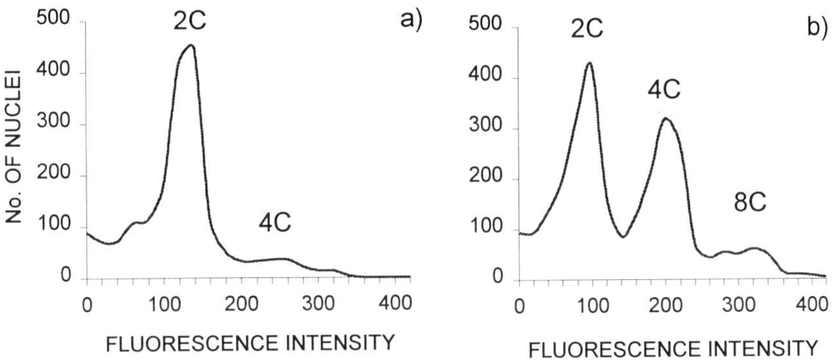

Figure 2. DNA histograms of intensity of DAPI fluorescence in: a)*Quercus canariensis* somatic embryo nuclei, and: b) *Q. suber* somatic embryo nuclei. The abscissas (fluorescence intensity in channel units) are in a logarithmic scale.

A clone of cork oak somatic embryos, Qs 2 74, has given a chimeric composition, with diploid and tetraploid nuclei present after 14 months in culture, which has not been the case in *Q. canariensis*. Variations in ploidy level also have been detected in *Picea abies* somatic embryos by chromosome counting (Fourré *et al*, 1995). *Q. canariensis* seems to be more stable *in vitro*. On the contrary, some cork oak somatic embryo cultures showed tetraploidy. At present, experiments are in progress to see if this is a general characteristic of this species. This alterations in the genetic stability of somatic embryos should lead us to be cautious with the plant material that is going to be regenerated, although somaclonal variations may also be valuable for genetic improvement. Flow cytometry provides a rapid and sensitive method to analyze the ploidy level of large numbers of clones of somatic embryos at an early stage of development, before regeneration of selected clones is attempted.

5. Acknowledgements

We thank Mr. Javier Vázquez and Miss Arancha González for providing the seeds of *Q. canariensis*, and to Mrs. Dolores Salvador and Mrs. Carmen García-Barriga for their technical assistance.

6. References

Bueno, M.A., Astorga, R. and Manzanera, J.A. (1992) Plant regeneration through somatic embryogenesis in *Quercus suber*. *Physiol. Plant.* **85**, 30-34.

Beckwar, M.R., Nagmani, R. and Kann, S.R. (1990) Initiation of embryogenic cultures and somatic embryo development in loblolly pine (*Pinus taeda*). *Can. J. For. Res.* **20**, 810-817.

Fourré, J.L., Berger, P. and André, P. (1995) Somatic embryogenesis and somaclonal variation in Norway spruce, in *Somatic Cell Genetics and Molecular Genetics of Trees. Joint Meeting of the IUFRO Working Parties S.04-07 and S.04-06. Gent, Belgium, Sept. 26-30* Kluwer Academic Publishers, Dordrecht (in press).

Greilhubert, J. (1988) "Self-tanning" - a new and important source of stoichiometric error in cytophotometric determination of nuclear DNA content in plants. *Pl. Syst. Evol.* **158**, 97-96.

Lee, S.K. (1977) On the asymptotic variances of u-terms in log-linear models of multidimensional contingency tables. *J. Am. Statist. Assoc.* **72**, 412-419.

Lloyd, G. and McCown, B. H. (1981) Commercially-feasible micropropagation of Mountain laurel, *Kalmia latifolia* by use of shoot tip culture. *Int. Plant Prop. Soc. Proc.* **30**, 421-427.

Manzanera, J.A., Astorga, R. and Bueno, M.A. (1993) Somatic embryo induction and germination in *Quercus suber* L. *Silvae Genetica*. **42**, 90-93.

Merkle, S.A., Wiecko, A.T. and Watson-Pauley, B.A. (1991) Somatic embryogenesis in American chestnut. *Can. J. For. Res.* **21**, 1698-1701.

Murashige, T. and Skoog, F. (1962) A revised medium for rapid growth and bio assays with tobacco tissue cultures. *Physiol. Plant.* **15**, 473-497.

Ruiz de la Torre, J. (1979) *Arboles y Arbustos de la España Peninsular,* Escuela Técnica Superior de Ingenieros de Montes, Madrid, 512 pp.

Sung, Z. R., Smith, J. M., Choi, J. H., Krauss, M., Borkird, C. and Liu, S. (1988) Gene expression in embryogenesis, *HortScience.* **23**, 513-515.

IDENTIFICATION OF AN EARLY MARKER OF EMBRYOGENESIS IN PINE

H. DAVID[1], J.M. DOMON[1], G. NEUTELINGS[1], B. DUMAS[2], Y. MEYER[3], A. DAVID[1]

[1]*Laboratoire de Biotechnologie et Physiologie Végétales, Faculté des Sciences , UPJV , 33 rue Saint-Leu 80039 Amiens Cédex France*

[2]*Centre de Physiologie Végétale, URA CNRS 1457, Université Paul Sabatier, 118 route de Narbonne 31062 Toulouse Cédex France*

[3]*Laboratoire de Physiologie Végétale, URA CNRS 565, Université de Perpignan Avenue de Villeneuve 66025 Perpignan France*

Abstract

Gymnosperm somatic embryogenesis undergoes several well-defined stages. Early stage somatic embryos, associated in an embryogenic tissue, consist of small closely packed cells referred to as the embryo-heads attached to highly vacuolated suspensor cells.

In our laboratory we have produced embryogenic tissues of pine from which embryos can undergo maturation and further conversion into plantlets (Laine and David 1990 ; Laine et al. 1992).

We have strong interest in the characterization of wall proteins specific of early embryo development. With respect to that, among the proteins ionically bound to the cell surfaces, we selected a cluster of three glycosylated polypeptides, representative of all the embryogenic lines (Domon et al. 1994).

N-terminal amino acid sequencing of one of them (Domon et al. 1995) showed homologies with sequences of germin precursors deduced from cDNA sequencing (*Arabidopsis*, ice plant, cereals) and with mature germins (wheat, barley). Moreover, all three were recognized by an antiserum raised against the non-glycosylated monomer of wheat germin expressed as a fusion protein in *E. coli* (Domon et al. 1995 ; David et al. 1995).

Germins have been extensively studied in germinated embryos of cereals (Lane 1994) and recently they have been found to be part of stress responses in dicots (Michalowski and Bohnert 1992 ; Delseny et al. 1993 ; Dumas et al. 1995).

For the first time though, this family of proteins has been shown to be associated with somatic embryogenesis. This was demonstrated in our pine

M. R. Ahuja et al. (eds.), Somatic Cell Genetics and Molecular Genetics of Trees, 29–38.
© 1996 *Kluwer Academic Publishers. Printed in the Netherlands.*

embryogenic system. Germins have also been found in very immature pine zygotic embryos but not in the corresponding female gametophytes. These proteins are still expressed in mature embryos. At this stage, an enzymatic activity (oxalate oxidase) has been detected within their oligomeric form, which disappears as soon as germination starts.

Functions and expression of germin genes will be studied through cloning of germin sequences.

1. Introduction

The Carribean pine, *Pinus caribaea* Morelet var. *hondurensis* Barret et Golfari, is a fast growing and high yielding species originating from central America and planted mainly in South America, West Africa and Australia. It is an important coniferous species for timber and pulp production.

Among the technologies leading to clonal propagation of genetically transformed plants in this species or related ones, a "cell/cell cluster to plant" system such as somatic embryogenesis is a powerful one.

Somatic embryogenesis has been achieved in our laboratory through the production and subculture of embryogenic tissue, the maturation and conversion of the embryos, the use of protoplasts as starting material and the cryopreservation of embryogenic cell lines (Lainé and David, 1990 ; Lainé and David, 1992 ; David et al., 1994). Growth of emblings regenerated from this material was similar to that of seedlings when established to soil.

This regeneration system is also suitable for developing fundamental studies such as the understanding of biochemical and molecular mechanisms occuring during embryo development. With respect to that, the conifer embryogenic system is worthwhile because 1) the embryogenic tissue derived either from zygotic embryos or from non-embryonic material is exclusively formed of young preglobular somatic embryos (stage 1) ; these embryos express different phenotypes such as small or large embryoheads attached to more or less numerous suspensor cells (Domon et al., 1994) ; this early multiplication step is controlled by the presence of an exogenous auxin, often 2,4-D ; 2) embryo maturation leading to globular and cotyledonary stages is controlled by the presence of ABA along with or without an osmotic stress.

The goal of our research was to characterize the early embryogenic structures at the extracellular protein level. The extracellular matrix (ECM) has been described by Knox (1992) among others as the subcellular compartment including walls, intercellular spaces and external plasmalemma surfaces. Due to the specific properties of its wall : plasticity and cross-linking of wall polymers, along with involvment of enzymatic/structural proteins in such processes, ECM controls morphogenesis through orientated cell enlargement. The pine model system developed in our laboratory has been used for that purpose i.e. the characterization of wall proteins specific of early embryo development, further identification of one glycopeptide and behaviour of this gene product in somatic/zygotic embryogenesis and germination. We raise the question of the possible biological as well as biochemical functions of such proteins during pine embryogenesis.

2. Materials and Methods

Source of plant material, electrophoretic analysis of extracellular proteins and detection of oxalate oxidase activity on blots will be quickly described.

Embryogenic and non-embryogenic lines of Carribean pine were maintained on proliferation medium as reported earlier (Domon et al. 1994). Zygotic embryos and female megagametophyte from very immature or dry seeds were collected from open-pollinated cones freshly received from Australia in January 1994. The embryos and megagametophytes were treated for extracellular protein sampling as described elsewhere (Domon et al., 1994). Due to the great number of immature zygotic embryos to be dissected out from the seeds it was not possible to resolve their extracellular protein extracts in a NEpHGE-PAGE system. Thus, we restricted the analysis to SDS-PAGE, western blots and enzymatic activity on blots. Except for the restrictions mentionned above, all the experiments were performed according to protocoles already published (Domon et al. 1994 ; Domon et al. 1995).

3. Results

3.1. PROTEIN PROFILES OF EMBRYOGENIC (E) VS NON-EMBRYOGENIC (NE) LINES

Five E lines originating from the same open-pollinated tree have been compared to two NE cultures, isolated from stem tissues of 3 year old embling-derived plants. One of those two lines (NE7) had the same genotype as one of the five E ones (E7).

Electrophoretic analysis (SDS-PAGE) showed very different E vs NE extracellular protein profiles. Moreover some minor differences being resolved as discrete bands were present or absent according to the line. On the contrary, no striking differences could be deduced from the comparison of 1D intracellular protein profiles of embryogenic vs non embryogenic samples. Using 2D NEpHGE-PAGE analysis that is known to be very resolutive (all basic proteins are maintained within the gel, none of them reaching its pHi) we numbered more than a hundred extracellular proteins in E as well as NE gels. Only a small number migrated at the same position and were considered having similar physico-chemical properties. Most of them were badly resolved such as the high molecular mass secreted proteins with high glycosylation rate while a few of them emerged clearly from all the E profiles and were absent from the NE samples (Domon et al. 1994). Thus the major differences consisted of a set of small, glycosylated, slightly acidic to neutral polypeptides strictly representative of the embryogenic lines (Fig. 1).

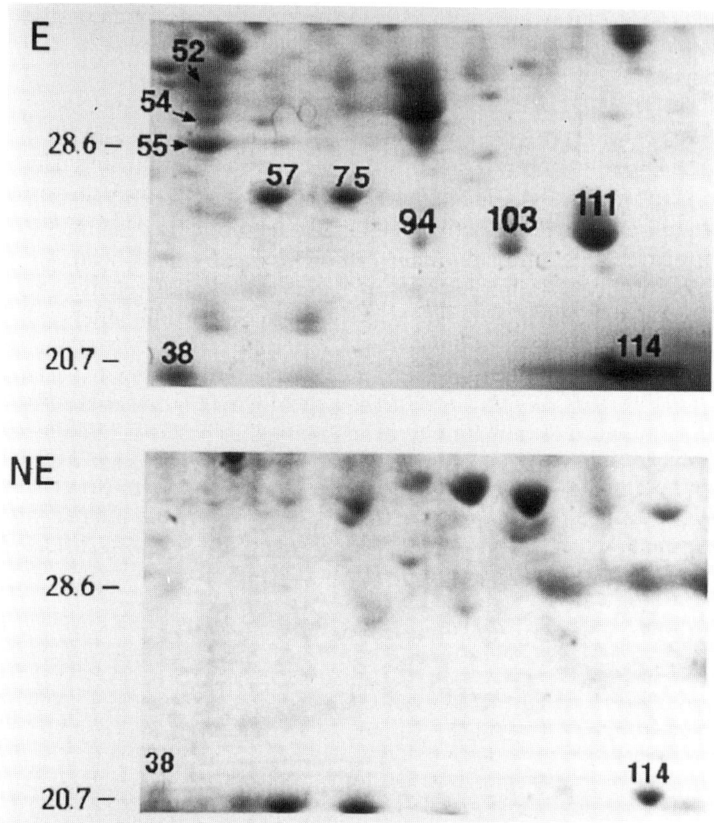

Figure 1. Partial electrophoretogram of extracellular proteins from a pine E (upper) or NE (lower) line (from Domon et al 1995).

These three spots : 111, 103 and 94 have a *M*r of 25-27000, were weakly stained with the concanavaline A peroxidase complex (detection of antenary N-linked glycans). Moreover they show cross-reactivity with immunsera directed against complex glycans with xylose and/or fucose residues (Domon et al. 1994). Their *in vivo* synthesis in the presence of tunicamycin, an antibiotic which inhibits the first step of glycosylation (transfer of NAcGlc-1P on Dolichol P) shows that the non glycosylated form of GP 111 (GP for glycopeptide) is around 21-22000. Moreover, *in vivo* protein synthesis in the presence of ^{35}S methionine shows that these 3 polypeptides are constantly synthesized and secreted between two subcultures of all the E lines maintained on multiplication medium (proliferation of stage 1 embryos) (Domon, 1995).

3.2. IDENTIFICATION OF GP 111 AND RELATED PROTEINS

3.2.1. *N-terminal Sequencing*

The GP 111 pine sequence was obtained for a single pine protein from one cell line only (E6) (Domon et al. 1995). The first series of alignments indicates that twelve aminoacids show strong similarity to corresponding N-terminal sequences in wheat germins and to open-reading frames found in germin-like coding elements isolated from other angiosperms (cereals, *M. crystallinum, A. thaliana*).

When comparing the N-terminal sequences deduced from microsequencing it appears that the mature protein starts at the same site, except one amino acid is absent. One major difference between these two groups of plants (cereals and pine) could be the site of excision of the signal peptide on the germin precursor, with, as a consequence, one amino acid less in the N-terminal part of the pine mature protein. The consensus sequence DPXXLQDFCVAD for these multiple alignments was deduced (Fig. 2). Since that, other germins or germin-like proteins have been cloned and sequenced in other cereals (rice) and crucifers (*Sinapis alba, Brassica napus* for example). Dowstream this consensus sequence, among the 12 other amino acids sequenced in the pine protein K, F, N and P still aligned on some germins or germin-like protein precursors.

```
1                                               S D P D P L Q D F C V A D L D G K - A V   S V N G H T C K
2    M  G Y S K T L A V S L F A V L L L A P A V L A - - - S D P D P L Q D F C V A D L D G K - A V   S V N G H T C K
3    M  G Y S K T L V A G L F A M L L L A P A V L A - - - T D P D P L Q D F C V A D L D G K - A V   S V N G H T C K
4    M  G Y S K T L V A G L F A M L L L A P A V L A - - - T D P D P L Q D F C V A D L D G K - A V   S V N G H T C K
5    M  G Y S K N I A S G M F A M L L L A S A V L S - - - S N P H P L Q D F C V A D L D G K - A V   S V N G H T C K
6    M  - - - - - - L R T I F L L S L L - - - - - - - F G L S N A S V Q D F C V A N L - - - K R A E   T P A G Y P C I
7    M  R V S Q S L V P - - F A I I A L V L S F V N A Y - - D P S P L Q D F C V A I D D L K - G V   F V N G H M C K
8    M  - - - - - - L R I I F L L S L L - - - - - - - F A L S N D S V Q D F C V A N L - - K R A E   T P A G Y P C I
9    M  - - - - K M R I Q F F I L S L F S S I S F A - - - S - - - - V Q D F C V A D P K G P Q N P   S - - G Y S C K
10   M  E A Y K - M F A - - F V V L - L A T T L Y Q A Y A T D P T Q L Q D F C V G V N K P N D G L   F V N G L F C K

11                                                D P E A L Q D F C V A D - - T K S T I   F M N X - P L N

12                                              A D P D P L Q D F C V A D L X D N - A V   X
13                                              Y D P S P L Q D F C V A D
14                                              Y D P S P L Q D F C I A D
```

Figure 2. Homologies between pine GP 111 (11) cereal mature germins obtained from microsequencing [12 (wheat) : Acc N° S27247, Lane et al. 1992 ; 13 (barley) : N° P28525, Hurkman et al. 1991 ; 14 (barley) : N° P526 Hurkman et al. 1991] and cereal and dicot germin precursors deduced from cDNAs [1 (barley) : Acc N° L15737, Lane et al. 1993 ; 2 (barley) : N° UO1963, Hurkman et al. 1994 ; 3 (wheat) : N° JO4592, Dratewka-Kos et al. 1989 ; 4 (wheat) : N° M63223, Lane et al. 1991 ; 5 (wheat) : N° M63224, Lane et al. 1991 ; 6 (thale cress) : N° Z30804, clone VBVBC10, Desprez et al. unpublished ; 7 (thale cress) : N° Z17674, clone YAP134T, Raynal et al. unpublished ; 8 (rape) : N° U21743, Saez-Vasquez et al. unpublished ; 9 (white mustard) : N° X84786, Heintzen et al. 1994 ; 10 (ice plant) : 99041, Michalowski & Bonhert 1992]. Note the consensus sequence within the N-terminal parts (First 12 amino acids of the mature proteins).

3.2.2. *Serological Affinity of GP111 with Antibodies against a Non-Glycosylated Germin Monomer*
Polyclonal antibodies were raised against the MBP-wheat germin monomer coding sequence expressed as a fusion protein in *E. coli*. The immunserum which recognizes unambiguously the germin apoprotein (Dumas et al. 1993) has been used to confirm the homologies with germin and glycoproteins deduced from the multiple alignments. Barley germin oxalate oxidase in its oligomeric form (125-130 KDa) partially resolved in a monomeric polypeptide of 25-27 KDa was used as a positive control and glycosylated horseradish peroxidase as a negative control. We could show that a cross-reactivity between anti wheat germin protein antibodies and a band around 25-27 KDa (1D) was detected on western blots of extracellular proteins of all the embryogenic lines analysed (Fig. 3A). No signal was seen either in the NE extracts or in the intracellular E/NE proteins (data not shown). Immunodetections on blots of 2D (NEpHGE-PAGE) gels showed that the 25-27 KDa band corresponds to the three spots GP 111, GP 103 and GP 94 (Fig. 3B).

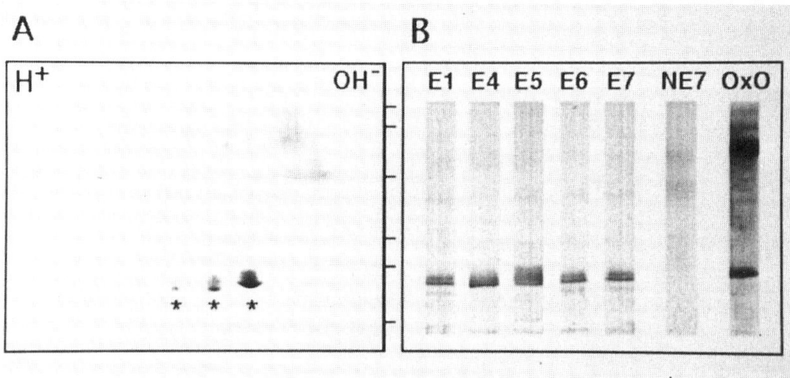

Figure 3. Serological relationships between pine proteins and wheat germin. Note (B) a band around 25-27 KDa restricted to E lines and resolved (A) in 3 polypeptides (GP 111, 103, 94 from right to left) (For details see Domon et al., 1995).

From the presence of these 3 proteins associated with preglobular stage somatic embryos of pine we considered these GLPs as early markers of embryogenesis in pine.

3.3. USE OF GLP MONOMERS IN DIAGNOSIS OF THE EMBRYOGENIC POTENTIAL IN PINE.

The markers identified as germin monomers have been proved to be powerful tools for diagnosis of embryogenic potential. The line NE7 that was one of both lines used as a negative control in detection of proteins specific of the embryogenic state was subjected to a series of drastic treatments in order to reverse its phenotype into an embryogenic one. The inductive treatment led to necrosis of most of the explants. Only twenty per cent of them exhibited some white cell masses protruding from the dying callus. Those masses could be subcultured separately. We could establish the embryogenic character of these calli by histological as well as biochemical control. Thus, the typical morphological traits (embryoheads on top of suspensor cells) as well as the presence of the three GLPs (GP 111, GP 103 and GP 94) in the extracellular protein extracts of the reversed culture, confirmed by serological reactivity with anti wheat germin antiserum, indicated unambiguously the embryogenic potential of this culture. These calli are currently under maturation treatments. In terms of forestry biotechnologies this is a breakthrough since our NE lines were established from stem tissues (brachyblasts) from a 3 year old plant and were subcultured for more than 3 years without loosing the specific morphological traits of a non-embryogenic culture. Let us mention that for early diagnosis of embryogenic potential, cell printing techniques that we developped in our laboratory might be preferred for very early detection since we could detect directly extracellular germins on a few preglobular embryos, under the microscope, using the same protocole as that used for western blots.

3.4. ENZYMATIC FUNCTION OF GLPs IN EMBRYO DEVELOPMENT AND GERMINATION

Cereal germin-oxalate oxidases are active in their oligomeric form (125-130 KDa) while their monomers (25-27 KDa) are not. This enzymatic activity has been detected during germination. Oxalate oxidase activity was directly assessed on blots after 1D electrophoresis in semi-native conditions (0,1% SDS, a concentration known to avoid total denaturation of the oligomeric form (Dumas et al., 1993 ; Lane et al., 1993). Figures 2A and B show immunodetection (A) of GLPs and oxalate oxydase activity (B) on blots of extracellular protein samples of pine. Preglobular somatic or zygotic embryos and their corresponding haploid megagametophyte, cotyledonary zygotic embryos and gametophyte from dry seeds, as well as germinating embryos, were analysed (Fig. 4).

Only the monomeric form of germins is present in immature somatic or zygotic embryos, either because the multimeric form of the embryo GLPs does not stand the semi-native conditions or because the oligomeric form does not assemble in the walls of immature embryos (Fig. 4A). As for cereals, this monomer does not express oxalate oxidase activity. On the other hands, as for cereal germins expressed during germination, the oligomeric form is maintained and expresses oxalate oxidase activity in the mature embryos from dry or imbibed seeds (Fig. 4B).

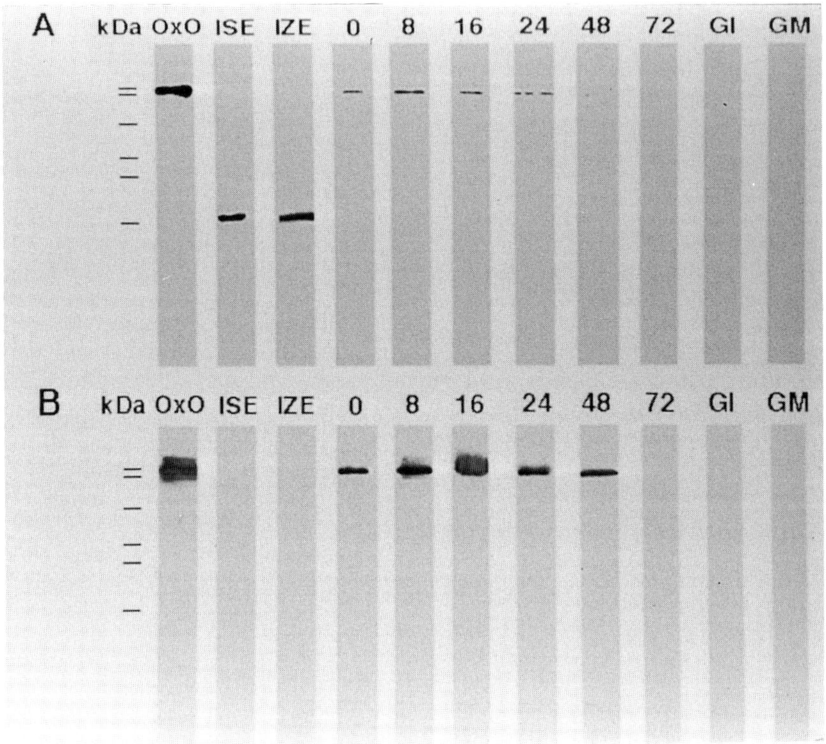

Figure 4. GLPs detection (A) and oxalate oxidase activity (B) during pine embryogenesis and germination ; 0, 8, 16, 24, 48, 72 : time (h) post imbibition.
(A) Note the presence of GLPs monomers in somatic/zygotic immature embryos and the presence of GLPs oligomers during germination.
(B) Only the oligomeric form expresses the enzymatic activity.

After 72h post imbibition, the expression as well as the enzymatic activity of pine GLPs disappear. This is concomitant with emergence of the radicle indicating the end of germination (Figs. 4a and B).

Germin association into an oligomeric form was checked on 2D NEpHGE-PAGE (2,3% SDS) derived immunoblots showing the resolution of the high molecular mass GLPs (band of 125-130 KDa) into 4-5 spots of 25-27 KDa (Data not shown).

4. Discussion

Germin oxalate-oxidases of cereals (for review see Lane 1994) are among the best characterized protein markers of cereal embryo development, their accumulation coinciding with the onset of growth in germinating wheat embryos. They catalyse H_2O_2 formation and Ca^{2+} release from Ca^{2+} oxalate degradation. Both H_2O_2 and Ca^{2+} are needed for several physiological purposes such as cross-linking reactions and signalling. Pseudogermin, an antigenically related isoform of germin is synthesized during embryogenesis, at the time of maximal cell enlargement in the immature embryo (Lane et al. 1992).

Germin-like proteins are also present in several dicotyledonous species such as the model species *Arabidopsis thaliana* and other crucifers. In that species a germin clone (Höfte et al. 1993) belongs to a cDNA library established from a mix of mRNAs of immature siliquae at different developmental stages. In our laboratory we have studied the developmental regulation of the corresponding gene and showed that, during the life cycle of the plant, its expression was restricted to the siliquae and, within the siliquae, to globular to cotyledonary embryos (David 1995).

The GLPs that we identified associated with early stages of pine embryo development are early markers of embryogenesis in this conifer system. These proteins, found in zygotic as well as somatic embryogenesis, represent a new group of proteins useful for the diagnosis of embryogenic potential. The physiological implications of such markers are still questionned since we have not detected any oxalate oxidase activity at that early stages of development yet. In agreement with the results on cereal germins, pine GLPs accumulating in mature embryos express oxalate oxidase activity until 48h post imbibition (radicle emergence). We observed that such proteins are secreted by somatic as well as zygotic embryos of pine but are absent from the corresponding female gametophyte.

Although recent papers reporting on this family of proteins showed their association with rhythmic phenomena, with biotic and abiotic stress response (Dumas et al., 1995 ; Michalowski and Bonhert, 1992) with embryogenesis and early development (Lane et al. 1993, Dumas et al 1993, Domon et al. 1995) a biochemical function has only been clearly demonstrated in cereal and pine germination (Lane, 1994, and this paper). We are currently isolating clones encoding GLPs from a library established from pine embryogenic tissue. From their sequences we hope to gain knowledge about regulation of GLPs expression and about biological significance of these proteins in our early pine embryogenic system.

References

David, A., Lainé, E., and David, H. (1995) Somatic embryogenesis in *Pinus caribaea*, in : S.M. Jain (ed.), *Somatic Embryogenesis of Woody Plants*, Kluwer Academic Publishers, Dordrecht, Netherlands, pp 145-182.

Delseny, M., Gaubier, P., Hull, G., Saez-Vasquez, J., Gallois, P., Raynal, M., Cooke, R. ang Grellet, F. (1993) in Basra AS (ed) *Stress induced gene expression* Harwood Academic Publishers, Reading, UK, pp.25-59.

Domon, J.M., Meyer, Y., Faye, L., David, A., and David, H. (1994) Extracellular (glyco)proteins in embryogenic and non-embryogenic cell lines of Caribbean pine. Comparison between phenotypes of Stage one somatic embryos, *Plant Physiol. Biochem* . **32**, 1-11.

Domon, J.M., Dumas, B., Lainé, E., Meyer, Y., David, A., and David, H. (1995) Three glycosylated polypeptides secreted by several embryogenic cell cultures of pine show highly specific serological affinity to antibodies directed against the wheat germin apoprotein monomer, *Plant Physiol.* **108,** 141-148.

Domon, J.M. (1995) Glycoprotéines sécrétées et embryogenèse chez le Pin. Identification d'un glycopeptide, Ph D Thesis of the University of Picardie Jules Verne, Amiens, France. (Plant Physiology and Development) 135 pages.

Dumas, B., Sailland, A., Cheviet, J.P., Freyssinet, G., and Pallett, K. (1993) Identification of barley oxalate oxidase as a germin-like protein, *C. R. Acad. Sci. Paris*, **316**, 793-798.

Dumas, B., Freyssinet, G., and Pallet, K.E. (1995) Tissue specific expression of germin-like oxalate oxidase during development and fungal infection of barley seedlings, *Plant Physiol.* 1091-1096.

Höfte, H., Desprez,T., Amselem, J., Chiapello, H., Caboche, M., Moisan, A., Jourjon, M.F., Charpenteau, J.L., Berthomieu, P., Guerrier, D., Giraudat, J., Quigley, F., Thomas, F., Yu DY, Mache, R., Raynal, M., Cooke, R., Grellet, F., Delseny, M., Parmentier, Y., De Marcillac, G., Gigot, C., Fleck, J., Philipps, G., Axelos, M., Bardet, C., Tremousaygue, D., and Lescure, B. (1993) An inventory of 1152 expressed sequence tags obtained by partial sequencing of cDNAs from *Arabidopsis thaliana,The Plant Journal* **4**, 1051-1061.

Heintzen, C., Fischer, R., Melzer, S., Kappeler, K., Apel, K. and Staiger, D. (1994) Circadian oscillations of a transcript encoding a germin-like protein that is associated with cell walls in young leaves of the long-day plant *Sinapis alba* L. *Plant Physiol.* **106**, 905-915.

Knox, J.P. (1992) Molecular probes for the plant cell surface, *Protoplasma* **167**, 1-9.

Lainé, E. and David, A. (1990) Somatic embryogenesis in immature embryos and protoplasts of *Pinus caribaea,Plant Science*, **69**, 215-224.

Lainé, E., Bade, P., and David, A. (1992) Recovery of plants from cryopreserved embryogenic cell suspensions of *Pinus caribaea*, *Plant Cell Rep.* **11**, 295-298.

Lane, B.G., Cuming, A.C., Fregeau, J., Carpita, N.C., Hurckman, W.J., Bernier, F., Dratewka-Kos, E., and Kennedy, T.D. (1992) Germin isoforms are discrete temporal markers of wheat development, *Eur. J. Biochem.* **209**, 961-969.

Lane, B.G., Dunwell, J.M., Ray, J.A., Schmitt, M.R. , and Cuming A.C. (1993) Germin, a protein marker of early plant development , is an oxalate oxidase *J. Biol. Chem.* **268**, 12239-12242.

Lane, B.G. (1994) Oxalate, germin, and the extracellular matrix in higher plants, *FASEB J* **7**, 294-301.

Michalowski, C.B. and Bohnert, H.J. (1992) Nucleotide sequence of a root-specific transcript encoding a germin-like protein from the halophyte *Mesembryanthemum crystallinum*, *Plant Physiol.* **100**, 537-538.

SOMATIC EMBRYOGENESIS AND SOMACLONAL VARIATION IN NORWAY SPRUCE: CYTOGENETIC AND MOLECULAR APPROACHES

J.-L. FOURRÉ, P. BERGER and P. ANDRÉ
University of Louvain, Unit of Water and Forestry
Place Croix du Sud 2, B-1348 Louvain-la-Neuve, Belgium

Abstract

Chromosome counting and RAPD (Random amplified polymorphic DNA) were used to study somaclonal variation in Norway spruce somatic embryogenesis. The cytogenetic approach revealed, for the first time in this species, three cases of mutated acclimated somatic plants: one totally trisomic and two chimeras with trisomic buds and diploid roots. Other cases of totally trisomic, double trisomic, tetraploid, or mixoploid 5-year-old embryogenic masses were also detected. The molecular approach (RAPD) revealed no somaclonal variation despite the large sample of DNA and primers used and the important interclonal variation observed.

1. Introduction

Four embryogenic clones of Norway spruce have been subcultivated and observed over several years to determine the evolution of production of mature embryos and to assess the quality of the produced embryos. A wide range of intraclonal quantitative and qualitative variability has been observed within this production. Certain morphologic deviations appeared at the immature stage and after maturation, such as immature embryos with a diffuse organization, entire or part albino mature embryos or acclimated somatic seedlings comparable to dwarf mutants. All these phenotypic variations could be the result of a modification of the genome itself or only of the expression of the genome. Two approaches, chromosome counting and RAPD (Random amplified polymorphic DNA), were choosen for their abilities to detect genotypic variations: respectively genomic and chromosomic or genic mutations. This study was done on embryogenic masses as well as on acclimated somatic plants.

2. Material and Methods

2.1. PLANT MATERIAL

Four embryogenic mass clones of Norway spruce (*Picea abies* (L.) Karst.), initiated from mature zygotic embryos, were selected out of several hundred clones for their ability to produce healthy mature embryos. The A, B and C clones were initiated on July 3, 1989 (from seeds collected in 1984 at seed stand "Cedrogne 031A", Belgium) and the V clone was initiated on March 5, 1990 (from a seed collected in 1984 at seed stand

M. R. Ahuja et al. (eds.), Somatic Cell Genetics and Molecular Genetics of Trees, 39–45.
© 1996 *Kluwer Academic Publishers. Printed in the Netherlands.*

"Jubilé St Jean", Belgium). Seeds were soaked in water for 24 hours, surface-sterilised for several seconds in 70% ethanol and aseptically dissected in a laminar flow cabinet. Embryos were placed in contact with the initiation medium (1M1) of Gupta *et al.* (1986), modified as follows : 10g/l of sucrose, 5 µM of 2,4-dichlorophenoxyacetic acid (2,4-D), 2 µM of kinetin (KIN), 2 µM of 6-benzylaminopurine (BAP) and gelified with 3 g/l of Gelrite. After initiation, the embryogenic masses were subcultivated every week on the multiplication medium (1M2), which was identical to the initiation medium, except for the concentrations of the growth regulators that were ten times lower. Each week, embryo maturation was obtained by transferring a part of the embryogenic mass to a multiplication medium deprived of auxin and cytokinin and containing 30 g/l of sucrose. After autoclaving, abscissic acid (ABA) was added to this medium (medium 1M3). During these successive stages, the Petri dishes remained in darkness at a temperature of 22°C.

The embryos were germinated under a 16-hour photoperiod, in Petri dishes on the modified "Murashige and Skoog" medium (Misson *et al.* 1988), containing a quarter of the NH_4NO_3 concentration, 20 g/l of sucrose, 4 g/l of activated charcoal and gelified with 3 g/l of Gelrite. Plantlets were acclimated in a growth chamber under a constant light using a hydroponic culture system. They were lodged within the holes of a 2 mm thick perforated expanded polystyrene plate placed in a plastic basin containing the constantly ventilated B+C nutrient solution of Ingestad (1971). The surviving individuals were planted in containers filled with a mixture of compost and Perlite in a proportion of 2:1.

2.2. CHROMOSOME SQUASHES

2.2.1. *Acclimated Plants*

Metaphase chromosome spreads were obtained from root tips and/or sprouting buds of acclimated somatic seedlings. The samples were collected when the mitotic index was higher: in the morning when there was no late frost for the root tips, and in the afternoon for the sprouting buds. The samples were pretreated in a saturated solution of 1-bromonaphthalen at 0 °C. Adequate condensation of chromosomes required 24-hour pretreatment. The material was then successively fixed in a 80% ethanol and glacial acetic acid solution (3:1) at 4 °C for 24 hours, hydrolysed in 5 N HCl solution at room temperature for 8 minutes, and rinsed with water. The samples were stained by the Feulgen method for one hour. The meristematic tissues were picked up onto glass slide using a needle with flattened tip and squashed with the coverglass in a drop of alun-hematoxylin. When chromosome preparation was judged satisfactory, it was sealed with the VALAP (1/3 vaseline, 1/3 lanolin, 1/3 paraffin).

2.2.2. *Embryogenic Masses*

Metaphase chromosome spreads of embryogenic masses were obtained according to the procedure of Nkongolo and Klimaszewska (1994).

2.3. DNA ISOLATION

DNA was extracted from flushing bud needles, embryogenic tissues, and somatic embryos according to Doyle and Doyle (1990). Approximately 50 mg of tissue were ground using a 1.5 ml Eppendorf tube and a plastic pestle in 450 µl of 2X CTAB buffer. The extraction buffer consisted of 2% (w/v) CTAB (Cetyl trimetylammonium bromide), 1,4 M NaCl, 20 mM EDTA (Ethylenediaminetetraacetic acid, UCB), 100 mM Tris pH 8, 1% Poly[1-vinyl-2-pyrollidone] (PM: 25000, Merck) and 0.2% (v/v) ß-mercaptoethanol. The homogenate was incubated at 60 °C for 30 min, extracted with 800 µl of chloroform:isoamylalcohol (25:1), and then centrifuged at 13000 g for 5 min. The chloroform:isoamylalcohol extraction was repeated twice more. DNA was precipitated from the aqueous phase by mixing with 350 µl isopropanol (- 20 °C). After centrifugation at 13000 g for 15 min at 4 °C, the DNA pellet was rinsed with 70% ethanol, vacuum dried, and resuspended in 40 µl distilled water. After fluorimetric measurement of DNA concentration, the sample was diluted with water at 50 ng/l.

2.4. DNA AMPLIFICATION

Ten nucleotides in length primers were used for PCR amplification following a protocol previously described by Williams *et al.* (1990). Amplification reactions were performed in volumes of 25 µl containing 50, 100 or 200 ng of genomic DNA, 100 µM of each dNTP (Pharmacia), 0.6 or 1 unit of thermostable DNA polymerase (Dynazyme from *Thermus brockianus*, Finnzymes or Taq DNA polymerase from *Thermus aquaticus*, Boehringer), 2.5 µl of 10 X reaction buffer supplied with the DNA polymerase and 0.4 µM of primer (Operon Technologies). The mixture was covered with a drop of mineral oil. For the DNA amplification, a Perkin Elmer (DNA Thermal Cycler 480) or a Prem (LEP scientific) Thermal Cycler was programmed for 35 cycles, each consisting of a denaturation step of 1 min at 94 °C, followed by an annealing step of 1 min at 37 °C, and an extension step of 3 min at 72 °C. The first cycle was preceded by 3 min at 94 °C, 1 min at 37 °C and 3 min at 72 °C, and the last cycle was followed by 10 min at 72 °C. After amplification was completed, the samples were loaded and electrophoresed on 1.2% agarose gel, followed by staining with ethidium bromide. In all cases lambda phage DNA digested with EcoRI and HindIII (Pharmacia) was used as size marker.

We only compared RAPD profiles obtained with DNA samples of the same clone tested with the same primer, DNA polymerase, thermal cycler, and reaction mixture (same primer, DNA, DNA polymerase concentrations). The results were obtained after three experiments using respectively 3, 10 and 29 primers.

3. Results

3.1. CHROMOSOME COUNTING

Chromosome counting of 7 substrains of 5-year-old embryogenic masses (table 1) revealed important genomic mutations such as monoploidy ($2n$-1=23), trisomy ($2n$+1=25), double trisomy ($2n$+2=26) and tetraploidy ($4n$=48). Some masses seemed to be completely mutated (one trisomic, one double trisomic and one tetraploid) while others were normal (diploid $2n$=24) or presented different ploidy levels: monosomy, trisomy and diploidy. The completely mutated embryogenic masses (Vn°1,Vn°2 and C1) were not able to regenerate mature embryos. On the other hand some genomically normal embryogenic masses were also unable to maturate (A1 and A2). The mixoploid substrains are able to produce mature embryos (B) or are only producing nodules (C).

A sample of 240 acclimated plants (on a total of 1000) regenerated during hundreds of weekly subcultures (3.8 to 28.5 months after initiation) were analysed by chromosome counting: 218 in root tip cells only, 13 in bud cells only and 9 in both types of cells (table 2). One plant with a quite normal phenotype was totally trisomic (Fourré 1992). Another phenotypically normal plant was a chimera, with trisomic buds and diploid roots. A third one, with a dwarf phenotype, was also a chimera with the same genomic anomaly. The three mutants were regenerated from three different clones, respectively 12.9, 13.2 and 23.3 months after initiation (Fourré 1995).

TABLE 1. Chromosome counting of 7 substrains of 5-year-old embryogenic masses. Distribution of the analysed metaphasic plates according to the chromosome number observed

Substrain	Behaviour on 1M3	Chromosome number observed 23	23+f	24	24+f	25	26	±48	Σ	% of anomalies
B	mse	4	1	21	2	2	-	-	30	30
C	nod	3	-	24	-	3	-	-	30	20
V n°1	nec	-	-	-	-	-	10	-	10	100
V n°2	nec	-	-	-	-	8	-	-	8	100
A1	prol	-	-	3	-	-	-	-	3	0
A2	prol	-	-	9	-	-	-	-	9	0
C1	prol	-	-	-	-	-	-	5	5	100

1M3 = maturation medium, mse = production of mature somatic embryos, nod = production of nodules, nec = necrosis, prol = proliferation of the immature embryos, f = presence of a chromosome fragment

TABLE 2. Chromosome counting of acclimated somatic seedlings. Distribution of the plants according to the ploidy level (diploidy (d) or trisomy (t)) observed in the roots and/or in the buds. Globally, 622 plant metaphasic plates were analysed: 580 from roots and 42 from buds. The mean number of metaphasic plates by plant was 2.6. Four metaphasic plates (3 from roots + 1 from buds) were scored for the A clone genomic chimera, 12 (8+4) for the B clone genomic chimera and 30 (25+5) for the trisomic C clone plant.

Clone	Root (d,?)	Root & Bud (d,d)	Root & Bud (d,t)	Root & Bud (t,t)	Bud (?,d)	Σ
A	19	-	1	-	-	20
B	112	-	1		7	120
C	8	-	-	1	-	9
V	79	6	-	-	6	91
Σ	218	6	2	1	13	240

t = trisomy, d = diploidy, ? = not analysed

3.2. RAPDs

RAPDs were used to assess the genetic integrity of our *in vitro* production. 2154 RAPD profiles were scored: 50 resulting from a combination of root extracted DNA of 5 somatic plants and 10 primers (revealing 133 bands), 1691 resulting from a combination of bud extracted DNA of 443 somatic plants and 3, 29 or 32 primers or amplification conditions (revealing 77, 285 or 362 bands), and 413 resulting from a combination of 56 DNA samples (extracted from embryogenic masses, mature somatic embryos or somatic seedlings) and 3, 10 or 29 primers (revealing 77, 133 or 285 bands)(Fourré 1995).

The analysed somatic plants were regenerated 3.8 to 27.1 months after initiation. The samples of embryogenic masses, mature embryos and somatic seedlings were taken 27 to 57 months after initiation. Some of the embryogenic masses, mature embryos or somatic seedlings were abnormal (unable to maturate, swollen or hyperhydric). Buds of the A and the B clone genomic chimeras were analysed respectively with 3 and 29 primers. Roots and Buds of the C clone trisomic plant were analysed respectively with 10 and 32 primers.

No true intraclonal variation was observed despite the large sample of DNA and primers used, but for many primers we were able to differentiate the four clones (figure 1).

FIGURE 1. Example of RAPD profiles. Gel electrophoresis of RAPD fragments obtained with primer OPA-11 or OPA-16 (Operon Technologies). Lanes l are size marker l digested with EcoRI and Hind III. For the 4 somaclones (A, B, C and V), lanes 1 to 5 show RAPD products of 5 different individuals.

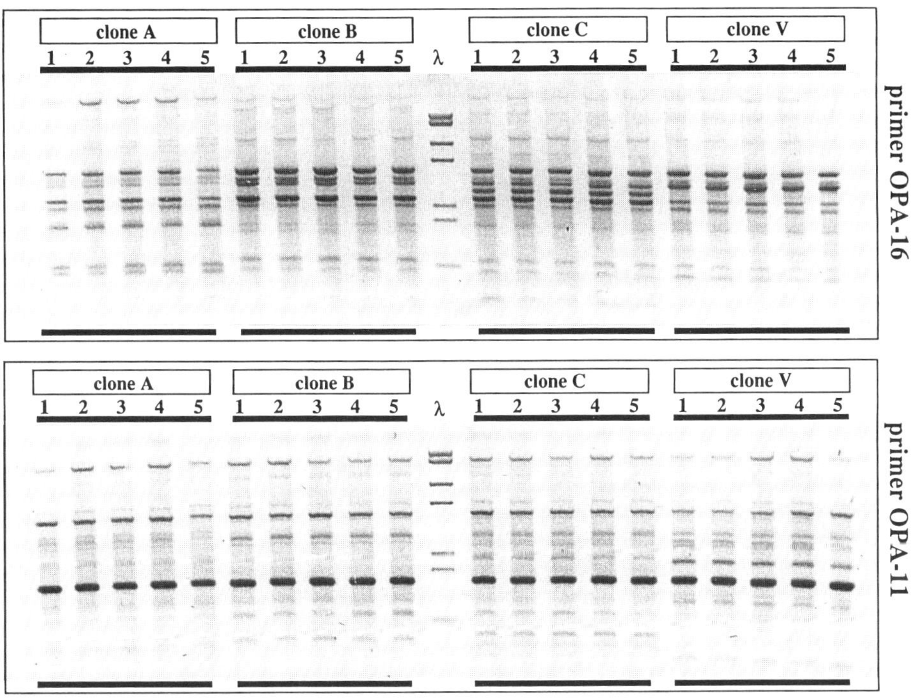

4. Conclusion

The cytogenetic approach revealed for the first time important mutations in embryogenic masses as well as in acclimated somatic plants of a conifer. Some substrains of embryogenic masses were totally mutated and at least 3 out of 218 acclimated plants (1.4 %) were totally or partially trisomic. Moreover, trisomy was not always correlated with a visible phenotypic modification such as dwarfism. All these genotypic somaclonal variations are *a priori* not desirable for a biotechnology of clonal propagation.

Lelu (1988) also reported two tetraploid globular somatic embryos in Norway spruce. In contrast, using cytometry and isozyme analyses of embryogenic masses or somatic embryos of Norway spruce or interior spruce, Mo *et al.* (1989) and Eastman *et al.* (1991) were not able to detect any genotypic variations.

The RAPD approach did not reveal intraclonal variations despite the presence of important phenotypic and genomic anomalies. Similarly, other authors failed to observe intraclonal variations using RAPD for ± 150 somatic plants (2 clones) and 15 somatic embryos (2 clones) of Norway spruce (Heinze 1993) or for 90 somatic embryos (3 clones) of *Picea mariana* (Isabel *et al.* 1993). On the other hand, De Verno *et al.* (1994), observed with RFLP some mitochondrial DNA modifications occuring in somatic embryogenesis of *Larix* species.

We therefore believe that it would be hazardous to use exclusively the RAPD method to study somaclonal variation. Nevertheless this method is very useful to identify our clones, to produce what we call fingerprints.

Acknowledgements

This research was supported by the "Région Wallonne", Belgium (convention GRAGEF). The authors are grateful to Prof. M. Briquet, University of Louvain, for providing equipment and advice for RAPD analysis and to Prof. J.-M. Kinet, University of Louvain, for his careful review of this manuscript and his helpful comments. The authors also thank O. Charlier and D. Verbois for technical assistance.

References

De Verno, L.L., Charest, P.J. and Bonen, L. (1994) Mitochondrial DNA variation in somatic embryogenic cultures of *Larix*, Theor. Appl. Genet. **88**, 727-732.

Doyle, J.J. and Doyle, J.L. (1990) Isolation of plant DNA from fresh tissue, BRL Focus **12**(1), 13-15.

Eastman, P.A.K., Webster, F.B., Pitel, J.A., Roberts, D.R. (1991) Evaluation of somaclonal variation during somatic embryogenesis of interior spruce (*Picea engelmannii* complex) using culture morphology and isozyme analysis, Plant Cell Reports **10**, 425-430.

Fourré, J.-L. (1992) Mise en évidence d'un chromosome surnuméraire chez un plant de *Picea abies* (L.) Karst. produit par embryogenèse somatique, in AFOCEL-IUFRO (eds.), *Mass Production Technology for Genetically Improved Fast Growing Forest Tree Species, Tome II*, AFOCEL Publishers, Nangis, France, pp. 485-487.

Fourré, J.-L. (1995) Embryogenèse somatique et variation somaclonale chez l'épicéa: approches morphogénétique, cytogénétique & moléculaire, PhD Thesis, Université catholique de Louvain, Belgium.

Gupta, P.K. and Durzan, D.J. (1986) Plantlet regeneration via somatic embryogenesis from subcultured callus of mature embryos of *Picea abies* (Norway spruce), In Vitro Cellular & Developmental Biology **22**, 685-688.

Heinze, B. (1993) Genetic stability in Norway spruce plants and embryos derived from somatic embryogenesis as analysed by randomly amplified DNA, PhD Thesis, Universität für Bodenkultur Wien, Austria.

Isabel, N., Tremblay, L., Michaud, M., Tremblay, F.M. and Bousquet, J. (1993) RAPDs as an aid to evaluate the genetic integrity of somatic embryogenesis-derived populations of *Picea mariana* (Mill.) B.S.P., Theor. Appl. Genet. **86**, 81-87.

Lelu, M.A. (1988) Variations morphologiques et génétiques chez *Picea abies* obtenues après embryogenèse somatique , Annales de recherches sylvicoles de l'AFOCEL de l'année **1987**, 35-47.

Misson, J.-P. (1988) Multiplication du *Thuja plicata* par culture *in vitro* de tissus juvéniles et âgés, Can. J. For. Res. **18**, 473-477.

Mo, L.H., von Arnold, S., Lagercrantz, U. (1989) Morphogenic and genetic stability in long term embryogenic cultures and somatic embryos of Norway spruce (*Picea abies* L. Karst.), Plant Cell Reports **8**, 375-378.

Nkongolo, K.K. and Klimaszewska, K. (1994) Karyotype analysis and optimization of mitotic index in *Picea mariana* (black spruce) preparations from seedling root tips and embryogenic cultures, Heredity **73**, 11-17.

Williams, J.G.K., Kubelik, A.R., Livak, K.J., Rafalski, J.A. and Tingey, S.V. (1990) DNA polymorphisms amplified by arbitrary primers are useful as genetic markers, Nucleic Acids Research **18**, 6531-6535.

TOWARDS WATER STRESS-TOLERANT POPLAR AND PINE TREES : MOLECULAR BIOLOGY, TRANSFORMATION AND REGENERATION

A. ALTMAN[1], D. PELAH[1], O. YARNITSKY[1], T. TZFIRA[1], A. YA'ARI[1], W-X. WANG[1], O. SHOSEYOV[1], A. VAINSTEIN[1], A. HÜTTERMANN[2], AND S. WANG[3]

[1]*The Kennedy-Leigh Centre for Horticultural Research and the Otto Warburg Center for Biotechnology in Agriculture, The Hebrew University of Jerusalem, Rehovot 76100, Israel* ; [2] *Institute of Forest Botany, University of Göttingen, 37077 Göttingen, Germany;* [3]*Experimental Center of Forest Biology, Beijing Forestry University, 100083 Beijing, China.*

ABSTRACT. Novel procedures for the improvement of current breeding, selection and forestation techniques can only be achieved by combining conventional breeding and selection procedures, rapid and efficient propagation, new planting and land use approaches, and novel biotechnological methodologies. An integrated biotechnological approach should be useful both for determining the physiological and molecular basis of tree tolerance, and for rapid selection and tree improvement. In the following, we present data from our studies on micropropagation, molecular analysis of water-stress tolerance, and transformation of *Populus* spp. and *Pinus halepensis*. Axillary bud break was induced in *Populus* stem and leaf cultures, and adventitious buds and roots were formed, resulting in acclimatized plants. Procedures for adventitious shoot formation in root cultures of aspen (*Populus tremula*) are currently under study. Explants of mature zygotic embryos, seedlings, and cytokinin-treated mature trees of *P. halepensis* are being studied for in vitro clonal propagation. Efficient procedures for the massive induction of first- and second-cycle adventitious shoots from embryos were established. Buds were also induced on pine seedling explants and on detached needles from long-term embryo culture. Fascicular buds were induced in 3-year-old pine trees, and they were cultured in vitro for further regeneration. Somatic embryogenesis is also being studied. Drought-specific proteins/genes were studied in several *Populus* species and in *P. halepensis*. A novel 66 kDa boiling-stable protein was highly expressed in *P. tremula* shoots, as early as 1 h after gradual water loss and ABA application, as found by Pelah et al. [19]. Using *Populus* clones which differed in their drought tolerance, we found a good correlation between the expression of dehydrin-like proteins and sucrose synthase and the degree of drought tolerance and ion leakage [21]. cDNAs of the 66 kDa and related proteins are being employed for additional molecular characterization. Here we present additional evidence for the effect of cold and osmotic stresses on protein expression. The 66 kDa boiling-stable protein was highly expressed in aspen callus cultures subjected to osmotic stress in a medium containing 6 or 12% mannitol, simultaneously with a decrease in cellular water potential. The same protein accumulated in response to cold stress, upon culture of aspen callus and shoots at 4°C. A highly efficient transformation and regeneration procedure of aspen, which does not require a pre-selection stage on antibiotics, was established. *P. halepensis* embryos, seedlings and

M. R. Ahuja et al. (eds.), Somatic Cell Genetics and Molecular Genetics of Trees, 47–56.
© 1996 *Kluwer Academic Publishers. Printed in the Netherlands.*

mature buds were also efficiently transformed, as monitored by root formation and/or GUS expression. These are fully described by Tzfira et al. [29, 31].

1. Introduction

The growing global awareness of environmental issues is coupled with a realization of the unique contribution of forestry to the well-being of this planet and to the conservation of its biodiversity. Thus, intensive, combined research and development efforts are being called for to increase forest area worldwide and to replace forests which are continuously being destroyed by industrialization and urbanization, deleterious changes in the global climate, and alterations in agricultural practices. Practically, three of the major threats to forest growth, depending on the region and on specific local conditions, include desertification in arid and semi-arid areas, biological pests, and pollution. Desertification, resulting from over-exploitation by the local inhabitants, is aggravated by regional climatic changes, and results in increased soil erosion and a decrease in land and agricultural productivity. Desertification, and the accompanying drought stress, is the single most detrimental factor limiting land and agricultural productivity in many countries. Water shortages, found in arid and semi-arid zones worldwide, have severe implications on our ecosystem, which go well beyond general scientific interest.

Ecologically sound management of the ecosystem can efficiently fight desertification and increase land productivity. Large-scale forestation, requiring many billions of trees annually, is a key factor in fighting desertification in many of the world's arid and semi-arid regions. Realization of this goal is dependent on forestation with tree species that are tolerant to the prevailing drought and salinity conditions, and that are at the same time fast-growing and multipurpose, as well as on novel practical methods for propagation, tree planting and water conservation and utilization. Selection and breeding of forest trees for better performance in marginal lands and climatic conditions requires many generations, making this a long-term goal. Novel procedures, aimed at shortening and improving the current selection and forestation techniques, are therefore much in need. These can only be achieved by combining conventional breeding and selection techniques, rapid and efficient propagation, new planting and land-use approaches and, where relevant, novel biotechnological methodologies. An integrated biotechnological approach should be useful both for determining the physiological and molecular bases of tree tolerance, and for rapid selection and tree improvement. In this respect, bud cultures have been used previously by us to elucidate the hormonal control of bud growth and development in trees of *Fagus sylvatica*, *Populus trichocarpa* and *P. tremula* [15, 16, 17].

In the following, we present data on micropropagation, molecular analysis of water-stress tolerance, and transformation of poplars and pines. Other aspects of these studies have been published elsewhere [2, 19, 20, 29, 30, 31]. The various integrated experimental stages required for any biotechnological approach of this type are indicated in Fig. 1.

2. Experimental

2.1. PLANT MATERIAL

Aspen (*Populus tremula*) and other *Populus* clones, as well as Aleppo pine (*Pinus halepensis*) were used throughout. Both in vitro-cultured plantlets and tissues, and greenhouse-grown plants were employed.

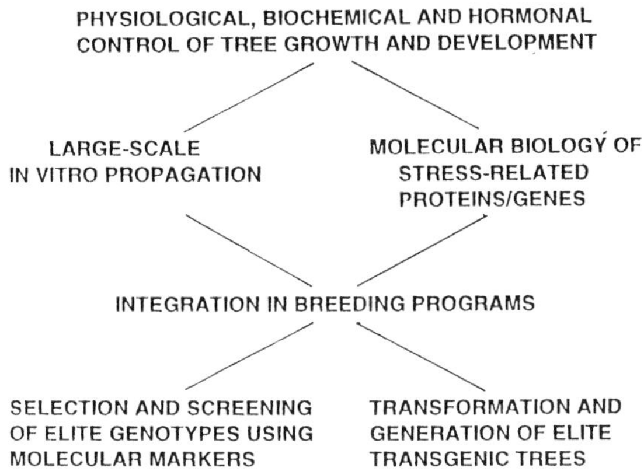

Fig. 1. The steps and key-methodolgies of an integrated biotechnological approach for the improvement of tree growth under stress conditions.

2.2. IN VITRO CULTURE AND REGENERATION

Shoot explants of aspen were routinely subcultured every 5-6 weeks on half-strength Murashige and Skoog basal medium (MS) supplemented with 500 mg/l casein enzymatic hydrolysate. The shoot cultures were maintained at 25^{0}C under a 16-h photoperiod using cool-white fluorescent light (50-60 mmol/m^2/sec). Callus was initiated from leaf sections and subcultured on a similar medium, supplemented with 1 mM 2,4-D, 4.4 mM BA, and 2.6 mM NAA.

For bud regeneration in *P. halepensis*, excised mature embryos were cultured in an inverted position on 0.5 MS medium supplemented with cytokinins and auxins, then placed in their normal position for further growth, and rooted (for additional details see Altman et al. [2]. Alternatively, the resulting adventitious buds served as a source for needles. Elongated needles were excised and placed horizontally on 0.5 MS medium with different concentrations of BA and/or ABA. In addition, 3-year-old pine trees were pruned and sprayed with BA or TDZ to induce fascicular buds, and the resulting elongated shoots were then used for in vitro culture and bud regeneration. For somatic embryogenesis, mature pine embryos were grown on DCR medium containing 500 mg/l casein hydrolysate and 50 mg/l L-glutamine [8], with 2,4-D at 1, 3, or 10 mg/l.

2.3. ANALYSIS OF WATER STRESS

Aspen shoots were subjected to water stress as described previously [19]. To induce osmotic stress in aspen callus, callus samples were cultured for 4 days in a callus induction medium containing 0, 6 or 12% mannitol, after which they were analyzed for stress proteins. Water potentials of the medium and the callus were determined using a vapor-pressure osmometer. For cold-stress treatments, callus cultures and shoot cultures were grown in the dark at 4^{0}C, for 1 or 4 days and for 4, 12 or 24 h, respectively. Shoot and callus cultures were frozen in liquid nitrogen, homogenized for protein extraction, and analyzed by gel electrophoresis of total and boiling-stable proteins and western blots, as described previously [19].

2.4. TRANSFORMATION

Both co-cultivation with *Agrobacterium rhizogenes* strain LBA 9402 (and other strains) and biolistics served for transformation of aspen and *P. halepensis*, as reported by Tzfira et al. [29, 31].

3. Results

3.1. IN VITRO CULTURE AND REGENERATION OF ALEPPO PINE

P. halepensis is highly recalcitrant to in vitro propagation, although some success has been achieved using mature embryos [11, 12]. We have previously studied explants of zygotic embryos, seedlings and cytokinin-treated mature trees of *P. halepensis* for their regeneration potential, and efficient procedures for massive induction of adventitious shoots from mature embryos, as well as for their rooting, were established [2]. Here we report additional results, including induction of a second cycle of adventitious buds from inverted mature embryos, regeneration from embryo and seedling explants, regeneration from elongated juvenile needles, growth and regeneration of cytokinin-induced fascicular buds, and attempts to induce somatic embryogenesis.

A second cycle of adventitious buds was successfully induced from the first cycle of buds formed on inverted mature embryos [2, 20]. Mature embryos and embryo explants were incubated on 0.5 MS medium with several combinations of NAA, 2,4-D, a phenoxy-auxin-glycine conjugate, and BA. Highest rates of adventitious shoots were induced on a medium containing 1 μM conjugate and 10 μM BA (Fig. 2). Buds were induced on this medium from both cotyledon explants and whole embryos, and from whole embryos only with several other combinations of growth regulators. Culture of explants from aseptically germinated seedlings resulted in adventitious shoot formation from apical explants only (containing the cotyledons), and at a higher rate on a medium containing 1 μM conjugate and 10 μM BA (Fig. 3).

Needles, available in large numbers on seedlings and trees, can serve as an alternative source for bud induction. This approach has been successfully utilized with mature *Picea abies* trees [10] and *P. sitchensis* [25]. Juvenile needles from *P. brutia* 12-months-old seedlings were also induced to form adventitious buds [1]. Needles were excised from embryo shoots that had been cultured for 8 months on 0.5 MS medium with 11 μM BA, and placed horizontally on 0.5 MS medium with different concentrations and combinations of BA and ABA. After 2 weeks, cell division and swelling occurred in the basal part of 70-80% of the needles, followed by callus growth. Although the survival rate of needles was low in several treatments, a considerable number of buds were induced from them, especially when cultured with 11 μM BA with or without 10 μM ABA (Table 1). One to four buds developed per needle, most of them during the 4th and 5th weeks of culture.

Cloning via micropropagation from mature and established pine trees has many advantages over the use of juvenile tissues, but is considerably more difficult. Three-year-old seedlings were pruned and sprayed several times with 200 mg/l BA or TDZ. A large number of fascicular shoots developed within 3-4 weeks after treatment. Spraying with either BA or TDZ induced morphologically different buds: shoots arising from the TDZ treatments were stunted and did not develop well, whereas BA induced well-developed and elongating shoots. These shoots are being examined for further growth in culture.

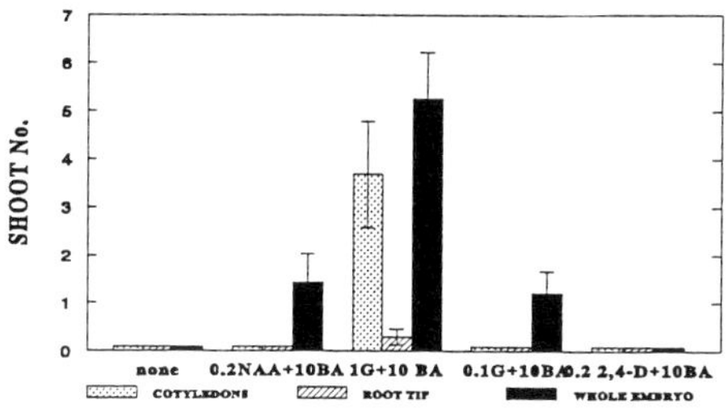

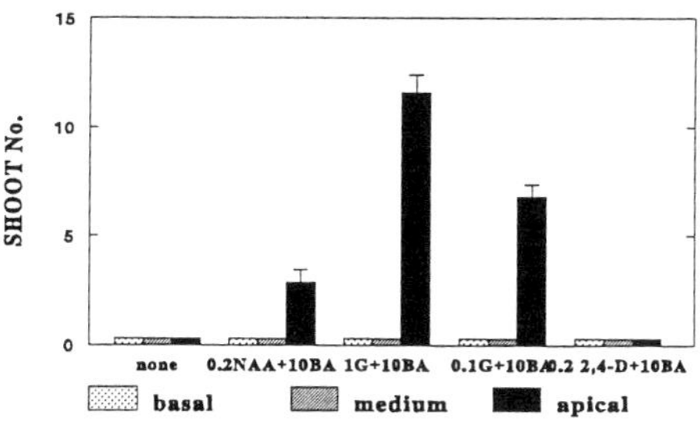

Fig. 2 (Left). Adventitious shoot formation in intact mature embryos and in embryo explants of *P. halepensis*. Whole embryos or the cotyledon and root half of the embryo were cultured for 30 days on media with the indicated concentrations (μM) of growth regulators. Data represent the average of four replicates, 30 embryos and explants each, ± SD.

Fig. 3 (Right). Adventitious bud development in seedling explants of *P. halepensis*. 21 day- old seedlings were excised into three types of hypocotyl explants: the basal part, the median part, and the upper part which also contained the whirl of cotyledons. All were cultured for 45 days on media with the indicated concentrations (μM) of growth regulators. Data represent the average of four replicates, 30 explants each, ± SD.

Table 1. Induction of adventitious buds on *P. halepensis* needles. Elongated needles were excised from embryos cultured in vitro for 8 months on 0.5 MS medium with 11 μM BA. The excised needles were cultured for 84 days on 0.5 MS media with the indicated concentrations (μM) of growth regulators. Necrosis and bud induction is expressed as percent of the total number of needles, and bud induction is further expressed as percent of surviving needles. Data are average of 60 needles per treatment.

GROWTH REGULATOS μM	None	1 ABA	10 ABA	11 BA	11 BA + 1 ABA	11 BA + 10 ABA
Necrosis, %	55	46.67	60	30	16.67	50
Buds, % induction	0	0	0	8.33	3.33	6.67
Buds, % of survived	0	0	0	12	4	13.33

In addition to the regeneration of shoot buds, we studied the conditions for somatic embryogenesis in Aleppo pine. Mature embryos were cultured on DCR medium containing 500 mg/l casein hydrolysate and 50 mg/l L-glutamine, and 2,4-D was added to the medium at 1, 3, or 10 mg/l. After 1 month in culture, embryos on 1 mg/l 2,4-D developed firm callus, whereas mucilaginous callus developed with 3 or 10 mg/l 2,4-D. Light-microscopic examination of the mucilaginous callus revealed many long vacuolated cells, typical of the preliminary stages of somatic embryogenesis in conifers [8]. Embryos with callus and separated calli were transferred to a medium without 2,4-D, but no further development was observed. The use of immature and mature embryos is being further studied.

3.2. ANALYSIS OF WATER-STRESS PROTEINS

We have previously shown that a novel 66 kDa boiling-stable protein is highly expressed in *P. tremula* shoots, as early as 1 h after gradual water loss and ABA application [19]. This protein accumulated during progressive water stress, and decreased upon rehydration. The N-terminal amino acid sequence of BspA was determined, and exhibited high homology to wheat germins. Moreover, in *P. popularis* and *P. tomentosa*, which differ in their drought tolerance, we found a positive correlation between the expression of dehydrin-like proteins (35 and 36 kDa) and sucrose synthase (90 kDa) and the degree of drought tolerance, as judged by immunological cross -reactivity and ion-leakage tests [21]. Similar boiling-stable proteins were differentially expressed in pine seedlings. In the following we present additional evidence for the effect of cold and osmotic stresses on protein expression.

When cells are subjected to osmotic stress, dehydration soon occurs, resulting in a decrease in the water potential of the cells. This is clearly indicated by the marked decrease in the water potential of aspen callus cultured for 4 days with 12% mannitol (Fig. 4). Medium containing 6% mannitol only slightly affected the osmotic potential of the cells, but this clearly corresponded to the negligible change in the water potential of the medium. Our previous studies, and others [3, 6, 22, 26] have suggested that boiling-stable proteins play a role in desiccation tolerance by protecting cell membranes and other cell proteins from water stress. Western blot analysis of the osmotically-stressed aspen callus, using anti-BspA antibodies, showed marked accumulation of the 66 kDa boiling-stable protein upon exposure to both severe (12% mannitol, -1.44 MPa) and mild (6% mannitol, -0.75 MPa) water stress (Fig. 5). This supports our previous results on the effect of drought (i.e. dehydration resulting from water loss from leaves) on expression of this unique protein in aspen shoots.

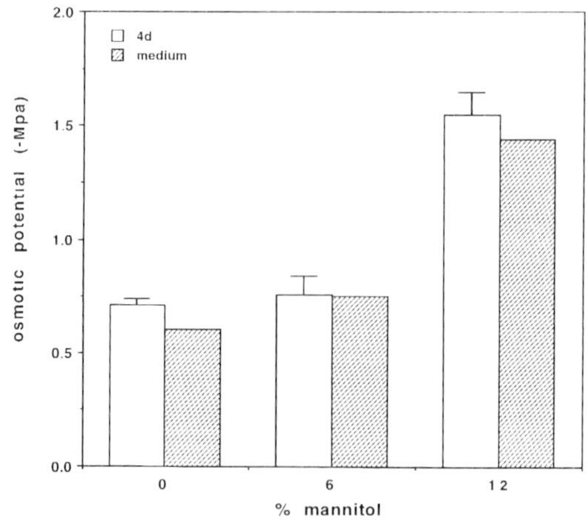

Fig. 4. Osmotic potential of the culture media supplemented with 6 or 12% mannitol, and of aspen callus cultured on these media for 4 days. Data represent the mean of three replicates, two calli (3 mm in diameter) each, per treatment , ±SD.

Cold stress may share common detrimental effects with drought, because both result in "physiological drought" due to water loss from cells, and in membrane damage. Plants vary widely in their responses to cold temperatures. Many tropical and subtropical species are "chilling sensitive", and they are injured or killed by exposure to low, non freezing temperatures. In contrast, plants from temperate regions generally acclimate to cold temperatures and may survive under these conditions. A number of plant genes are upregulated in response to low temperature, as shown by Cattivelly et al. [5].

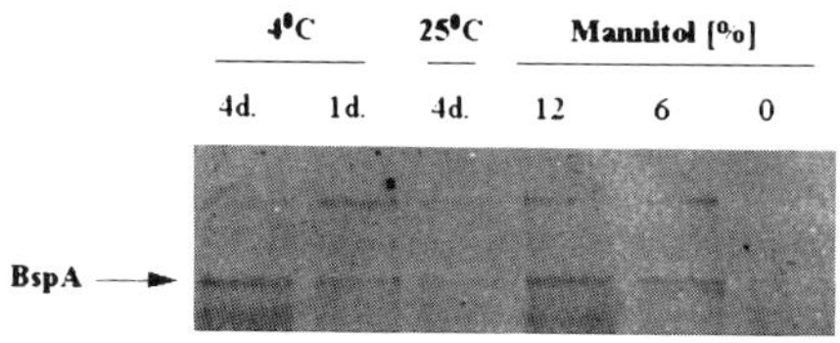

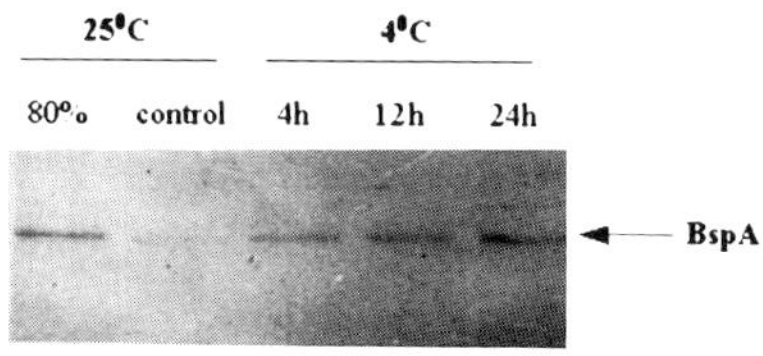

Fig. 5 (Left). Expression of the 66 kDa boiling-stable protein in aspen callus cultures in response to osmotic and cold stresses. Western blots, using BspA polyclonal antibodies (1:5000), of SDS gels of total proteins extracted from callus cultured with 0, 6, or 12% mannitol for 4 days, and from callus cultured at 25 or 4^0C for 1 or 4 days.

Fig. 6 (Right). Expression of the 66 kDa boiling-stable protein in aspen shoots in response to wilting and cold stress. Western blots, using BspA polyclonal antibodies (1:5000), of SDS gels of total proteins that were extracted from shoots wilted to 80% of their initial fresh weight and held in plastic bags for 4 h (lane 1), and from intact aspen plantlets that were kept at 4^0C for 4 (lane 3), 12 (lane 4), or 24 h (lane 5). Control explants were kept at 25^0C for 24 h (lane 2).

The water stress-responsive 66 kDa protein clearly accumulated in callus cultures subjected to 1 and 4 days of cold treatment (Fig. 5). Aspen shoots that were cold-stressed for 4, 12 or 24 h showed a gradual accumulation with time of this protein, similar to its expression in wilted aspen shoots (Fig. 6). The expression of stress proteins has been studied mainly in herbaceous plants, and a few proteins have been found to accumulate due to osmotic, drought and cold stress [18, 23]. It is thus of interest that similar proteins are expressed in woody plants in response to water stress, cold stress, and ABA application, as found in *Populus* spp. in our studies.

3.3. POPLAR AND PINE TRANSFORMATION

As indicated in the introduction (see also Fig. 1), an efficient transformation procedure and the establishment of transgenic plants constitute a necessary part of comprehensive breeding and engineering programs for forest trees, or for that matter any plant. This novel avenue is currently more routinely exploited in herbaceous plants, and it still presents great difficulties in most trees. It is thus of interest to note that *Populus* is becoming the model tree of choice for genetic engineering [7, 14, 24]. Nevertheless, improved procedures for poplar transformation are being sought, and efficient methods have to be developed for other forest tree species. Our studies on transformation of *P. tremula* and *P. halepensis* are discussed by Tzfira et al. [31] and elsewhere [29].

4. Discussion and concluding remarks

Pinus and *Populus* contain many species and clones, comprising a large proportion of world forests and adapted to many regions and climatic conditions. Hence these trees are important, both economically for timber production and as major components of the ecosystem. Their very wide distribution range, together with their wealth of species, subspecies, ecotypes and clones, offer a large gene pool for coordinated breeding programs. Enhanced tree growth under abiotic and biotic stress conditions is one of the major goals for poplar and pine improvement. Thus, the combination of conventional breeding and selection techniques, rapid clonal propagation, and novel molecular and biotechnological methodologies can become useful both for determining the physiological performance of the trees, and for rapid selection and tree improvement. In vitro clonal propagation, the study of drought-specific proteins and genes, and genetic transformation of *Populus* species and *P. halepensis* have been the focus of our studies.

Micropropagation of many, but not all *Populus* species, is relatively easy, mainly by enhancement of axillary bud growth and adventitious bud formation from various types of shoot explants. The remaining problems include: rapid in vitro clonal propagation of recalcitrant *Populus* species which also do not root well from cuttings, and establishment of economically efficient, large-scale in vitro propagation systems using alternative procedures and explant types. The latter is being studied in our laboratory, employing root cultures in liquid media and transformed phenotypes (Carmi, Vinocur, Ziv and Altman, unpublished data). *Pinus* species, including *P. halepensis*, are much more recalcitrant to in vitro propagation, especially from non-juvenile tissues. Even the use of juvenile explants, including those of embryos and seedlings, may present difficulties as far as a high propagation coefficient is concerned. Cloning of a large number of plantlets from individual mature zygotic seeds is one way to alleviate these difficulties. Our studies show a high rate of adventitious bud induction from mature embryos of *P. halepensis* and their rooting, resulting in the potential establishment of 100-1000 plantlets from a single embryo, when serial bud induction cycles are employed. The use of elongated needles for bud induction, and the possibility of using fascicular buds from cytokinin-treated trees, further increase the efficiency of in vitro clonal propagation of the Aleppo pine.

Water loss from plant tissues under drought conditions results in growth inhibition and a number of metabolic and physiological changes. These include accumulation of ABA [32], stomatal closure resulting in reduced transpiration rates, a decrease in the water potential of plant tissues, and decreased photosynthesis [4, 9, 28]. Changes in gene expression in response to water deficits, as evidenced by the synthesis of new proteins and mRNAs, have been recently studied, and a number of genes responsive to water stress and ABA treatments, identified as LEAs (late embryogenesis abundant proteins), RABs (responsive to ABA) and dehydrins, have been characterized [3, 22]. These proteins are highly hydrophilic (>55% hydrophilic residues), remain soluble after boiling, and contain a low amount of C and W residues with the presence of repeating motifs. These proteins are postulated to function in desiccation tolerance [26]. Similar proteins accumulate during cold stress [13, 27], suggesting that a major component of freezing tolerance involves tolerance to dehydration. These studies are based solely on studies with herbaceous plants, and the molecular drought-avoidance characteristics of many commercially important woody plants have not been documented. Efforts to improve growth and productivity of forest trees under stress conditions could benefit from an understanding of the expression of specific drought-induced proteins, which may contribute to drought tolerance. Further to our previous studies, additional evidence is presented here on the accumulation of this protein in response to osmotic and cold stresses. Furthermore, a good correlation between the expression of dehydrin-like proteins and sucrose synthase, and the degree of drought tolerance and ion leakage in *Populus* clones differing in their drought tolerance, was found. Further analysis of the expression of these proteins and their mode of action will allow us to better understand its role in drought tolerance of trees. Cloning and sequencing of the relevant gene has already been achieved in our laboratory, and its function(s) and expression patterns are being studied.

The improved procedures for *Populus* transformation and establishment of transgenic tree clones, as well as transformation of *P. halepensis*, are described elsewhere in this volume [31]. This will form the basis for additional studies on expression of the drought-specific gene, and perhaps other genes of interest, in transformed poplar and pine species.

5. Acknowledgments

This research was supported in part by the German-Israeli Agricultural Research Agreement for the benefit of the Third World (GIARA), by the Land Development Authority (Keren Kayemet LeIsrael), and by a grant from the G.I.F., the German-Israeli Foundation for Scientific Research and Development.

6. References

1. Abdullah A.A. and Grace J. (1987). Regeneration of calabrian pine from juvenile needles. Plant Sci. 53, 147-155.
2. Altman, A., Ya'ari, A., Pelah, D., Gal, A., Tzfira, A., Wang, W-X. , Shoseyov, O., Vainstein, A. and Riov, J. (1995). In vitro organogenesis, transformation and expression of drought-related proteins in forest tree cultures. In: M. Terzi et al., eds., Current Issues in Plant Molecular and Cellular Biology, Kluwer Academic Publ., pp. 87-94.
3. Baker, J.C., Steele, C. and Dure, L III. (1988). Sequence and characterization of 6 lea proteins and their genes from cotton. Plant Mol. Biol. 11, 277-291.
4. Boyer, J.S. and Bowen, B.L. (1970). Inhibition of oxygen evolution in chloroplasts isolated from leaves with low water potentials. Plant Physiol. 45, 612-615.
5. Cattivelli, L. and Bartels, D. (1990). Molecular cloning and characterization of cold-regulated genes in barley. Plant Physiol. 93, 1504-1510.
6. Close, T.J., Kortt, A.A. and Chandler, P.M. (1989). A cDNA based comparison of dehydration-induced proteins (dehydrins) in barley and corn. Plant Mol. Biol. 13, 95-108.
7. Fillatti J.J., Sellmer J., McCown B., Haissig B, and Comai L. (1987). Agrobacterium mediated transformation and regeneration of Populus. Mol. Gen. Genet. 206, 192-199.
8. Gupta P.K and Durzan D.J. (1986). Somatic polyembryogenesis from callus of mature sugar pine embryos. Bio/Tech. 4, 643-645.
9. Hsiao, T.C. (1973). Plant responses to water stress. Annu. Rev. Plant Physiol. 24, 519-570.
10. Jansson E. and Bornman C.H. (1981). In vitro initiation of adventitious structures in relation to the abscission zone in needle explants of Picea abies: anatomical considerations. Physiol. Plant. 53, 191-197.
11. Lambardi M., Sharma K.K. and Thorpe T.A. (1991). In vitro regeneration of shoot buds and plantlet formation from mature embryos of Pinus halepensis Mill. Acta Hort. 289, 123-124.
12. Lambardi M., Sharma K.K. and Thorpe T.A. (1993). Optimization of in vitro bud induction and plantlet formation from mature embryos of Aleppo pine. In Vitro Cell. Dev. Biol. 29, 189-199.

13. Lin, C., Gau, W.W., Everson, E. and Tomashow, M.F. (1990). Cold acclimation in Arabidopsis and wheat. A response associated with expression of related genes encoding "boiling-stable" polypeptides. Plant Physiol. 94, 1078-1093.
14. McCown, B.H., McCadeD.E., Russell D.R., Robinson, D.J., Barton, K.A., Raffa, K.F. (1991). Stable transformation of *Populus* and incorporation of pest resistance by electric discharge particle acceleration. Plant Cell Rep. 9, 590-594.
15. Nadel, B.L., Altman, A., Pleban, S. and Huttermann, H. (1991a). In-vitro development of mature Fagus sylvatica L. buds. I. The effect of medium and plant regulators on bud growth and protein profiles. J. Plant Physiol. 138, 596-601.
16. Nadel, B.L., Altman, A., Pleban, S., Kocks, R. and Huttermann, A. (1991). In vitro development of mature Fagus sylvatica L. buds. II. Seasonal changes in the response to plant growth regulators. J. Plant Physiol. 138, 136-141.
17. Nadel, B.L., Hazan, G., David, R., Altman, A. and Huttermann, A. (1992). In vitro propagation of Populus species:responses to growth regulators and medium composition. Acta Horticulturae 314, 61-68.
18. Neven, L.G., Haskell, D.W., Hofig, A., Li, Q-B. and Guy, C.L. (1993). Characterization of a spinach gene responsive to low temperature and water stress. Plant Mol. Biol. 21, 291-305.
19. Pelah, D., Shoseyov, O. and Altman, A. (1995). Characterization of BspA, a major boiling-stable, water-stress-responsive protein in aspen (Populus tremula). Tree Physiol. 15, 673-678 .
20. Pelah, D., A. Ya'ari, A. Altman, O. Shoseyov and Riov, J. (1995). Growth, in vitro propagation and desiccation-specific proteins of Populus and Pinus cultures. In S. Wang and X. Jiang (eds.), Growth and Development Control and Biotechnology in Woody Plants, Proc. IUFRO Workshop, Beijing 1993, China Forestry Publishing House, Beijing, pp 75-89.
21. Pelah, D., Wang, W-X., Altman, A., Shoseyov, O., and Bartels, D. (1996). Differential accumulation of water-stress related proteins and soluble sugars in Populus genotypes which differ in their water-stress tolerance (submitted).
22. Piatkowsky, D., Schneider, K., Salamini, F. and Bartels, D. (1990). Characterization of five abscisic acid-responsive cDNA clones isolated from the desiccation-tolerant plant Craterostigma plantagineum and their relationship to other water-stress genes. Plant Physiol. 94, 1682-1688.
23. Plant, A.L., Cohen, A., Moses, M.S. and Bray, E.A. (1991). Nucleotide sequence and spatial expression pattern of a drought and abscisic acid-induced gene of tomato. Plant Physiol. 97, 900-906.
24. Pythoud F., Sinkar V.P., Nester E.W. and Gordon M.P. (1987). Increase virulence of Agrobacterium rhizogenes conferred by the vir region of pTiBo542: application to genetic engineering of poplar. Bio/Tech. 5, 1323-1327.
25. Selby C. and Harvey B.M.R. (1990). The influence of composition of the basal medium on the growth and morphgenesis of cultured sitka spruce (Picea sitchensis) tissue. Ann. Bot. 65, 395-407.
26. Skriver, K. and Mundy, J. (1990). Gene expression in response to abscisic acid and osmotic stress. Plant Cell 2, 503-512.
27. Thomashow, M.F. (1990). Molecular genetics of cold acclimation in higher plants. Adv. Genetics 28, 99-131.
28. Turner, N.C. (1986). Adaptation to water deficits: A changing perspective. Aust. J. Plant Physiol. 13, 175-190.
29. Tzfira, T., Ben-Meir, H., Vainstein, A., and Altman, A. (1995). Highly efficient transformation and regeneration of aspen plants through shoot-bud formation in root cultures. Plant Cell Rep. (in press).
30. Tzfira, T., Yarnitzky, O., Vainstein, A., and Altman, A. (1996). Agrobacterium rhizogenes-mediated DNA transfer in Pinus halepensis Mill. (in preparation)
31. Tzfira, T., Yarnitzky, O., Vaistein, A., and Altman, A. (1996). Highly efficient transformation and regeneration of transgenic aspen plants through shoot-bud formation in root culture, and transformation of Pinus halepensis. (this volume).
32. Zeevaart, J.A.D. and Creelman, R.A. (1988). Metabolism and physiology of abscisic acid. Annu. Rev. Plant Physiol. Plant Mol. Biol. 39, 439-473.

MOLECULAR APPROACHES TO MATURATION-CAUSED DECLINE IN ADVENTITIOUS ROOTING ABILITY IN LOBLOLLY PINE (*Pinus taeda L.*)

Carmen Diaz-Sala[1], Patricia B. Singer[1], Keith W. Hutchison[1] and Michael S. Greenwood[2]

[1] Department of Biochemistry, Microbiology and Molecular Biology, University of Maine, Orono 04469-5735 USA and [2] Department of Forest Ecosystem Science, University of Maine, Orono 04469-5755 USA.

ABSTRACT

New approaches towards understanding the control of adventitious root formation include studies at the level of gene expression. We discuss attempts to determine the identity of as yet unknown genes which are specific to formation of roots, as well as studies on the expression of genes known to be involved in the wound response. A number of genes appear to be specific to adventitious root formation in several species including loblolly pine, and the expression of these genes changes with maturation.
Keywords: adventitious rooting, gene expression, loblolly pine

1. Introduction

Loss of adventitious rooting ability with maturation in woody plants is a long-standing problem, since vegetative propagation of mature, select trees becomes increasingly difficult as the tree develops. There are currently several industrial pilot programs for the vegetative propagation of such species as loblolly pine, Douglas fir and black spruce using rooted cuttings, and interest in the biology of adventitious root formation remains high. Recently, there have been several reports of the use of molecular methods to examine gene expression during adventitious root formation.

Useful insights into the role of gene expression in the adventitious rooting process have been obtained during the induction of adventitious roots by auxin in 6-day-old sunflower (*Helianthus annuus* L.) seedlings. Using 2-dimensional gel electrophoresis Oliver et al. (1994) demonstrated that after a cutting is made, at least 50 novel proteins are synthesized at the site of root formation within 10 hours. It is not known if the presence of these proteins is a function of regulation at the transcriptional or translational level.

Further work comparing cDNA libraries from cuttings at time 0, 10 and 24 hours after the cutting is made has resulted in detection of about 39 genes which may be uniquely expressed during the rooting process (Oliver et al. 1994). Of these, about 8 appear to be specific to the formation of adventitious roots or phloem fibers, and some are in the process of being characterized. All but one of the genes is expressed

M. R. Ahuja et al. (eds.), Somatic Cell Genetics and Molecular Genetics of Trees, 57–61.
© 1996 *Kluwer Academic Publishers. Printed in the Netherlands.*

in maturing primary root tissues, and are probably associated with the later stages of root formation. The remaining gene is a putative peroxidase, and is expressed specifically in primary and secondary root tips. Identifying genes that are specific to the rooting process is only a first step in the process of trying to understand how they and their products affect the process of root regeneration. Since many more genes may be identified, some of unknown function, this approach will be laborious and time-consuming.

Comparing gene expression in tissues from the same plant which have lost or retain the competence to form roots is another promising approach to the long-standing problem of decline in adventitious rooting ability with maturation. This problem has recently been approached at the molecular level by comparing gene expression in hard-to-root mature cuttings with easy-to-root juvenile cuttings from the petioles of English ivy from the same plant. Woo et al.(1995) screened 200,000 cDNA clones prepared from juvenile and mature petioles with and without auxin (NAA, α-naphthalene-acetic acid) treatment, and found 2 genes which were differentially expressed between juvenile and mature tissues. The mRNA from one of the genes (HW101, similar to a chlorophyll a/b binding protein) is expressed at higher levels (without auxin treatment) in juvenile petioles than mature petioles. NAA treatment suppressed expression in juvenile but not mature petioles. The mRNA from the other gene (HW103, similar to a cell wall proline rich protein gene) is expressed at a higher level in mature than juvenile petioles. NAA suppressed production only in chronologically young mature petioles. The location of expression of the HW103 gene is consistent with an inverse relationship between expression of this gene and the ability to form adventitious roots (Sanchez et al., 1995). We can tentatively conclude that failure to root in English ivy may involve only a small number of genes, or that the genes involved are expressed at levels too low to be detected by differential hybridization.

Using 50-day-old seedlings of loblolly pine, we have compared gene expression in hypocotyl cuttings, which root in 20-30 days only when auxin treated, with that of epicotyl cuttings, which do not root at all in this time period, with or without auxin (Greenwood and Weir, 1995, and unpublished data). A similar-sized cutting can be obtained from both the hypocotyl and epicotyl of each 50-day-old seedling.

2. Pal Expression and the Decline of Adventitious Rooting Potential

Wound response genes are among the first genes to respond after the cutting is made. Also, additional wounding at the site of the cut stimulates the rooting response (Diaz-Sala and Greenwood, unpublished data). Therefore, we have compared the wound response in terms of the expression of a PAL (phenylalanine ammonia-lyase) gene in both epicotyl and hypocotyl cuttings. Northern blots were performed, and PAL expression was detected with a loblolly pine cDNA probe. These data were normalized relative to the expression of ribosomal RNA.

Both types of cutting exhibit PAL expression as a consequence of wounding. However, the level of PAL expression is different depending on the cutting type and whether there is additional wounding at the cutting base. Without additional wounding the level of PAL expression is similar for both types of cutting but the peak level is reached sooner in epicotyls (3-6 hours) than in hypocotyls (6-12 hours). Higher levels of PAL are seen in both cutting types when additional wounding is performed, but the response is less in epicotyls. The patterns of expression were similar whether or not the cuttings were treated with the auxin IBA (indole-3-butyric acid), without which hypocotyl cuttings do not root. In addition, the phytotropin NPA (N-(1-napthyl)pthalamic acid) completely blocks rooting in hypocotyls when applied in combination with auxin at equimolar concentrations, but does not have an effect on PAL expression. These results indicate that rooting competent and incompetent cuttings do respond somewhat differently to wounding in terms of PAL expression. However, since IBA and NPA both affect rooting without changing PAL expression patterns, we conclude that there is probably no direct correlation between the loss of rooting competence and the wound response 24 hours after wounding.

3. Actin Expression and the Decline of Adventitious Rooting Potential

The determination of division planes and the establishment of cell elongation axes are crucial features in the coordination of the organization of a new meristem. In this respect, the cytoskeleton is of particular interest since any signal inducing a polarized developmental event, such as root regeneration, will very likely affect cytoskeletal organization and composition. Actin is a major component of cytoskeleton filaments, so we have examined actin expression using methods similar to those used for PAL, to see if its expression is correlated with competence to form root meristems in response to auxin.

Expression of an actin gene was followed using an homologous probe obtained by RT-PCR, and shows a closer correlation with rooting competence than that shown by PAL. No differences in the expression were detected in hypocotyl cuttings from 50-day-old seedlings during the first 24 hours of treatment. However, a higher level of expression was observed in hypocotyl cuttings from 20-day-old seedlings that root faster than cuttings from 50-day-old seedlings. Expression of actin is slightly higher in the presence of auxin in hypocotyl cuttings from 20-day-old seedlings, but the increase is blocked by the addition of NPA. In contrast, epicotyls showed a decrease in the level of the gene expression with and without auxin. Our results suggest that actin gene expression is not only a function of wounding but may be associated with competence to regenerate adventitious roots. Studies are currently underway with 2 other cytoskeletal genes, tubulin and myosin.

4. Identification of Differentially Expressed Genes

The technique of differential display permits the identification of species of mRNAs that are expressed at different levels in different tissue types, or in response to different treatments(Liang and Pardee 1992). mRNA is reverse transcribed into cDNA using an anchored oligo-dT-primer containing 2 additional bases at its 3' end. Theoretically, one twelfth of the population of mRNAs will be transcribed into cDNA with each different anchored oligo-dT primer. The cDNA is then amplified using either one or two random 10mer primers and the addition of a radioactive dNTP. The products of the amplification are electrophoresed. Differentially expressed sequences will be present in one lane but not in others (Lang and Pardee 1992).

RNAs isolated from loblolly pine hypocotyls and epicotyls at time zero and at 24 hours after treatment with 10μM IBA were analysed. Also included were untreated samples at 24 hours. Typically 80-90 bands per lane were observed with most reactions. However, taking into account the possibility of multiple bands for one product due to separate strands running at different locations as well as each strand potentially containing an additional A, the number decreased to about 50-60 bands per lane. In comparing the RT-PCR products from hypocotyl and epicotyl RNA at time zero, about 90% of the bands were present in both samples.

Thus far we have isolated 35 PCR products that were produced off RNA from specific treatments or tissues. Initially, sequences found in the treated hypocotyls at 24 hours but not in epicotyls at 24 hours were selected for further study. The fragments of interest were cloned, sequenced and analyzed by differential display southern blots. We have now begun to analyze the clones by the RNase protection assay in order to get a more quantitative measure of expression between the RNA samples.

5. CONCLUSIONS

There are significant changes in gene expression associated with the loss of rooting ability in both English ivy and loblolly pine. More work is needed to determine whether or not these changes are the primary cause of loss of rooting ability, and whether they are generally applicable to other changes associated with maturation.

6. ACKNOWLEDGEMENTS

This work was supported by Grants from the Maine Agriculture and Forestry Experiment Station and the North Carolina State Loblolly and Slash Pine Rooted Cutting Project to MSG and KWH. CD-S was supported by a fellowship from Comite Conjunto para la Cooperacion Cultural y educativa (Spain-USA) and Instituto de fomento para la Investegacion cientifica y technica (Principado de Asturias, Spain). The authors wish to thank Ross Whetten for the loblolly pine PAL cDNA probe. MAFES publication #1967.

LITERATURE CITED

Greenwood, M.S., and R.J. Weir 1995. Genetic variation in rooting ability of loblolly pine cuttings: effects of auxin and family on rooting by hypocotyl cuttings. Tree Physiology 15: 41-45.

Liang, P. and A.B. Pardee 1992. Differential display of eukaryotic messenger RNA by means of the polymerase chain reaction. Science 257:967-971.

Oliver, M.J., J.H. Liu, I. Mukherjee, and D.M. Reid 1994. Molecular investigations into adventitious root initiation in sunflower seedlings. Plant Physiol. 105: 186.

Oliver, M.J., I. Mukherjee, and D.M. Reid 1994. Alteration in gene expression in hypocotyls of sunflower (*Helianthus annuus*) seedlings associated with derooting and formation of adventitious root primordia. Physiol. Plant. 90: 481-489.

Sanchez, M.C., A.G. Smith, and W.P. Hackett 1995. Localized expression of a proline-rich protein gene in juvenile and mature ivy petioles in relation to rooting competence. Physiol. Plant. 93: 207-216.

Woo, H.H., W.P. Hackett, and A. Das 1994. Differential expression of a chlorophyll *a/b* binding protein gene and a proline rich protein gene in juvenile and mature phase English ivy (*Hedera helix*). Physiol. Plant. 92: 69-78.

A COMPARATIVE STUDY OF EMBRYOGENIC AND NON EMBRYOGENIC CELL CULTURES IN *PICEA ABIES* (L.) KARST.

Identification of Mitochondrial DNA Polymorphisms and Detection of Differentially Expressed Polypeptides

P. DU JARDIN[1], A. CHANDELIER[1], F. CHAUVEAU[2], J. BERCETCHE[2], M. PAQUES[2]

[1] *Unité de Biologie végétale, Faculté des Sciences agronomiques de Gembloux, Avenue Maréchal Juin, Bât 48, B-5030 Gembloux, Belgium;* [2] *Station de Biotechnologies, AFOCEL, Domaine de l'Etançon, F-77370 Nangis, France*

Abstract

Embryogenic cell cultures (or 'ESMs', for embryonal-suspensor masses) and non embryogenic cultures (or 'NEC', for non embryogenic calli) have been initiated from young cotyledons in Picea abies *(L.) Karst. and separately multiplied for over two years. In order to better understand their physiological peculiarities, two molecular approaches were followed. First, bidimensional electrophoreses were used for comparing the polypeptide patterns in ESMs and NEC. Multiple extracellular polypeptides appeared specific to the embryogenic cell cultures, while several intracellular polypeptides were only observed in the non embryogenic calli, cultivated in the light or in darkness. Second, a mitochondrial DNA library was constructed in* Picea abies *and 168 plasmid clones were used as RFLP probes for comparing the structure of the mitochondrial genome in the two types of cell material. Two probes detected extensive and qualitative DNA polymorphisms between embryogenic and non embryogenic cell cultures. Interestingly, the mtDNA pattern in ESMs was distinct from that of the corresponding mother plants (as defined in buds and needles), but apparent reversion to the mother plant pattern was observed upon embryo maturation. When comparing ESMs or NEC between genotypes, the two mtDNA probes detected extensive polymorphism, while the protein markers seemed to be genotype-independent.*

1. Introduction

In *Picea abies*, embryogenic cell cultures can be initiated from zygotic embryos (Hakman et al., 1985; Chalupa, 1985; Gupta and Durzan, 1986; von Arnold, 1987) and cotyledons (Lelu *et al.*, 1987; Bercetche, 1989; Ruaud *et al.*, 1992) with a good efficiency. However, in order to fully integrate this method of genotype amplification within breeding programs, it is necessary to obtain somatic embryogenesis from non-juvenile material. The genetic and physiological bases of explant competence to initiate embryogenic cell cultures still

M. R. Ahuja et al. (eds.), Somatic Cell Genetics and Molecular Genetics of Trees, 63–68.
© 1996 *Kluwer Academic Publishers. Printed in the Netherlands.*

remains unclear. In *Picea abies*, the long generation time and the large genome size prevent the genetic dissection of embryogenic competence by mutant isolation and mapping, but cellular and molecular studies are susceptible to provide markers for assessing genotype and explant capacity to embryogenesis.

In this work, embryogenic and non embryogenic cell cultures have been established from cotyledons of zygotic plantlets and analysed at both the protein and mitochondrial DNA levels. Both types of cultures were initiated from the same cotyledon and the starting seedling could be preserved. Hence, markers of the embryogenic state of the cultured cells could be searched for by comparing embryogenic and non embryogenic cultures from the same genetic background, and their expression investigated in the corresponding mother- and regenerated plants.

2. Materials and Methods

2.1. PLANT MATERIAL

The protocol used for obtaining somatic embryos from spruce seedlings and regenerated plants was decribed elsewhere in more details (Bercetche, 1989 ; Ruaud *et al.*, 1992). Briefly, cotyledons from 5-day-old plantlets are placed on induction media and produce two types of cell cultures : (1) embryonal-suspensor masses » (ESMs), which contain immature embryos equivalent to the initial developmental stages of zygotic embryos, and (2) non embryogenic calli (NEC). The ESMs differentiate on the primary calli which give rise to NEC. Later, ESMs and NEC are separated and cultivated independently for extended periods (Chauveau *et al.*, 1995). Based on this protocol, two types of genotypes (seedlings) are defined : a genotype for which at least one cotyledon differentiated embryogenic tissues is termed embryogenic ; in the opposite case, the genotype is termed non embryogenic. ESMs can be induced to mature into cotyledonary embryos, which can germinate and be converted into plantlets, as described in Bercetche *et al.* (1993). In the present study, five genotypes (provenance Gerarmer, Vosges, France) have been used to initiate *in vitro* cell cultures (table 1). Two embryogenic genotypes produced ESMs, cotyledonary embryos and regenerated plants. The three others are non embryogenic (produced NEC only).

2.2. PROTEIN ANALYSIS

Total proteins were extracted by TCA-acetone precipitation after grinding the plant material in liquid nitrogen (Granier, 1988). Extracellular proteins were recovered from ESMs and NEC essentially as described by Domon *et al.* (1994). Both protein extracts were solubilized in urea/ampholine buffer and subjected to two-dimensional electrophoresis using standard procedures.

2.3. MITOCHONDRIAL DNA ANALYSIS

A mitochondrial DNA library was previously constructed in Norway spruce, after purification of mitochondria from embryogenic cultures (Chandelier and du Jardin, unpublished). Southern-blot analysis was performed by probing the DNA samples, extracted from tissues, cells or isolated mitochondria, using a modified CTAB isolation protocol, with the ^{32}P-labelled mtDNA inserts from the plasmid library.

Table 1. *Plant and cell material used in the present study*

GENOTYPE	M	E S1	E S2	E S3	NE S1	NE S2	NE S3	R37	R100
30701	+	+	+	+	+	+	+	+	+
20719	+	+	+	+	+	+	+	+	-
30601	+	-	-	-	+	+	+	-	-
10603	+	-	-	-	+ (a,b)	+ (a,b)	+ (a,b)	-	-
10504	+	-	-	-	+	+	+	-	-

M, mother plant ; E, embryonal-suspensor mass ; NE, non embryogenic callus, R37 and R100, plants regenerated from subcultures 37 and 100 of ESMs. a, b indicate two phenotypically distinct non embryogenic calli in genotype 10603. S1, S2, S3 correspond to three different long-term subcultures. +/- means that the material is /is not available in the corresponding genotype.

3. Results

3.1. EMBRYOGENIC AND NON EMBRYOGENIC CELL CULTURES CAN BE DISTINGUISHED BY THEIR PROTEIN PATTERNS.

When total soluble proteins from ESMs and non embryogenic calli were compared by two-dimensional gel electrophoresis (figure 1A), most of the steady-sate proteins appear to be common to the two culture types. However, at least two polypeptides with apparent molecular masses of 35 kDa and with close pI values were specific to the non embryogenic cultures. Interestingly, these polypeptides were observed independently of the genotype, the subculture and the light conditions (not shown). They thus constitute valuable markers of the non embryogenic state of the cultures.

Polypeptides specific to the ESMs were evident from the 2-D separation of the extracellular proteins (figure 1B). It is worth noting that these proteins were recovered in a saline buffer and thus include proteins ionically bound to the cell surface. Multiple sets of polypeptides with identical apparent molecular masses and close pI values were detected. These polypeptides were found in the two genotypes for which ESMs were obtained, as well as in distinct subcultures. They thus constitute interesting markers of the embryogenic state of the cell cultures.

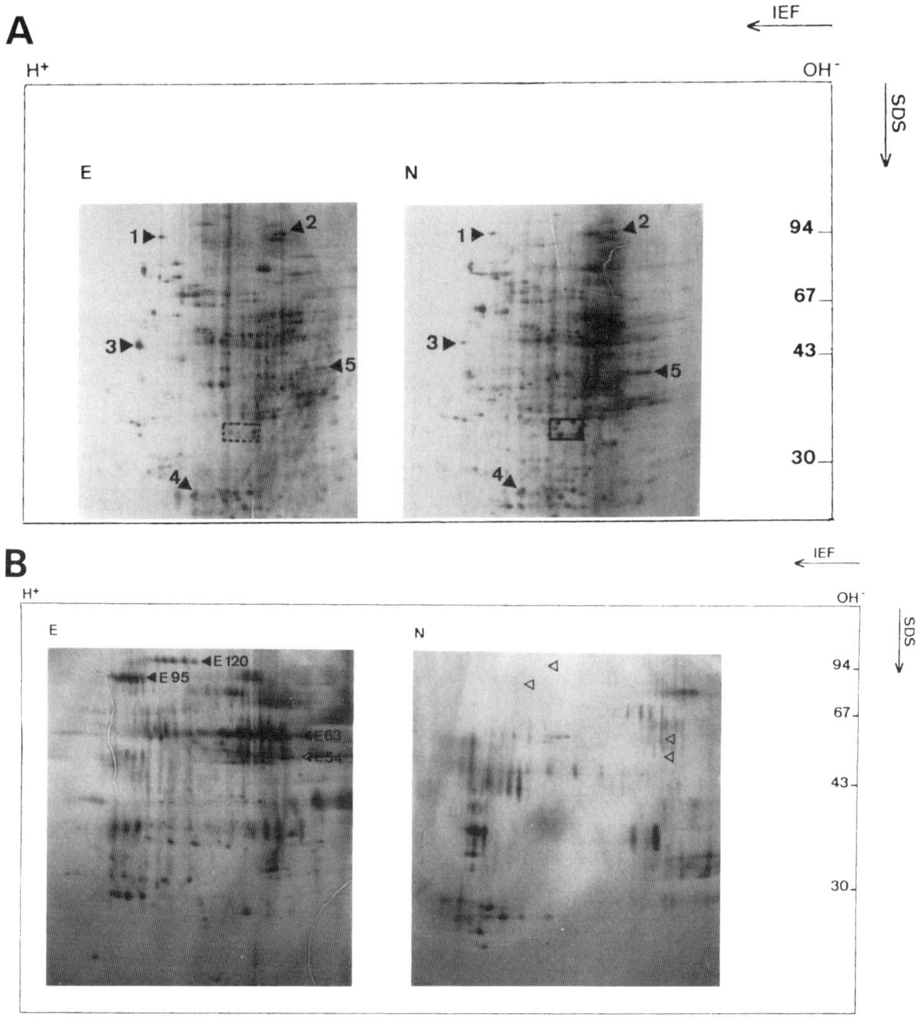

Figure 1. A. Two-dimensional electrophoretic separaration of total proteins from embryogenic (E) and non embryogenic (N) cell cultures. The gel area with the differentially expressed polypeptides is included in the rectangle. Numbered spots are not commented in the text. **B.** Two-dimensional electrophoretic separaration of extracellular proteins from embryogenic (E) and non embryogenic (N) cell cultures.The positions of the polypeptides specifically found in the embryogenic cultures are indicated by the filled triangles, and their corresponding positions in the non embryogenic pattern are indicated by the open triangles. A molecular mass scale (kDa) is indicated at the right of each panel.

3.2. EMBRYOGENIC AND NON EMBRYOGENIC CELL CULTURES CAN BE DISTINGUISHED BY THEIR MTDNA PATTERNS.

Total DNA samples were extracted from embryogenic and non embryogenic cultures and searched for restriction fragment length polymorphisms, using 168 mtDNA probes from a plasmid library previously constructed in *Picea abies*. Most of the probes did not show any RFLP between the two cell cultures, incidentally indicating that the digestibility of the DNA was equally good with both extracts. Two probes however, called mh44 (figure 2) and mp6 (not shown), showed qualitative RFLPs between embryogenic and non embryogenic cultures (compare adjacent lanes marked E and NE in figure 2). The mtDNA patterns proved to be stable when three subcultures of the same cell line were compared, the first and last subculture being spaced by 13 weeks. Interestingly, the pattern characterizing the ESMs was shown to be different from the mother plant pattern, as defined in needles and buds (figure 2 and data not shown). The question as to whether the plants regenerated from the ESMs contain the ESM mtDNA or the mother plant mtDNA has thus to be raised. For the two embryogenic genotypes, the regenerate plants were shown to contain the mother plant pattern (figure 2), indicating that the mtDNA modifications detected by the two probes in the ESMs are reversed when maturation, germination and conversion are achieved. Qualitative changes in band intensities were found however, which deserves further analysis. Furthermore, in genotype 30701, cotyledonary embryos were Southern-analyzed and shown to contain the mother plant type mtDNA, indicating that the apparent reversion of the mtDNA changes occured during embryo maturation.

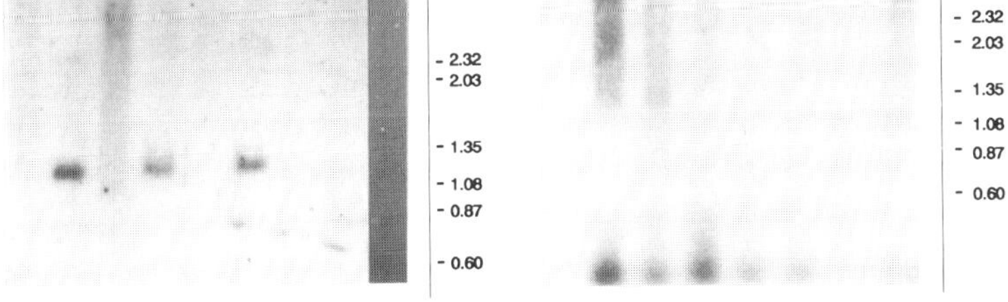

Figure 2. Southern hybridization of total DNA from embryonal-suspensor masses (E), non embryogenic calli (NE), mother plants (M), regenerated plants (R) and cotyledonary embryos (ES) in genotypes 30701 and 20719, with the mtDNA probe mh44. S1, S2 and S3 correspond to three different long-term subcultures. Molecular weight scales are positioned at the right of each panel. The DNA samples were digested to completion by ClaI and run in 0.8 % agarose.

4.Conclusion

The present study was aimed at identifying molecular polymorphisms between embryogenic and non embryogenic cultures in *Picea abies*. Both approaches - the description of the steady-state protein contents and that of the mtDNA structures - were successful in that respect. Polypeptides specific to the embryogenic and non embryogenic cultures were identified, and proved to be independent of the genotype, the subculture number and the light treatment. In contrast, the mtDNA polymorphisms identified were genotype-specific. The highly dynamic nature of the mtDNA during somatic embryogenesis prompts us to further study the conditions for the true-to-type amplification of spruce genotypes with respect to their mitochondrial genome, as well as to define the physiological significance of the mtDNA modifications observed in both the embryogenic and non embryogenic cultures.

Acknowledgements: This work was supported in part by EU (FORESTcontract n°MAE2B CT91 0039).

5. References

1. Bercetche J. (1989) Optimisation des conditions d'obtention de plantules à partir de cals embryogènes chez *Picea abies*. *Annales AFOCEL* **1988**, 97-115.
2. Bercetche J., Reymonds I. and Pâques M.(1993) Conversion des embryons somatiques de *Picea abies* en plantes : influence du support de cultures. *Annales AFOCEL* **1992**, 5-28.
3. Chalupa (1985) Somatic embryogenesis and plantlet regeneration from cultured immature and mature embryos of *Picea abies* (L.) Karst. *Commun. Inst. For. Cech.* **14**, 57-63.
4. Chauveau F., Brachon N.and Pâques M. (1995) Marqueurs protéiques associés aux propriétés embryogènes des cultures cellulaires de *Picea abies*. *Annales AFOCEL* **1993-1994**, 5-22.
5. Domon J.-M., Meyer Y., Faye L., David A.and David H. (1994) Extracellular (glyco)proteins in embryogenic and non embryogenic cell lines of Caribbean pine. Comparison between phenotypes of stage one somatic embryos. *Plant Physiol. Biochem.* **32**, 137-147.
6. Granier F. (1988) Extraction of plant proteins for 2-D PAGE. *Electrophoresis* **9**, 712-718.
7. Gupta P.K. and Durzan D.J. (1986) Plantlet regenration *via* somatic embryogenesis from subcultured callus on mature embryos of *Picea abies* (Norway spruce) In vitro *Cell Dev. Biol.* **22**, 685-688.
8. Hakman I., Fowke L.C., von Arnold S. and Eriksson T. (1985) The development of somatic embryos in tissue culture initiated from immature embryos of *Picea abies* (Norway spruce). *Plant Sci.* **38**, 53-59.
9. Lelu M.A., Boulay M. and Arnaud Y. (1987) Formation of embryogenic calli from cotyledons of *Picea abies* (L.) Karst. collected from 3 to 7-day-old seedlings. *C.R. Acad. Sci. sér. 3* **305**, 105-109.
10. Ruaud J.N., Bercetche J., Pâques M. (1992) First evidence of somatic embryogenesi from needles of 1-year-old *Picea abies* plants. *Plant Cell Rep.* **1,** 363-366
11. von Arnold S (1987) Improved efficiency of somatic embryogenesis in mature embryos of *Picea abies* (L.) Karst. *J. Plant Physiol.* **128**, 233-244.

PRIORITIES FOR APPLICATION OF BIOTECHNOLOGY TO PLANTATION IMPROVEMENT

Priorities in Forest Biotechnology

R.D. TEASDALE
ForBio Research Pty Ltd,
50 Meiers Road, Indooroopilly, Brisbane, Australia. 4068

The development of forest biotechnology is now at a stage where application is a reality. We have placed considerable emphasis on the use of molecular markers for efficient production of genetic maps and detection of correlations with quantitative trait loci. These studies are complicated by the use of families that are small, unbalanced, or contaminated by rogue trees. Nevertheless, a growing number of correlations are being detected between markers and major genes controlling valuable polygenetic traits including wood density, volume growth, branching patterns, and insect resistance. The direct benefits of marker assisted selection are significant and will continue to increase through guided breeding to fix genes, and through further selection. Such genetic gains will inevitably exceed the potential of rejuvenation and cryogenic storage strategies, so that pursuit of these goals is now of dubious benefit. The efficient use of markers depends not only on the value of the correlated traits, but also on the cost of screening which increases exponentially with the number of loci, and the ability to efficiently propagate elite genotypes as large clones. We are using robotic automation of both marker analyses and micropropagation processes to reduce costs and meet massively increasing demands in both these areas.

We have programmes at advanced stages to produce trees that exhibit total reproductive incompetence and consequently enhanced vegetative growth through removal of competitive sinks for photosynthate. Reproductive incompetence is an example of an *intrinsic* trait in contrast to genetic engineering targets such as resistance to insects, pathogens or herbicides, which are extrinsic traits susceptible to external factors such as alternative insects, or new selective herbicides. Available transformation methods generally represent an uncontrolled art which can suffer from variable and unstable levels of expression, and genotype dependency of the regeneration route. With progress in molecular biology yielding an increasing number of genes for valuable traits, it is becoming more important that these limitations are overcome. The integration of robust transformation-regeneration methods with marker-aided selection

M. R. Ahuja et al. (eds.), Somatic Cell Genetics and Molecular Genetics of Trees, 69–78.
© 1996 *Kluwer Academic Publishers. Printed in the Netherlands.*

and genotype-independent propagation methods is essential for operational deployment of these and other exotic genes.

Introduction

The development of biotechnology has now advanced to the stage where its benefits can be directly applied to forestry, perhaps more quickly than in plant agriculture, but almost certainly with greater potential. This is timely in view of the need to find solutions to the pressures now emerging in world forestry as a result of a three-fold squeeze of inability of present production forests to meet present demands indefinitely, increasing demands from countries with emerging economies (particularly in Asia), and considerable pressures to reduce harvesting from native forests. With *ca.* 90% of the world's wood requirements presently provided by native forests, a reduction to 80% of present supply of this source will involve essentially doubling of supply from plantation forests without any change in total demand. Biotechnology is probably the main hope for realistic change to the imperatives of this reality. Developments in three principal areas provide the basis for confidence that biotechnology can make a significant impact on ameliorating this supply-demand dilemma. These are in the area of molecular markers for improvement of natural characteristics, the judicious use of genetic engineering methods, and efficient clonal propagation methods to quickly deliver these benefits.

Molecular Marker Methods

A variety of molecular marker methods have appeared in recent years, but in our view by far the most important of these for forestry is the introduction of RAPDs. Originally developed by Williams *et al.* in 1990, as a general marker for use in plant and animal agriculture, the use of RAPDs in forestry was pioneered by the NCSU Forest Biotechnology group (Grattapaglia *et al.,* 1991) due to their particular advantage for preparation of simple linkage maps using the haploid megagametophyte tissue of conifers for three important reasons. First, the dominant nature of RAPDs was inconsequential due to the haploid character of this tissue, which corresponds to the maternal haplotype of the new embryo; secondly, the amount of haploid tissue was enough for up to 1000 reactions, so that sufficient information could be generated for a complete linkage map from a single tree; thirdly, an essentially unlimited number of random markers can be generated without any prior DNA sequence or other information to permit a saturated map to be constructed. The further use of megagametophytes for identification of QTL involves other considerations, irrespective of the marker system employed. While dominant markers, such as RAPDs in general, are less informative than co-dominant markers, the impact of this is often exaggerated. An appropriate set of dominant markers from both phases will provide the samc information as co-dominant markers - the difference in information yield is less than a factor of two.

However, despite excellent mapping results from several laboratories, RAPD markers have not been universally accepted, with a number of workers claiming lack of reproducibility. Our experience has been that RAPD markers are highly reliable; in fact, the repeatibility of over 2000 duplicate RAPD marker assays with the same 40 DNA preparations was 99.57%; when 55 independent foliage preparations were employed in over 3000 assays the reproducibility was almost identical at 99.44%. We are unaware of any other marker system which has been shown to exhibit such a high level of reliability, so that RAPDs should be considered one of the very best marker systems for reproducibility. The information yield of RAPDs is also intrinsically high. On agarose gels resolution permits up to 40 loci to be sampled, with a mean of approximately 15, depending on the reaction and gel separation conditions. If acrylamide gels are used then this can increase to over 60 loci. However, the effort required to produce one set of reactions and run these on acrylamide gels probably exceeds the effort for perhaps ten such sets run on agarose gels. On a strict theoretical basis, we have optimised the efficiency for pre-selecting the most informative reactions. Such pre-selection regimes also increase efficiency when using acrylamide gels, but the relative gain is substantially less when greater number of loci are sampled — a result of the averaging effect of marker distributions when larger numbers of loci are involved. Accordingly, the information yield of (for example) ten "random" primer reactions with agarose gels can be obtained on perhaps one or two agarose gels after judicious preselection of primers, which equals the information yield of one such acrylamide gel, but with much less effort.

It has been argued that RAPD markers will not be transferable to other pedigrees because they are not poly-allelic. However, our experience has been that a significant proportion of those markers that segregate in one family are transferable. In the case of *Pinus radiata*, 30% of markers from one pedigree are generally found to segregate in another. By simply producing "super maps" with typically three-fold the "standard" marker density, sufficient markers are also available for any new pedigree. Quick pre-screening will identify those that segregate in a target pedigree. We have found that marker order and approximate linkage distances are generally conserved between independent maps of *Pinus radiata* (i.e. synteny has been confirmed), so that RAPDs are quite transferable. Moreover, the ability to commence mapping using RAPDs with little prior information and effort makes these markers particularly valuable in forestry where most pedigrees are only a small number of generations from the wild with little preceding genetic analyses. Accordingly, all of the prevailing criticisms of RAPDs are found to be invalid in our laboratory, and are presently the method of choice for linkage and QTL analysis in forestry. We have now produced linkage maps of five tree species, with over 800 markers and 1000 markers for the "super maps" of *Pinus radiata* and *Eucalyptus grandis* respectively.

Marker Analyses in Forestry

The reliability of RAPDs has proved to be important in detection of rogues in field trials. The progeny of the defined crosses being used for our marker studies were generally established by other workers for different purposes. Although these were established by a number of different organisations, essentially every pedigree we have studied using molecular markers has been found to contain contaminant trees. Faith in the reliability of marker assays is clearly necessary to confidently declare a tree a contaminant. Entire rows of trees have been found to be unrelated to either of the declared parents; in two pedigrees one of the parents was a full-sib to the nominal parent; contamination rates of 25% are not unusual, and on average 4% of the progeny are contaminants. These results raise concerns about estimates of heritability and other genetic parameters obtained on the assumption of high fidelity in establishment of genetic trials. Moreover, if rogues are not eliminated, which can be difficult in species like *Pinus radiata* for which the breeding and production populations seem to exhibit a narrow genetic base, analysis will be complicated, resulting in map expansion and difficulty in assigning best marker order.

For forestry species commercial traits such as volume growth are essentially always quantitative, and the natural variation across progeny of a full-sib cross is generally very wide. It is illustrative to consider the extent of such variation as a baseline for what we may want to improve upon. For example, in a *Pinus radiata* pedigree with a mean (μ) diameter at breast height (DBH) of 15.8 cm, one standard deviation (σ) is 3.71 cm (which almost corresponds to a Poisson distribution where $\sigma = \sqrt{\mu} = 3.97$). Such large variance corresponds to a volume differential of approximately 75% for an increase of one σ unit above the mean. Two of the best individuals in such a pedigree have DBH values of nearly 25cm corresponding approximately to 2.5xσ above the mean. Crossing these individuals yields progeny with $\mu = 16.8$ cm, which is only 1.0 cm or 0.27xσ above the mean. The quantum of such gains on breeding will depend on the trait and the pedigree, but improvements of 0.25 to 0.4x σ would not be atypical.

QTL have now been identified for a number of traits within several families with several commercial plantation species. For example, in one *E. grandis* x *E. urophylla* hybrid we have found a single major locus for which selection yields a 70% increase in volume over the mean; in another family an increase of 18% was provided by a single locus; QTL have been identified in other eucalypt families for bark thickness, wood density, height, limonene (turpene) biosynthesis and insect resistance. Essentially every family we have examined has resulted in QTL identification. Rather than a large number of loci of small effect, we have generally found a few loci of major effect. This should not be surprising; in the absence of any standardising influence a random distribution of size of effects may be expected, and those loci most readily detected (as well as having most value) will be those of greatest effect.

These QTL were identified in full-sib families containing in the vicinity of 100 progeny. Such large full-sib families are not common in forestry trials established before the relatively recent ascent of molecular marker technology as a robust tool for forestry. New trials are being established which will be well-suited to QTL analysis, but these will require field growth to a suitably advanced stage for valid phenotype assessment — the greatest benefit of markers is as predictive tools to select for traits that are not measurable at an early stage. At present, there is a paucity of pedigrees that are ideal for QTL detection. Nevertheless, existing pedigrees can be made informative, if not with the cost-effectiveness and confidence that would be provided if they had been established in numbers appropriate for analyses using molecular markers. For *Pinus radiata* we have studied a set of six small full-sib families (11 to 58 trees) which share a common parent, thereby creating a half-sib array of 181 trees. These were planted on three different sites, with the planting on the third site of different age to the other two. Many markers that are heterozygous in the common parent will not be informative in those families where the second parent is homozygous, however, the availability of a "super map" of *Pinus radiata* provided alternative markers in these regions so that sufficient population coverage could be obtained in all regions of the genome. Despite the complexity of this case, it was possible to standardise phenotype data and to adequately genotype all individuals to permit QTL analyses for a number of traits. Analyses across the entire half-sib array identified significant QTL for a number of quantitative traits. Of particular interest were three controlling wood density. Analyses of individual full-sib families revealed that in some of these one or two of the QTL had a significant effect, whereas the remaining loci clearly had no effect. In other families similar behaviour was observed for some other loci. The lack of trait segregation at a particular locus is simply explained by the second parent being homozygous positive for the trait, so that all progeny of this cross carry the allele for the trait. This illustrates how detection of QTL in such half-sib arrays requires not only that genes of major effect are involved, but also that these are not obscured by non-segregation in too many families of the array. Where QTL are identified, such a defined half-sib array has the advantage of providing information on multiple parents, and thereby recognition of how to most effectively apply this information. Most important, however, is the recognition that many existing families may allow valid QTL analyses without waiting many years for the new crosses to grow to an assessable stage. We have now confidently identified 16 QTL over 11 traits in various eucalypt and pine pedigrees, and have tentatively identified another 49 putative loci for 9 traits in an F2 pedigree of a *Pinus elliottii* x *Pinus caribaea* hybrid, and expect this list to extend substantially in the next few years.

Application of QTL Information

For gymnosperms there is a long delay, generally two years, in production of full-sib seed, whereas open-pollinated seed is generally available immediately; in addition, selection for maternal markers is simplified by the availability of megagametophytes which correspond to the maternal haplotype of the new embryo in each seed. For such seed the choice of QTL to use in selection is based on the respective improvements their selection will provide. This can be estimated by assuming that the parents to the half-sib array are random with respect to their effects at these loci. The benefit of selection for particular combinations of QTL was assessed with the progeny used for the QTL analyses. When all three loci are used in selection across all progeny of the half-sib array, the mean of selected individuals was 0.88x σ above the mean for wood density. When one of the full-sib families was similarly analysed but with selection only for the two segregating QTL, the improvement in the mean was 1.01xσ, that is, a similar benefit was obtained for fewer loci and therefore lower selection intensity than with the half-sib progeny.

Such improvements of approximately one standard deviation above the mean are far greater than improvements of perhaps 0.3xσ expected from conventional breeding. Moreover, there was significant distribution about these means — there needs to be clear understanding from those applying marker-aided selection that although the mean will be improved, not all progeny will be superior. Sufficient progeny need to be selected to ensure random mixing of alleles at other loci. With appropriate recognition and accommodation of such requirements, it is clear that not only can we undertake valid QTL analyses, but that the benefits of this information can be applied immediately. Greater gains will be obtained by judicious selection of parents to combine desirable traits in the most efficient manner. It is relevant to note that the half-sib array approach used here also identifies those parents for which a desirable allele is fixed, not only those that are segregating, which will be preferred in any guided breeding program, allowing selection for segregating traits in addition to those indicated to be fixed, without combined selection intensity if both were segregating.

Where valuable traits are segregating, the impact of selection intensity needs to be considered. For purposes of simplicity the effect of eliminating individuals of uncertain genotype due to recombinations close to the QTL can be ignored. The number of QTL (n) used in selection will increase the selection intensity according to $1/(2^n)$; because a distribution of performance is expected due to the randomisation of other genes, it is necessary to select (say) 20 individuals for clonal propagation; accordingly, if eight loci are used, at an intensity of 1/256 there will be need to screen 5120 progeny to obtain the target of 20 genotypes with the desired marker combination. While this is of itself not unreasonable, it will also be necessary to accommodate any constraints of clonal propagation systems. In general, embryogenesis systems are very genotype dependent — a responsiveness of 1% of genotypes to embryogenic protocols is not uncommon with many species — in combination with selection for eight loci this would involve screening of over 500,000 full-sib progeny (if available) to obtain the target of 20 individuals which have both the correct genotypc and arc cmbryogenic. The lack of a suitable tissue culture system combined with selection for more than three or four loci clearly leads to a screening task of daunting scale, which is unlikely to be justified except for traits of particularly high value.

Is There a Role for Genetic Engineering of Plantation Trees?

Our laboratory has a strong program in genetic engineering of trees to be totally incapable of forming reproductive structures. Such trees will obviously be sterile, thereby providing gene containment for environmental and commercial protection, but this is a secondary benefit. Avoidance of the reproductive burden is estimated to provide vegetative growth increases of up to 15%. There will be additional benefits including saving of the management cost of removing regrowth seedlings, avoidance of wood quality damage in some species by stem cones, and large losses of nutrient rich pollen. While several strategies hare been pursued in parallel, we have favoured a tissue ablation approach whereby a promoter to an early flower-specific gene is used to express a lethal gene such as a RNase, protease, nuclease, anti-sense to a central function, a cytotoxin, etc. These specific studies cannot be adequately discussed here and will be the subject of a separate presentation.

Exotic traits such as herbicide resistance and insect resistance also have value, but must be considered against alternative strategies, such as selection for natural insect resistance, or carefully localised application of available herbicides; the development of new selective herbicides is also a possibility to be considered. There appear to be no exotic traits which alone are likely to exceed the value of reproductive incompetence for intrinsic enduring benefit with a wide range of species. It is noted that some of the targets of genetic engineering can be better addressed through marker methods. For example, quantitative adjustment of the levels of lignin in trees will be difficult using the blunt instrument of transformation with an anti-sense or other knock-out strategy. Marker methods seem far more appropriate for quantitative adjustments within the natural range of levels — excessive reduction of lignin will yield aberrant trees such as those seen under copper-deficiency.

Moreover, the benefits of reproductive incompetence and other exotic traits are generally expected to provide value improvements in the vicinity of 10 to 15%, which is much less than the improvements which seem attainable through marker-aided selection and guided breeding. This is not to suggest that exotic traits are not worth pursuing; clearly, such exotic traits should not be developed as alternatives to marker-aided selection, but as additional benefits. That is, it is somewhat pointless to transform otherwise poor material to elevate it to a value that remains far below that more simply obtained by selection; technologies need to be integrated so that only germplasm with the superior natural genes through selection is transformed. However, as for propagation, genotype dependence of transformation-regeneration can introduce an additional constraint to be combined with marker selection. Moreover, with what are considered responsive genotypes, the efficiency of transformation-regeneration nevertheless is generally low on an explant basis, often below 1/1000, so that it is necessary to ensure amplification of the target material following trait selection using markers. If embryonic or other materials that occur in limited numbers with each genotype are used for transformation then the numbers of genotypes to be screened and the cost of marker assays will be amplified inordinately.

Much greater efforts are needed to develop transformation systems to a far more robust level than generally prevails. There is too much art and insufficient science in many of the present approaches to transformation. There is need to better understand and control gene insertion to increase frequency of transformation, and to understand position effects on expression. Development of a robust capability for targeted insertion into sites of known stability would allow repeated transformation without production of chimaeras, and provide confidence that expression would not decline over time — an important consideration with trees. This would then herald the use of gene transfer not just for exotic genes but also for native genes to shorten the time constraints of breeding.

Propagation Systems

To capture the genetic gains of selected and/or transformed material, it is necessary to clonally propagate using tissue culture in order to deliver elite material in a short time frame — if improved genotypes perform, say, 20% better than alternative material, each year of delay leads to a corresponding loss of production opportunity. While harvesting of trees may not occur for many years, plantings are generally made to meet production requirements, so that use of improved material will allow lower plantings (and lower establishment costs) for the same targets — the benefits should therefore be realised immediately and not at harvest age. Tissue culture is generally the only means to provide large quantities of improved material very quickly. However, the total demand needs to be spread across numerous genotypes, perhaps 20 genotypes for each of five families. With germplasm improvements expected to be continuous and dynamic, individual clones will become progressively obsolete, so that an average clone life is likely to be only five years, or less. Thus, if a tissue culture facility is to produce (say) five million plants annually, this will correspond to 250,000 plants per clone over five years or 50,000 annually. Inordinately high screening costs would not be easily recovered with such numbers — any genotype dependency of the propagation system will amplify all selection costs, which may not be justifiably recovered across such clone sizes. It is important that propagation occur in a robust, cost-efficient manner with minimum genotype dependency and little risk of generating aberrant plants, such as through physiological stress or somaclonal variation.

While embryogenesis systems are of interest to us and may prove to be valuable in the future, at present they are costly to initiate and highly genotype dependent with most species. To improve the cost-effectiveness of micropropagation we have developed a robust robotic system which is capable of producing 3 million plants annually. Using potato plants as the initial target species, the machine automatically opens a container of approximately 40 plants, then harvests single plants and transfers these to a turntable surface where each is stereoscopically imaged. The image is processed with advanced software to identify nodal segments, which are then individually dissected and transferred to new media. Tips are placed in separate trays to other nodal segments, and trays are closed automatically, with the entire operation carried out in HEPA-filtered air. Storage buffers hold and automatically transfer sufficient input and output stocks for several hours of operation. One operator can thereby operate three machines which are compact for delivery through a standard doorway and operate using standard power supply. These machines have now been proven with potatoes and are being developed for application to eucalypts, acacias and pines in 1996. This will then allow very reliable and cost-effective micropropagation of large numbers of improved planting stock for plantation use.

Concluding Remarks

Our approach to the application of biotechnology to forestry has been to first make informed judgments of what is needed for forest improvement, and to then develop strategies to attain these goals. Molecular marker methods have surpassed expectations in the size of the genetic gains that can be identified and clearly will have greater impact on forest tree improvement than genetic engineering for the foreseeable future. However, these benefits are limited without the ability to quickly and reliably deliver elite material; our development of an efficient robotic system for micropropagation has therefore effectively addressed this need. Acceptance of these gains in the generally conservative forestry sector may be a greater impediment to field implementation than the technical factors.

Genetic engineering must be integrated with marker aided selection methods as well as with propagation — otherwise there will be no role for it. This imposes constraints on tissue culture systems that are used so as to avoid inordinate expansion of the numbers of plants used in screening. Moreover, following the same rationale, different exotic traits in general should be combined into cassettes and introduced through a single transformation process.

The developments of the last few years in forest biotechnology now have the potential to change the nature of plantation forestry. We are at the beginning of what should be an exciting time which will see increasingly sophisticated breeding plans being implemented. In the not too distant future perhaps gene transfer through breeding will be replaced, at least in part, with direct transfer of natural genes.

References

Grattapaglia, D., Wilcox, P., Chaparro, J.X., O'Malley, D.M., McCord, S., Whetten, R., McIntyre, L., and Sederoff, R. (1991) A RAPD map of loblolly pine in 60days. International Society for Plant Molecular Biology International Congress, Tucson, AZ, Abstract #2224.

Williams, J.G.K., Kubelik, A.R., Livak, K.J., Rafalski, J.A. and Tingey, S.V. (1990) DNA polymorphisms amplified by arbitrary primers are useful as genetic markers, *Nucleic Acids Research* **18**, 6531 - 6535.

Section II.

Transformation and gene expression

GENETIC ENGINEERING OF LIGNIN BIOSYNTHESIS IN POPLAR

W. BOERJAN[1], H. MEYERMANS[1], C. CHEN[1], J.-C. LEPLÉ[1,2], J.H. CHRISTENSEN[1], J. VAN DOORSSELAERE[1], M. BAUCHER[1], M. PETIT-CONIL[3], B. CHABBERT[4], M.-T. TOLLIER[4], B. MONTIES[4], G. PILATE[2], D. CORNU[2], D. INZÉ[5], L. JOUANIN[6], and M. VAN MONTAGU[1]

[1] Laboratorium voor Genetica, affiliated to the Flanders Interuniversity Institute for Biotechnology, and [5] Laboratoire Associé de l'Institut National de la Recherche Agronomique (INRA), Universiteit Gent, K.L. Ledeganckstraat 35, B-9000 Gent, Belgium; [2] Station d'Amélioration des Arbres Forestiers, INRA, F-45160 Ardon, France; [3] Centre Technique du Papier, F-38004, Grenoble Cedex 09, France; [4] Laboratoire de Chimie Biologique, INRA, F-78850 Thiverval-Grignon, France; [6] Laboratoire de Biologie Cellulaire, INRA, F-78026 Versailles Cedex, France

1. Abstract

The removal of lignin in paper manufacture is a toxic and energy-requiring process. For the paper industry and for the environment, it would be very beneficial to process trees with less lignin or with lignin that is more easily extractable. To achieve this goal, we are down-regulating the activity of a number of lignin biosynthesis enzymes in poplar by genetic engineering and studying their role in determining the quantity and quality of lignin. Here, our recent results obtained in this field are summarized with emphasis on bispecific caffeic acid/5-OH-ferulic acid-*O*-methyltransferase, cinnamoyl-CoA-reductase, caffeoyl-CoA-*O*-methyltransferase, and peroxidase.

2. Introduction

One of the main obstacles in the production of pulp and paper is the removal of lignin (Whetten and Sederoff, 1991, Dean and Eriksson, 1992). Lignin is a complex aromatic polymer that is deposited mainly in the walls of cells that constitute the wood and represents between 15 and 35% of the dry weight of wood, depending on the tree species (Higuchi, 1985). The removal of lignin in paper manufacture is a toxic and energy-requiring process. For example, during Kraft pulping, which is the mostly used chemical delignification process, 800 kg of NaOH and 300 kg of Na_2S are required to remove one ton of lignin (Chiang *et al.*, 1988). After this extraction process, a series of bleaching steps are applied (using ClO_2, O_3, H_2O_2, O_2) to remove the residual lignin that is strongly bound to the cellulose fibers (Biermann, 1993). These lignin extraction procedures also have the disadvantage of reducing the length of the cellulose fibers and,

M. R. Ahuja et al. (eds.), Somatic Cell Genetics and Molecular Genetics of Trees, 81–88.
© 1996 *Kluwer Academic Publishers. Printed in the Netherlands.*

hence, the strength of the paper. If the lignin is not removed, the paper turns yellow upon exposure to sunlight. Given the value of the wood market worldwide, the increasing demand for paper and the increasing concern for a clean and green environment, the genetic engineering of trees to obtain a lignin that is more easily extractable during the paper making process is a worthwhile objective.

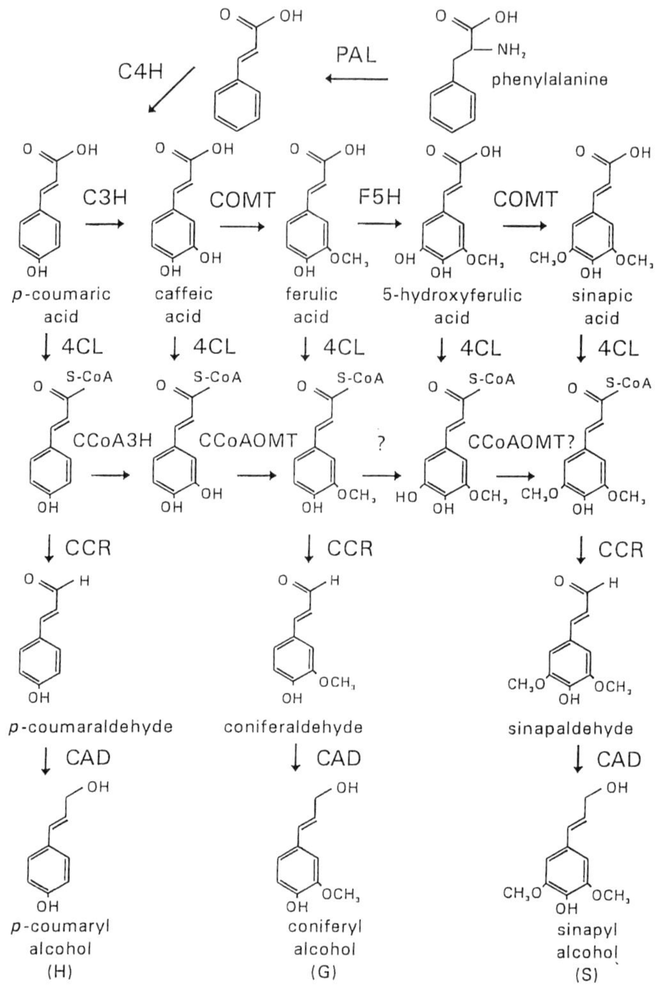

Figure 1. The Lignin Biosynthesis Pathway.
PAL, phenylalanine ammonialyase; C4H, cinnamate-4-hydroxylase; C3H, coumarate-3-hydroxylase; COMT, bispecific caffeic acid/5-OH-ferulic acid-*O*-methyltransferase; F5H, ferulate-5-hydroxylase; CCoA3H, coumaroyl-CoA-3-hydroxylase; CCoAOMT, caffeoyl-CoA-*O*-methyltransferase; 4CL, 4-hydroxy-cinnamate-CoA ligase; CCR, cinnamoyl-CoA-reductase; CAD, cinnamyl alcohol dehydrogenase.

To obtain this goal, a fundamental insight into the biosynthesis of lignin is needed. Lignin is a polymer that is built by the polymerization of three alcohols, the monolignols *p*-coumaryl alcohol, coniferyl alcohol, and sinapyl alcohol (Figure 1), which differ only by their degree of methoxylation and which give rise to the *p*-hydroxyphenyl (H), guaiacyl (G) and syringyl (S) units in the polymer, respectively. Two broad categories of linkages can be distinguished in the polymer, the carbon-carbon (C-C) linkage and the ether linkage (Figure 2). The latter is easily broken during Kraft pulping whereas the C-C linkages are not. Because C-C linkages can be formed between the C-5 positions of two G units but not between two S units, it is logic that the relative amount of the three types of monolignols determines, at least partially, the ease for delignification.

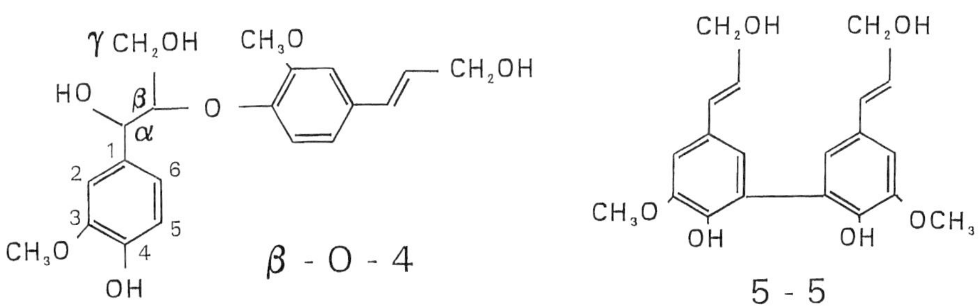

Figure 2. Schematic representation of a ß-O-4 ether linkage and a C-C linkage between two coniferyl alcohol units. For an overview of the different linkages in the lignin polymer, the reader is referred to Adler (1977).

To unravel the *in vivo* role of the lignin biosynthesis enzymes in determining the amount and composition of lignin, we have cloned the corresponding genes of four of these enzymes from poplar and made transgenic trees containing antisense genes. Here, we present the state of the art of our research, that has been part of European Union-funded projects and involved collaboration with laboratories throughout Europe. At the Laboratory of Genetics (Gent, Belgium), we are focusing on bispecific caffeic acid/5-OH-ferulic acid-*O*-methyltransferase (COMT), cinnamoyl-CoA-reductase (CCR), cinnamyl alcohol dehydrogenase (CAD), caffeoyl-CoA-*O*-methyltransferase (CCoAOMT) and peroxidases (POx). The results obtained by down-regulation of CAD are presented by Baucher *et al.* (this volume).

Poplar (*Populus tremula* x *P. alba)* is used as a model system. This hybrid can be efficiently transformed (Leplé *et al.*, 1992). Poplars are among the most planted tree species in Europe because of their high economic interest. It is fairly easy to generate inter- and intraspecific hybrids. In this way, trees with superior growth characteristics have been obtained. Most poplars are readily vegetatively propagated and hence large clonal plantations can be made in a short time. In Europe, poplar wood is mainly processed to veneer, particle board, soft board and pallets, whereas in the U.S., poplars are mainly cultivated for pulp production.

3. Results and Discussion

3.1. DOWN-REGULATION OF COMT IN POPLAR

COMT is a bifunctional enzyme that catalyses the conversion of caffeic acid to ferulic acid and 5-OH ferulic acid to sinapic acid *in vitro* (Figure 1). COMT has been one of the attractive enzymes to down-regulate because it was already known from natural mutants of maize and sorghum, the brown-midrib (*bm* and *bmr*) mutants, that they had reduced levels of COMT activity (or COMT and CAD activity) and simultaneously also reduced amounts of lignin. It has recently been demonstrated that the *bm3* maize mutant has a defective *COMT* gene (Vignols *et al.*, 1995). Therefore, it seemed straightforward to obtain the same results, less lignin, in trees through down-regulation of COMT by genetic engineering.

At the Laboratory of Genetics (Gent, Belgium), we have purified and cloned a cDNA for COMT from *P. trichocarpa* x *P. deltoides* (Dumas *et al.*, 1992). The 3' and 5' parts of the cDNA were cloned in antisense orientation behind the CaMV 35S promoter and transferred into the poplar hybrid *P. tremula* x *P. alba*. All transgenic lines, 15 for each construct, were screened for a reduced steady-state *COMT* mRNA level. In four transformants (ASB4B, ASB6B, ASB2B, and ASB10B, all containing the 3' part of the *COMT* cDNA in antisense direction), a reduced level of *COMT* mRNA was detected with the lowest levels in ASB2B and ASB10B. COMT activity analysis showed that ASB4B and ASB6B had a reduction in COMT activity by about 50% whereas in ASB2B and ASB10B plants, the COMT activity was reduced by 95%. These plants were propagated in the greenhouse to provide sufficient plant material (wood) for lignin analysis and Kraft pulping simulation experiments. The data obtained from the lignin analyses are presented in Table 1.

Table 1. Klason lignin content and lignin composition of xylem from control and antisense *COMT* plants[a]

	KL	G	S	5-OH-G	S/G
JUNE 1994					
PBI121	19.0	133	189	-	1.42
ASB2B	16.8	150	34	16	0.23
ASB10B	17.1	160	51	13	0.31
OCTOBER 1994					
PBI121	19.6	132	207	-	1.57
ASB2B	19.3	197	50	16	0.25
ASB10B	19.2	179	84	6	0.47

[a] Data are expressed as μmole·g^{-1} of cell wall residue (CWR) (the residue that is left after successive extractions of the xylem with toluene ethanol, ethanol, and water). The data are the mean of two independent measurements of one or two plants. Standard deviations for KL and for G, S, and 5-OH-G were 3% and 6% at the most, respectively. KL, Klason lignin content (expressed as percent CWR). Lignin composition was analyzed by thioacidolysis. G, guaiacyl unit; S, syringyl unit; 5-OH-G, 5-OH-guaiacyl unit; -, not detected. PBI121 is the control, ASB2B and ASB10B are plants with a 95% reduced COMT activity.

Despite the reduction of 95% in COMT activity, no reduction in the amount of lignin was observed (Klason method; Effland, 1977). However, when the composition of the lignin was analyzed, significant differences were detected. The antisense plants were characterized by a reduction of the S/G ratio (due to a decreased level of S units and an increased level of G units) when they were analyzed by thioacidolysis (Lapierre *et al.*, 1986). Thioacidolysis determines the level of S and G units that are linked by ß-O-4 ether linkages in the lignin. In addition, a novel lignin unit, 5-OH-guaiacyl, corresponding to the non-methylated precursor of the S unit, was detected in ASB2B and ASB10B trees. This unit has previously also been detected in brown-midrib mutants (Lapierre *et al.*, 1988).

Wood of 2-year-old ASB2B and ASB10B trees was analyzed for its lignin extractability by simulated Kraft pulping experiments. Based on the reduced S/G ratio in these lines, one might expect that the lignin would be more difficult to extract, because of the relative increase in the level of G units and, thus, probably also the number of C-C linkages between the C-5 positions of the lignin units. In agreement with this hypothesis, wood of gymnosperms, of which the lignin consists mainly of G units, is more difficult to delignify than the wood of angiosperms, of which the lignin consists of G as well as S units. On the other hand, the 5-OH-guaiacyl units that had appeared in the lignin of the transgenic lines might render the lignin more soluble, due to the OH group instead of the methoxy group at the C-5 position (Boudet *et al.*, 1995).

Simulated Kraft pulping experiments of ASB2B and ASB10B trees resulted in a higher kappa number for the lignin of the transgenic plants. The kappa number is a measure for the residual amount of lignin that is still present after the pulping step. Therefore, the proposed beneficial effect of the 5-OH-guaiacyl unit on the delignification process did not counterbalance the negative effect of having less S units. The results however, support the hypothesis that the S/G ratio and, hence, the degree of methoxylation of the polymer is an important parameter determining the extractability of the lignin from the wood.

Thus, there are four important conclusions that can be drawn from the antisense-*COMT* experiments. First, a 95% reduction in COMT activity does not reduce the amount of lignin but affects the number of S units that is incorporated into ß-O-4 ether linkages in the lignin polymer. Second, a new unit type, the 5-OH-guaiacyl unit, is incorporated into the lignin. Third, the level of G units is not reduced but had even even increased [These results have been published by Van Doorsselaere *et al.* (1995). Similar results have been obtained in tobacco by Atanassova *et al.* (1995)]. Fourth, a lower S/G ratio results in a more difficult delignification process.

3.2. CLONING OF *CCoAOMT* FROM POPLAR

The fact that the level of G units was not reduced in the antisense-*COMT* poplars raises the possibility that this level is controlled by an alternative pathway that bypasses the conversion of caffeic acid to ferulic acid catalysed by COMT (Figure 1). This bypass might occur at the level of the cinnamoyl-CoA esters, by the conversion of caffeoyl-CoA to feruloyl-CoA, catalysed by CCoAOMT. Evidence for this hypothesis was also generated by Ye *et al.* (1994) and Ye and Varner (1995), who showed by

Northern analysis that in *Zinnia*, the expression of *CCoAOMT* but not of *COMT* was induced during tracheary element formation. Furthermore, by tissue printing they showed that in young internodes of *Zinnia*, *CCoAOMT* mRNA was associated with differentiating xylem, whereas *COMT* mRNA was associated with differentiating phloem fibers. These authors suggested that CCoAOMT might play a role as important as COMT in lignification. To investigate this hypothesis, we have cloned cDNAs for CCoAOMT from poplar (*P. trichocarpa*). Four cDNAs were obtained through hybridization of a xylem cDNA library with the *CCoAOMT* cDNA from parsley as a probe (Schmitt *et al.*, 1991). Based on restriction maps and DNA sequence analysis, we have tentatively concluded that these four cDNAs correspond to two genes, each of which is represented by two allelic forms. All four cDNAs code for a protein of 247 amino acids. One of these cDNAs is being used to make sense and antisense constructs under the control of the p70 promoter (CaMV 35S with double enhancer).

As a first step towards exploring the regulatory mechanisms controlling the expression of *CCoAOMT*, a genomic library from *P. trichocarpa* was screened. Two genes corresponding to two of the cDNAs were cloned. The coding sequence of both genes is interrupted by four introns. Both promoters have been fused to the ß-glucuronidase gene to investigate whether expression of CCoAOMT colocalizes with lignifying cells.

3.3. DOWN-REGULATION OF *CCR* IN POPLAR

CCR catalyses the reduction of the cinnamoyl-CoA thioesters to the corresponding cinnamyl aldehydes. CCR is the first enzyme in the lignin branch pathway and therefore hypothesized to play a key regulatory role in lignin biosynthesis (Goffner *et al.*, 1994). CCR has been purified from *Eucalyptus gunnii* xylem (Goffner *et al.*, 1994; see also Pettenati *et al.*, this volume). This group has isolated a corresponding cDNA (Boudet *et al.*, 1995) which our laboratory has used to screen a xylem cDNA library from *P. trichocarpa*. One *CCR* cDNA was isolated. The corresponding amino acid sequence showed 83 % identity and 93 % similarity to the one deduced from the eucalyptus cDNA. Sense and antisense constructs have been made. These constructs have been transferred to wild-type poplar, but also to poplars that have previously been transformed with antisense-*COMT* and antisense-*CAD* constructs. Over 300 putative primary transformants are currently being analyzed for a reduced CCR activity.

3.4. THE INVOLVEMENT OF PEROXIDASES IN LIGNIFICATION IN POPLAR

Laccase and peroxidase are the enzymes supposed to be involved in the polymerization of the lignin monomers. However, the precise role of these enzymes in the polymerization process is unclear. In the case of peroxidase, this is largely due to the fact that there are many isoforms, and that all these isoforms are very similar in sequence. Most of these isoforms catalyse the polymerization of coniferyl alcohol, one of the lignin building units, *in vitro*. Because they are so similar in sequence, the role of one particular isoform in lignification is not easily resolved by an antisense approach. Lagrimini (1991) has tried this approach by transforming tobacco with an antisense construct made from the cDNA coding for the most abundant peroxidase isoform from tobacco xylem. Overexpression of this isoform resulted in an increased amount of lignin, but down-regulation of this isoform did not result in a significant reduction of the amount of lignin (Chabbert *et al.*, 1992).

To demonstrate the role of a particular peroxidase isoform in lignification, its primary sequence should be determined, it needs to be demonstrated that this isoform catalyses the polymerization of lignin monomers *in vitro*, and that the expression of this isoform correlates temporarily and spatially with lignification (Lewis and Yamamoto, 1990). Our strategy is to purify the different isoforms from poplar xylem, to sequence them, to isolate the corresponding genes, and to investigate their cellular and subcellular location by inserting a small DNA fragment into the coding sequence that will give rise, after expression in poplar, to a chimeric peroxidase that can be specifically recognized by an antibody (tag fusion).

From xylem of *P. trichocarpa*, six peroxidase isoforms have been purified to homogeneity. On activity gels, four of them seem to be specific for the xylem. From these four isoforms, several peptides were microsequenced (Bauw *et al.*, 1987) and corresponding degenerated oligonucleotides were synthesized to screen a xylem cDNA library and a genomic library from *P. trichocarpa*. We have cloned four cDNAs that were more than 95% identical at the nucleotide level. From the deduced amino acid sequence, we concluded that one of these most probably corresponds to one of the purified xylem-specific peroxidases. The uncertainty in this conclusion is due to the very high level of homology between the different peroxidase cDNAs and the fact that only a limited amount of sequence data can be obtained from the purified protein. Using primers that could discriminate between the different cDNAs, a genomic clone that corresponded to this cDNA was identified. This gene, *glpox26*, is currently being sequenced.

4. Acknowledgments

The authors thank Prof. Dr. U. Matern for providing the parsley *CCoAOMT* probe and Prof. Dr. A. Boudet for providing the eucalyptus *CCR* probe. This work was carried out in the framework of the European Union Research Programs, ECLAIR Project number AGRE-0021-C, AIR2-CT93-1661, and FAIR-PL 95424. J.V.D. is indebted to the Vlaams Instituut voor de Bevordering van het Wetenschappelijk-Technologisch Onderzoek in de Industrie for a post-doctoral fellowship. D.I. is a Research Director of the Institut National de la Recherche Agronomique (France).

5. References

Adler, E. (1977) Lignin chemistry - past, present and future. *Wood Sci. Technol.* **11**, 169-218.

Atanassova, R., Favet, N., Martz, F., Chabbert, B., Tollier, M.T., Monties, B., Fritig, B., and Legrand, M. (1995) Altered lignin composition in transgenic tobacco expressing *O*-methyltransferase sequences in sense and antisense orientation. *Plant J.* **8**, 465-477.

Bauw, G., De Loose, M., Inzé, D., Van Montagu, M., and Vandekerckhove, J. (1987) Alterations in the phenotype of plant cells studied by NH_2-terminal amino acid-sequence analysis of proteins electroblotted from two-dimensional gel-separated total extracts. *Proc. Natl. Acad. Sci. USA* **84,** 4806-4810.

Biermann, C.J. (1993) *Essentials of Pulping and Papermaking*, Academic Press, San Diego.

Boudet, A.M., Lapierre, C., and Grima-Pettenati, J. (1995) Biochemistry and molecular biology of lignification. *New Phytol.* **129,** 203-236.

Chabbert, B., Monties, B., Liu, Y.T., and Lagrimini, M. (1992) Lignin content and composition in transgenic tobacco plants with altered peroxidase activity, in M. Kuwahara and M. Shimada (eds.), *Proceedings 5th International Conference on Biotechnology in Pulp and Paper Industry*, Uni Publishers Co., Tokyo, pp. 481-485.

Chiang, V.L., Puumala, R.J., Takeuchi, H., and Eckert, R.E. (1988) Comparison of softwood and hardwood kraft pulping. *Tappi J.* **71,** 173-176.

Dean, J.F.D., and Eriksson, K.-E.L. (1992) Biotechnological modification of lignin structure and composition in forest trees. *Holzforschung* **46,** 135-147.

Dumas, B., Van Doorsselaere, J., Gielen, J., Legrand, M., Fritig, B., Van Montagu, M., and Inzé, D. (1992) Nucleotide sequence of a complementary DNA encoding *O*-methyltransferase from poplar. *Plant Physiol.* **98,** 796-797.

Effland, M.J. (1977) Modified procedure to determine acid-insoluble lignin in wood and pulp. *Tappi* **60,** 143-144.

Goffner, D., Campbell, M.M., Campargue, C., Clastre, M., Borderies, G., Boudet, A., and Boudet, A.M. (1994) Purification and characterization of cinnamoyl-coenzyme A:NADP oxidoreductase in *Eucalyptus gunnii*. *Plant Physiol.* **106,** 625-632.

Higuchi, T. (1985) Biosynthesis of lignin, in T. Higuchi (ed.), *Biosynthesis and Biodegradation of Wood Components*, Academic Press, Orland, pp. 141-160.

Lagrimini, L.M. (1991) Wound-induced deposition of polyphenols in transgenic plants overexpressing peroxidase. *Plant Physiol.* **96,** 577-583.

Lapierre, C., Rolando, C., and Monties, B. (1986) Thioacidolysis of poplar lignins: identification of monomeric syringyl products and characterization of guaiacyl syringyl-lignin fractions. *Holzforschung* **40,** 113-118.

Lapierre, C., Tollier, M.-T., and Monties, B. (1988) Mise en évidence d'un nouveau type d'unité constitutive dans les lignines d'un mutant de Maïs bm3. *C.R. Acad. Sci. Paris*, Série III, **307,** 723-728.

Leplé, J.C., Brasileiro, A.C.M., Michel, M.F., Delmotte, F., and Jouanin, L. (1992) Transgenic poplars: expression of chimeric genes using four different constructs. *Plant Cell Rep.* **11,** 137-141.

Lewis, N.G., and Yamamoto, E. (1990). Lignin: occurrence, biogenesis, and biodegradation. *Ann. Rev. Plant Physiol. Plant Mol. Biol.* 41, 455-496.

Schmitt, D., Pakusch, A.-E., and Matern, U. (1991) Molecular cloning, induction, and taxonomic distribution of caffeoyl-CoA 3-*O*-methyltransferase, an enzyme involved in disease resistance. *J. Biol. Chem.* **266,** 17416-17423.

Van Doorsselaere, J., Baucher, M., Chognot, E., Chabbert, B., Tollier, M.-T., Petit-Conil, M., Leplé, J.-C., Pilate, G., Cornu, D., Monties, B., Van Montagu, M., Inzé, D., Boerjan, W., and Jouanin, L. (1995) A novel lignin in poplar trees with a reduced caffeic acid/5-hydroxyferulic acid *O*-methyltransferase activity. *Plant J.* **8,** 855-864.

Vignols, F., Rigau, J., Torres, M.A., Capellades, M., and Puigdomènech, P. (1995) The *brown midrib3* (*bm3*) mutation in maize occurs in the gene encoding caffeic acid *O*-methyltransferase. *Plant Cell* **7,** 407-416.

Whetten, R., and Sederoff, R. (1991) Genetic engineering of wood. *Forest Ecol. Manag.* **43,** 301-316.

Ye, Z.-H., and Varner, J.E. (1995) Differential expression of two *O*-methyltransferases in lignin biosynthesis in *Zinnia elegans*. *Plant Physiol.* **108,** 459-467.

Ye, Z.-H., Kneusel, R.E., Matern, U., and Varner, J.E. (1994) An alternative methylation pathway in lignin biosynthesis in *Zinnia*. *Plant Cell* **6,** 1427-1439.

STABILITY AND EXPRESSION OF CHIMERIC GENES IN *POPULUS*

M.R. AHUJA and M. FLADUNG
BFH, Institute of Forest Genetics
Sieker Landstrasse 2
22927 Grosshansdorf, Germany

Abstract

We have employed *Populus* as a model system to investigate the stability and expression of transgenes at the morphological, physiological, and molecular levels. Four clones from European aspen (*Populus tremula*) and hybrid aspen (*P. tremula* x *P. tremuloides*) clones were genetically transformed, using an *Agrobacterium* binary vector system, with different gene constructs, in particular *rol*C. Two types of promoters, namely 35S from cauliflower mosaic virus, and rbcS from potato, were employed for control of *rol*C gene expression. An improved transformation method employing leaf discs was developed, and putative transformants were regenerated on a kanamycin-containing medium. More than 1000 transgenic aspens have been raised in the greenhouse during the last two years. The morphology and physiology of the transgenic plants seems to be influenced by the type of promoter employed. Transgenic aspens carrying the 35S-*rol*C gene construct exhibited much smaller leaves, as compared to the untransformed controls. On the other hand, rbcS-*rol*C carrying transgenic aspens showed only slightly smaller leaves, as compared to the controls. However, transgenic plants carrying the chimeric gene 35S-*rol*C or rbcS-*rol*C exhibited yellowish-green leaves, as compared to dark green leaf color of the untransformed controls. A number of leaf abnormalities, chimeras, revertants to normal state have been observed in the transgenic aspens. In order to silence the *rol*C gene expression, two constructs carrying the transposable element *Ac* from maize, along with *rol*C, were employed for genetic transformation in aspen. Transgenic aspens carrying the transgenes 35S-*Ac*-*rol*C and rbcS-*Ac*-*rol*C were morphologically similar to the untransformed aspens, in terms of leaf size, shape and color, indicating that at least the phenotypic expression of the *rol*C gene has been repressed by *Ac*. The *Ac* excision events were monitored by variegated spots on the leaves. There were significantly more *Ac* excisions in the transgenic plants carrying 35S-*Ac*-*rol*C than in rbcS-*Ac*-*rol*C transgenes. These observations suggest that the type of promoter and/or gene construct seems to influence the phenotype and physiology in the transgenic plants. Transformation was confirmed by PCR amplifications of the *rol*C and *Ac* coding regions, and Southern blot analysis was carried out to determine the copy number of the transgenes and their relationship to the morphogenetic response, as well as genetic stability in the transgenic aspens.

1. Introduction

Traditional breeding and selection methods have been and will continue to be employed for the genetic improvement of forest trees. However, these approaches may have limited application in forest trees because of long generation cycles. Genetic engineering, on the other hand, offers prospects for tree improvement at an accelerated rate in the tree species. A major concern regarding gene transfer in plants and animal systems is how to regulate or direct where and when a transgene would be expressed. Although model promoters have been investigated in a number of herbaceous plant species, very little is known about their expressive control in the woody plants. A survey of published reports on transgenic annual crops has revealed that expression of transgenes is less stable than had been originally thought. Transgene inactivation has been observed in annual plants in the form of position effects, gene silencing and cosuppression (Meyer et al. 1993; Finnegan and McElroy, 1994; Van Blokland et al. 1994). That

M. R. Ahuja et al. (eds.), Somatic Cell Genetics and Molecular Genetics of Trees, 89–96.
© 1996 *Kluwer Academic Publishers. Printed in the Netherlands.*

might suggest that it is hard to predict if a transgene would be stably expressed in plants, whether an annual or a perennial.

Forest trees are long-lived and have extended vegetative phase ranging from one to several decades. They are constantly exposed to changing environments over a long period of time, which may influence their physiology and complex morphogenetic processes. Therefore, trees may require special considerations when genetically transformed. It is relevant to ask if foreign genes will be stably integrated and expressed in the forest trees (Ahuja, 1988a, 1988b, 1988c). Are the newly introduced genes expressed immediately in a specific tissue or in all tissues, or some remain silent for a long time, or they are lost during the long vegetative phase of the trees, or perhaps cause genetic changes in the host genome by position effect, or by genetic rearrangements?

We have employed *Populus* as a model system to determine the stability/instability of chimeric gene expression. Trees of this genus are fast-growing, have short-rotation cycles, and can be cultivated on marginal sites. In addition, well developed regeneration systems (prerequisite for gene transfer) are available in many *Populus* species and hybrids (see Ahuja, 1986, 1987, 1993; Ernst, 1993; Chun, 1993). Aspens and poplars have been transformed by *Agrobacterium* vector system (Fillatti et al. 1987; Klopfenstein et al. 1991,1993; Olsson et al. 1992; Confalonieri et al. 1994). We have employed *rol*C gene from *Agrobacterium rhizogenes* under the control of two different promoters, and transformed aspens by *Agrobacterium*-mediated gene transfer. *Rol*C is a dominant, pleiotropic gene, with very pronounced phenotypic and physiologic effects, as has been shown in transgenic tobacco (Spena et al. 1987; Schmülling et al. 1988), and transgenic potato (Fladung, 1990; Fladung and Ballvora, 1992). In other words, *rol*C gene affects whole plant development in transgenics by altering the steady-state levels of several endogenous growth regulators (Schmülling et al. 1993; Nilsson et al, 1993). Since *rol*C gene is expressed at the morphological level, it can be employed as a selectable marker to monitor its phenotypic and molecular expression, as well as genetic stability in the transgenic plants. A morphological marker has a distinct advantage, when compared to the time-consuming analysis of biochemical markers at specific intervals, because the phenotype can be observed at all times during the life cycle of a transgenic plant.

In this report we describe the phenotypic effects of the *rol*C gene in transgenic *Populus* under the control of two different promoters: 35S from the cauliflower mosaic virus and rbcS from potato. In addition, we have also observed transposition events caused by the activator (*Ac*) element from maize, when transferred along with the *rol*C gene in aspen. Transformation and genetic stability events were monitored in the transgenic aspen at the molecular level.

2. Materials and Methods

Three clones (W52, W3, and Braunal1) from European aspen (*Populus tremula* L.) and one clone (Esch5) from hybrid aspen (*P. tremula* L. x *P. tremuloides* Michx.) were genetically transformed by a leaf disc-*Agrobacterium* cocultivation method (Horsch, 1985), and transgenic plants were regenerated on a previously established micropropagation method in aspen (Ahuja, 1986, 1987). Four chimeric gene constructs (35S-*rol*C, rbcS-*rol*C, 35S-*Ac*-*rol*C, and rbcs-*Ac*-*rol*C) (Spena et al. 1987; Spena et al. 1989; Jones et al. 1992; Schmülling et al. 1993) were introduced into aspen by an *Agrobacterium tumefaciens* binary vector system. The transformed leaf discs were cultured on a shoot-induction medium containing 0.5 mg/l BAP, 0.02 mg/l NAA, 500 mg/l Cefotaxim and 50 mg/l kanamycin sulphate. The shoots were rooted on a medium supplemented with 0.5 mg/l IBA, 0.1 mg/l NAA, 500 mg/l Cefotaxim and 50 mg/l, and then transferred to a hormone-free medium, but containing the antibiotics.

DNA and RNA were isolated from leaf tissues according to previously described methods (Edwards et al. 1991; Doyle and Doyle, 1987; Logemann et al. 1987). PCR amplifications for the *rol*C and *Ac* coding sequences were carried out as described by Fladung et al.(1996). The number of integrated copies of the transgenes were determined by a non-radioactive Southern hybridization method (Fladung and Ahuja, 1995) , by employing two restriction enzymes (EcoRI and HindIII). The detailed protocols for genetic transformation, bacterial strains carrying the chimeric genes, and molecular analysis are described elsewhere (Fladung et al. 1996).

3. Results and Discussion

3.1 THE ROLC SYNDROME

We have raised a fairly large population (more than 1000) of putative transgenic plants (all micropropagated on a kanamycin-containing medium) in the greenhouse during the last two years. A total of 124 different independent transformants were investigated following genetic transformation of the four aspens and hybrid aspen clones with the four gene construct (Fladung et al. 1996). In this report only data on the transgenic plants derived from hybrid aspen clone (Esch5) would be presented. The morphology and physiology of the transgenic plants seems to be influenced by the type of promoter employed: 35S-*rol*C transgenics plants showed much smaller leaves and reduced height (Figure 1) as compared to those in the rbcS-*rol*C transgenic aspens; the latter approaching the leaf size and height of the untransformed controls (Figure 2). However, all genotypes transformed with 35S-*rol*C and rbcS-*rol*C chimeric genes showed pale-green coloration of the leaves as compared to dark-green leaf color in the controls. In addition, a strong promotion of root growth was observed in the aspen plantlets in culture carrying the 35S-*rol*C gene. Most clones morphologically characterized as *rol*C transgenic tested positive in the PCR analysis (Figure 3). However, one transgenic clone did not show the phenotypic characteristics of the *rol*C gene, but revealed an amplification product specific for the *rol*C primer by a PCR analysis.

Figure 1. Comparison of untransformed hybrid aspen clone Esch5 (right) with the 35S-*rol*C transgenic plants of the same clone. Note phenotypic variability between 35S-*rol*C transgenic plants.

A total of 17 independently regenerated transformants were analysed for the rbcs-*rol*C gene construct. Of these 15 showed pale-green leaves, typical of the *rol*C syndrome, but two clones showed dark green leaves. Although all the transformants were isolated on a kanamycin selectable medium, these two clones may either have lost the *rol*C gene, or there was transgene inactivation. From these 17 putative rbcS-*rolC* transgenic clones, 16 were positive for the *rol*C gene in the PCR analysis. Obviously, one clone was deficient in the *rol*C gene. However, the second clone phenotypically *rol*C negative, was positive for the *rol*C gene in the PCR amplification analysis. We are examining questions on transgene inactivation in *rol*C gene under the control of two types of promoters.

The phenotype of the transgenic aspen and hybrid aspen conditioned by the *rol*C gene under the expressive control of cauliflower mosaic 35S promoter, as well as the light inducible rbcS promoter from potato, resembled, to some extent, that observed in transgenic potato, but differed from that in tobacco. In the 35S-*rol*C transgenic potato (Fladung, 1990; Fladung et al. 1993) and aspen, smaller leaves having normal shape or a slight tendency towards lanceolate leaf shape were produced. On the other hand, in 35S-*rol*C transgenic tobacco, mainly lanceolate leaves were induced (Schmülling et al 1988). However,

Figure 2. Leaves from control (Esch5) clone, and 35S-*rol*C transgenics (A), and rbcS-*rol*C transgenics (B) of the same clone Esch5. In addition to reduction in leaf size, the leaf color was pale-green in the transgenics conditioned by the 35S-*rol*C and rbcS-*rol*C gene constructs.

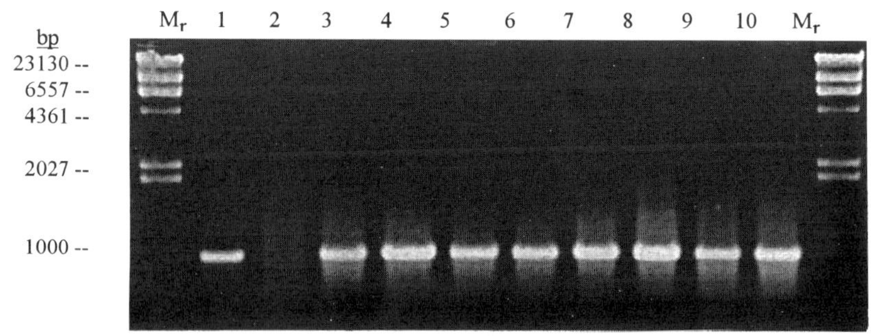

Figure 3. PCR amplification of the *rol*C coding sequence (lane 1), negative amplification in untransformed Esch5 clone (lane 2), and *rol*C positive, independent regenerants, transgenic plants (lanes 3-10). Molecular weight markers are included on left and right.

lanceolate-type leaves were also observed in the 35S-*rol*C transgenic plantlets *in vitro*. Regardless of the promoter, one of the main effect of the *rol*C gene was on the leaf color, that is pale-green, which was the same in the two herbaceous plant species, tobacco, potato, and a woody perennial, aspen.

3.2 THE TRANSPOSABLE ELEMENT AC

*Rol*C as a phenotypic marker has another advantage for studies in developmental biology, as its expression may be altered/silenced by position effect when a transposable element, *Ac*, is inserted between the promoter and the *rol*C coding sequences in the chimeric gene construct. However, after the transposition of *Ac*, the original *rol*C phenotype would be restored. This has been observed in tobacco (Spena et al 1989) and tomato (Jones et al. 1992). Whether the type of promoter would influence the transposion events in woody plants in not known.

We have tested the transposition events of *Ac*, in *rol*C transgenic aspens carrying two types of promoters: 35S-*Ac*-*rol*C and rbcS-*Ac*-*rol*C. The transgenic aspens carrying both types of gene constructs grow normally and are not distinguishable from the untransformed controls, in terms of height, leaf shape, leaf size, and leaf color. However, transgenic plants showing pale-green sectors on the green background of the leaf, indicate that transposition of *Ac* might have occurred in the leaf cells, thus allowing the *rol*C expression in the pale-green sectors. The number of pale-green spots on the leaves varied from one transgenic plant to another, suggesting that *Ac* excisions are random (Figure 4). In the Esch5 clone transgenic for 35S-*Ac*-*rol*C construct, 15 of 27 clones tested showed variegated leaves, and 12 transgenic clones exhibited dark green leaves. The PCR amplification analyses revealed that 23 out of 27 transgenic plants carried the *rol*C gene, while 21 out of 27 were tested positive for the *Ac* coding sequence. That means that some of the plants were either lacking in *rol*C and/or *Ac* genes.

Figure 4. Pale-green sectors of different sizes on the leaves from 35S-*Ac*-*rol*C transgenic Esch5 plants (A-D). These sectors are probably caused by the excision events of the *Ac* transposon from the 35S-*Ac*-*rol*C gene construct.

In the Esch5 transgenic containing the rbcS-*Ac*-*rol*C gene construct, only 2 out of 19 clones showed variegated leaves, and the remaining 17 exhibited normal green leaves. The PCR analysis revealed that 18 out of 19 carried the *rol*C as well as the transposable element *Ac*. The frequency of *Ac* transpositions were considerably lower in the rbcS-*Ac*-*rol*C transgenics, as compared to those in the 35S-*Ac*-*rol*C transgenics. Whether this difference is due to the promoter employed is not clear at this stage. However, the gene construct rbcS-*Ac*-*rol*C yielded a high frequency of *Ac* transpositions in tomato (Jones et al. 1992). This differences in the frequency of transpositions between aspen vs. tomato may possibly be attributed to the genetic background in which the gene construct is tested.

Although earlier studies (Howe et al. 1991, 1994) have indicated that it is possible to transfer the transposable element *Ac* from maize into poplars and demonstrate its presence and excisions at the molecular level, transpositions events have not been investigated at the phenotypic and histogenesis level in the transgenic aspen so far. We believe this is the first report which describes the *Ac* transpositions events in transgenic woody plants.

3.3 INTEGRATION AND EXPRESSION OF CHIMERIC GENES

Southern blot hybridization was carried out to confirm the PCR results and to determine the number of integrated copies of the transgenes in the transgenic aspens. Such analysis revealed that most of the independent primary transgenic clones contained one integrated copy of the *rol*C or *Ac*; however, some transgenics carried 2 or more copies (Figure 5). For example, out of 19 transgenic clones tested, 15 clones contained one copy each, three clones contained two copies each, and one clone had more than 3 integrated copies of the *rol*C gene. In other words, all transgenic clones tested contained at least one copy of the *rol*C gene. On the other hand, in the 35S-*Ac*-*rol*C transgenics, there were several plants which did nor reveal the presence of the transgenes. Out of 19 transgenics (35S-*Ac*-*rol*C) analysed, three plants did not carry the *rol*C gene, whereas 12 contained one copy each, three carries 2 copies each, and one had more than 3 copies. In the same transgenic sample of 19, five did not contain the *Ac* gene, while 11 carried one copy each, one had 2 copies, and 2 carried more than 3 copies each of the *Ac* gene. It would appear that some of the transgenics that had been selected on the kanamycin medium, may have lost of the transgene during the vegetative growth of the aspen trees.

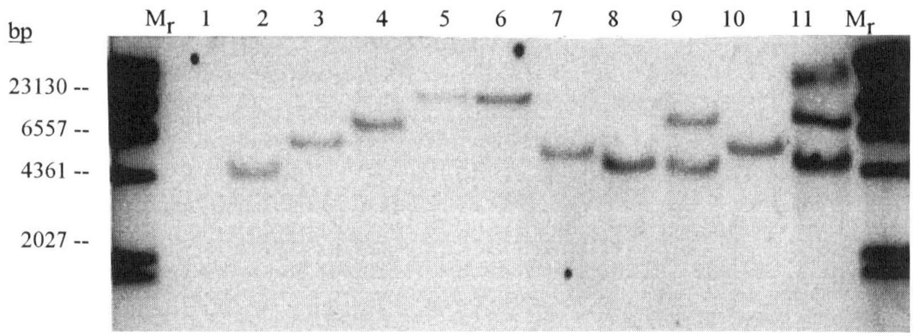

Figure 5. Non-radioactive Southern blot analysis of 35S-*rol*C transgenic hybrid aspen plants showing 1-3 integrated copies of the transgene (lanes 2-11). Untransformed Esch5 (lane 1) and molecular weight markers are shown on left and right.

The effect of the copy number on the transgenic plants is not well understood. In our experiments with transgenic aspen, we observed that plants with one or two integrated copies of the *rol*C gene did not appreciably deviate from the *rol*C phenotype. However, those transgenic plants carrying more than two copied of the integrated *rol*C did not survive after one year growth in the greenhouse. These plants overwintered in the greenhouse, but the buds did not break dormancy, and the plants died. Whether this phenomenon associated with the integrated copy number of *rol*C gene is specific to aspen is not known. It is possible that effects of high copy number of *rol*C as compared to other genes may be different. Alternatively, it may not be the copy number of the transgene, but the site of integration of the transgene(s), which might be important in determining the developmental course of a transgenic plant.

Northern blot analysis of the 35S-*rol*C transgenic plants revealed that the *rol*C gene is well expressed in the leaves and weakly in the stems of the greenhouse-grown plants. On the other hand, in those transgenics where *rol*C gene is controlled by the light-inducible rbcS promoter, the *rol*C expression was only detected in the leaves. Although, we have not checked the *rol*C expression in the roots, but predict that in those transgenics in which *rol*C expression is controlled by the constitutive 35S promoter, the *rol*C gene would be expressed, but very likely not in the roots of the transgenic plants in which *rol*C expression is governed by a tissue-specific rbcS promoter.

4. Concluding Remarks

It is clear from the above discussion that *rol*C gene can be expressed constitutively or only in a specific tissues, depending on the type of promoter employed. In addition, *rol*C gene expression can be expunged by the transposable element *Ac* when it is inserted next to the *rol*C in the chimric gene construct. However, following excision of *Ac*, *rol*C activity seems to be restored in the leaf tissues as monitored by the presence of pale-green spots on the green leaf tissue. We are continuing the Southern hybridization and northern blot analysis on the transgenic plants showing growth abnormalities, including chimeric plants. Further, we are investigating the mechanism of transgene inactivation in the transgenic aspens.

5. Acknowledgements

This research was in part suppoprted by a project AIR-2-CT94-1571 "Stability and expression of foreign genes in *Populus*" from the European Union, Brussels. We thank Mr. Olof Nowitzki for the technical assistance, and Hans Seuthe and Willi Holzwart for the greenhouse work.

6. References

Ahuja, M. R. (1986) Aspen, in Evans, D.E., Sharp, E.R. and Ammirato, P.J. (eds.), *Handbook of Plant Cell Culture,* vol. 4. Macmillan Publishing Company, New York, pp. 626-651.

Ahuja, M.R. (1987) In vitro propagation of poplar and aspen, in Bonga, J.M. and Durzan, D.J. (eds.), *Cell and Tissue Culture in Forestry*, vol. 3, Martinus Nijhoff Publishers, Dordrecht, pp. 207-223.

Ahuja, M.R. (1988a) Gene transfer in forest trees, in Hanover, J.E. and Keathley, D.E. (eds.), *Genetic Manipulation of Woody Plants,* Plenum Press, New York, pp. 25-41.

Ahuja, M.R. (1988b) Gene transfer in woody plants: problems and perspectives, in Ahuja, M.R. (ed.), *Somatic Cell Genetics of Woody Plants*, Kluwer Academic Publishers, Dordrecht, pp. 83-101.

Ahuja, M.R. (1988c) Molecular genetics of transgenic plants, in Hällgren, J.E. (ed.), *Molecular Genetics of Forest Trees*, Swedish University of Agricultural Sciences, Umeå, pp. 127-145.

Ahuja, M.R. (1993) Regeneration and germplasm preservation in aspen, in Ahuja, M.R. (ed.), *Micropropagation of Woody Plants*, Kluwer Academic Publishers, Dordrecht, pp. 187-194.

Chun, Y.W. (1993) Clonal propagation in non-poplar aspen poplar hybrids, in Ahuja, M.R. (ed.), *Micropropagation of Woody Plants*, Kluwer Academic Publishers, Dordrecht, pp. 209-222.

Confalonieri, M., Balestrazzi, A. and Bisoffi, S. (1994) Genetic transformation of *Populus nigra* by *Agrobacterium tumefaciens*, *Plant Cell Reports* 13, 256-261.

Doyle, J.J. and Doyle, J.L. (1987) A rapid DNA isolation procedure for small quantities of fresh leaf tissue, *Phytochemical Bull.* 19,11-15

Edwards, K. Johnstone, C. and Thompson, C. (1991) A simple and rapid method for the preparation of plant genomic DNA for PCR analysis, *Nucl. Acids Res.* 19,1349.

Ernst, S.G. (1993) *In vitro* culture of pure species non-aspen poplars, in Ahuja, M.R. (ed.), *Micropropagation of Woody Plants*, Kluwer Academic Publishers, Dordrecht, pp. 195-207.

Fillatti, J.J., McCown, B.H., Selmer, J., Haissig, B.E. and Comai, L. (1987) *Agrobacterium*-mediated transformation and regeneration of poplar, *Molec. Gen. Genetics* 206, 192-199.

Finnegan, J. and McElroy, D. (1994) Transgene inactivation: plants fight back, *Bio/Technology*, 12, 127-145.

Fladung, M. (1990) Transformation of diploid and tetraploid potato clones with the *rolC* gene of *Agrobacterium rhizogenes* and characterization of transgenic plants, *Plant Breeding* 104, 295-304.

Fladung, M. and Ahuja, M.R. (1995) 'Sandwich' method for non-radioactive hybridization, *Biotechniques* 18, 3-5.

Fladung, M. and Ballvora, A. (1992) Further characterization of *rolC* transgenic tetraploid potato clones, and influence of daylength and level of *rolC* expression on yield parameters, *Plant Breeding* 109, 18-27.

Fladung, M., Ballvora, A. and Schmülling, T. (1993) Constitutive or light regulated expression of the *rolC* gene in transgenic potato plant has different effects on yield attributes and tuber carbohydrate composition, *Plant Mol. Biol.* 23, 749-757.

Fladung, M., Kumar, S. and Ahuja, M.R. (1996) Genetic transformation of *Populus* genotypes with different chimeric gene constructs: Transformation efficiency and molecular analysis, *Transgenic Res.* 5, 1-11.

Howe, G.T., Strauss, S.H. and Goldfarb, B. (1991) Insertion of the maize transposable element *Ac* into poplar, in Ahuja, M.R. (ed.), *Woody Plant Biotechnology*, Plenum Press, New York, pp. 283-294.

Howe, G.T., Goldfarb, B. and Strauss, S.H. (1994) *Agrobacterium*-mediated transformation of hybrid poplar suspension cultures and regeneration of transformed plants, *Plant Cell Tissue & Organ Culture* 36, 59-71.

Jones, J.D.G., Bishop, G., Carrol, B., Dickson, M., English, J., Harrison, K., Jones, D., Scofield, S. and Thomas, C.M. (1992) Prospects of establishing a tomato gene tagging system using the maize transposon activator (Ac). *Proc. Royal Soc. Edinburgh*, 99B, 107-119.

Klopfenstein, N.B., Shi, N.Q., Kernan, A., McNabb, H.S., Hall, R.B., Hart, E.R. and Thornburg, R.W. (1991) Transgenic *Populus* hybrid expresses a wound inducible potato proteinase inhibitor II - CAT gene fusion, *Can. J. For. Res.* 21, 1321-1328.

Klopfenstein, N.B., McNabb, H.S., Hart, E.R., Hall, R.B., Hanna, R.D., Heuchelin, S.A., Allen, K., Shi, N.Q. and Thornburg, R.W. (1993) Transformation of *Populus* hybrids to study and improve pest resistance, *Silvae Genetica* 42, 86-90.

Logemann, J., Schell, J. and Willmitzer, L. (1987) Improved method for the isolation of RNA from plant tissues, *Anal. Biochemistry* 163, 16-20.

Meyer, P., Heidemann, I. and Niedenhof, I. (1993) Differences in DNA methylation are associated with paramutation phenomenon in transgenic petunia, *The Plant J.* 4, 89-100.

Nilsson, O., Moritz, T., Imbault, N., Sandberg, G. and Olsson, O. (1993) Hormonal charaterization of transgenic tobacco plants expression the *rolC* gene of *Agrobacterium rhizogenes* TL-DNA, *Plant Physiol.* 102, 363-371.

Olsson, O., Nilsson, O., Sundberg, B., Little, C.H.A. and Sandberg, G. (1992) *RolC* biosynthesizing transgenic *Populus* plants goes bonzai, *Fifth Workshop of the IUFRO WP Molecular Genetics*, Carcans, France, p. 3.6.

Schmülling, T., Schell, J. and Spena, A. (1988) Single genes from *Agrobacterium rhizogenes* influence plant development, *EMBO J* 9, 2621-2639.

Schmülling, T., Fladung, M., Grossmann, K. and Schell, J. (1993) Hormonal content and sensitivity of transgenic tobacco and potato plants expressing single *rol* genes, *The Plant J.* 3, 587-598.

Spena, A., Schmülling, T., Koncz, C. and Schell, J. (1987) Independent and synergistic activity of *rol A, B* and *C* loci in stimulating abnormal growth in plants, *EMBO J.* 6, 3891-3899.

Spena, A., Aalen, R.B. and Schulze, S.C. (1989) Cell autonomous behavior of the *rolC* gene of *Agrobacterium rhizogenes* during leaf development: a visual assay of transposon excision in transgenic plants, *The Plant J.* 1, 1157-1164.

Van Bokland, R., Van der Geest, N., Mol, J.N.M. and Kooster, J.M. (1994) Transgene-mediated suppression of chalcone synthase expression in *Petunia hybrida* results from increase inn RNA turnover, *The Plant J.* 6, 861-877.

STABLE GENETIC TRANSFORMATION IN BLACK SPRUCE AND TAMARACK AND THE TRANSGENIC EXPRESSION OF CONIFER GENES

P.J. CHAREST, D. LACHANCE, Y. DEVANTIER, C. JONES, L. TIAN, B. XUE, A. SEGUIN, K. KLIMASZEWSKA, AND B. RUTLEDGE
Molecular Genetics and Tissue Culture Group, Petawawa National Forestry Institute, Canadian Forest Service, Natural Resources Canada, Chalk River, Ontario, Canada, K0J 1J0.

1. Abstract

Stable genetic transformation of black spruce (*Picea mariana*) and tamarack (*Larix laricina*) was obtained via microprojectile bombardment of tissues from two stages of somatic embryogenesis (mature cotyledonary somatic embryos and suspension from embryonal masses) using the Biolistic PDS-1000/He device. A total of 14 transgenic lines were obtained with the two species with the vectors pRT99GUS and pBI426 using low level kanamycin selection. The integration of the foreign genes was confirmed by Southern hybridization and complex patterns were observed. Transgenic seedlings were regenerated for both species and expression of the GUS gene observed in needles. Two chimeric genes containing components from black spruce genes were used for transient gene expression into conifer tissues; a cDNA (MBF1) coding for a *myb* C1 related gene and a chalcone synthase promoter. Both components were proven functional when reintroduced into their original genome and in *L. laricina* tissues. Furthermore, the chimeric MBF1 gene was functional when introduced into corn cell suspension. These results indicate the evolutionary conservation of the genes involved in the anthocyanin pathways in gymnosperms and angiosperms. Additionally, the Green Fluorescent Protein gene from a jellyfish *(Aequorea victoria)* was evaluated as a marker for genetic transformation in conifer tissues.

2. Introduction

Recent advances in the characterization of conifer genes have shown the need for efficient genetic transformation procedures in this class of plants. Genes for wood quality, photosynthesis, and developmental pathways have been isolated and characterized (Charest et al. 1993a, Charest and Michel 1991). To be able to properly study gene structure and function, the reintroduction of the isolated gene back into its original genome is required to allow for identification of gene components and factors involved. Some of this work can be done in heterologous systems such as *Arabidopsis* and tobacco but this does not provide definitive data on gene regulation mechanisms present in gymnosperms.

M. R. Ahuja et al. (eds.), Somatic Cell Genetics and Molecular Genetics of Trees, 97–104.
© 1996 *Kluwer Academic Publishers. Printed in the Netherlands.*

Successful regeneration of transgenic conifers is relatively recent. Transgenic *Larix decidua* (European larch) trees have been produced following *Agrobacterium rhizogenes*-mediated transformation of aseptically grown seedlings (Huang et al. 1991) and transgenic *Picea glauca* (white spruce) have been regenerated following electric discharge particle bombardment of mature cotyledonary somatic embryos (Ellis et al. 1993). In both cases, the frequencies of transformation were low; only a few trees have been produced with *L. decidua* (Huang et al. 1991) and less than three dozen transgenic lines have been produced with *P. glauca* so far (Ellis, personal communication). In previous attempts, conifer transgenic tissues without tree regeneration were produced using microprojectile bombardment of somatic embryos in *Picea abies* (Norway spruce; Robertson et al. 1991) and in *P. glauca* (Bommineni et al. 1993), and tumor tissues via *Agrobacterium* transformation in several conifer species (reviewed by Charest and Michel 1991). Other experiments have led to the successful introduction of foreign genes in conifers via particle bombardment in various tissues (Charest et al. 1993a, 1993b, Seguin et al. 1995) and via electroporation of protoplasts (reviewed by Charest and Michel 1991, Seguin et al. 1995) without transgenic tissue recovery.

We report here results on stable genetic transformation and regeneration of transgenic trees for two conifers, *Picea mariana* and *Larix laricina*. This was achieved by using somatic embryogenesis combined with microprojectile bombardment with the helium Biolistic™ device. Furthermore, we present data on the expression of two chimeric black spruce genes (*myb* homologue and chalcone synthase) reintroduced back into conifer tissues by microprojectile bombardment.

3. Material and Methods

3.1 TREE TISSUES

Two embryogenic cell lines of *Picea mariana* (Mill.) B.S.P. (R4F14 and 119794-014) and one line of *Larix laricina* (Du Roi) K. Koch (L2) were used. These lines were cultured as described by Cheliak and Klimaszewska (1991) and Klimaszewska et al. (1995). Maturation and germination of somatic embryos were done according to Lelu et al. (1993).

3.2 PLASMID VECTORS

The plasmids for stable transformation were pRT99GUS which carried both the GUS (β-glucuronidase) gene and the NPT II (neomycin phosphotransferase II) gene under the control of 35S promoters from Cauliflower mosaic virus (Topfer et al. 1988), and pBI 426 which carried a translational fusion between the GUS coding region and the NPT II coding region (Datla et al. 1991) under the control of a double 35S promoter (Kay et al. 1987) linked to a translational enhancer from alfalfa mosaic virus (Datla et al. 1993). Plasmids for transient gene expression were pPCV8145, which carried a quadruple 35S promoter with the TMV translational enhancer linked to the firefly luciferase gene (O. Olsson, unpublished), pR and pC1, each of which carried a 35S promoter with the coding regions for the R and C1 genes, respectively (Bodeau and Walbot 1992), pBz2IL which carried the Bronze gene promoter with the alcohol dehydrogenase 1 intron linked to the firefly luciferase gene (Bodeau and Walbot 1992), pMBF1 which carried a black spruce *myb* gene homologue (cDNA) driven by the 35S promoter (Xue and Rutledge, in preparation) and pCHS-3 which carried a GUS gene linked to a black spruce chalcone synthase promoter (Xue and Rutledge, unpublished). Plasmid SLJ4D4 carried the GUS gene linked to a 35S promoter and a TMV translational enhancer (Jones et al. 1992) and was used to construct pCHS3 and pMBF1. Plasmid pBlueScript II KS-mGFP4 was made by taking the HindII-EcoRI fragment from pBIN 35S-mGFP4 (Dr. Jim Haseloff, MRC Lab. Molecular Biology, Addenbrookes Hospital, Cambridge, England) containing the modified Green Fluorescent Protein (GFP) gene from a jellyfish (*Aequorea victoria*) linked to a 35S promoter and a NOS (nopaline synthase) terminator.

3.3 MICROPROJECTILE BOMBARDMENT

The microprojectile bombardment procedures using the PDS-1000/He Biolistic™ device were carried out according to Charest et al. (1995), Seguin et al. (1995), and Klimaszewska et al. (1995) for stable genetic transformation, and according to Charest et al. (1993b) and Seguin et al. (1995) for transient gene expression.

3.4 β -GLUCURONIDASE ASSAYS

The β-glucuronidase fluorescent and histochemical assays were performed according to Jefferson et al. (1987).

3.5 NEOMYCIN PHOSPHOTRANSFERASE ELISA ASSAY

The NPTII ELISA assay was performed according to the manufacturer's recommendations (5 Prime --> 3 Prime Inc., Co, USA).

3.6 LUCIFERASE ASSAY

The Luciferase Assay System (Promega Product E1500, Fisher Scientific, Ottawa, Canada) was used to detect transient gene expression of the firefly luciferase, following the manufacturer's recommendations.

3.7 SOUTHERN HYBRIDIZATION

Southern hybridization experiments were performed as described by Charest et al. 1995.

4. Results and Discussion

4.1 STABLE GENETIC TRANSFORMATION IN BLACK SPRUCE AND TAMARACK

Recovery of stabley transformed embryogenic cell lines of black spruce and tamarack was possible by using the helium-driven particle gun (Biolistic PDS-1000/He™) to accelerate gold microparticles into cotyledonary somatic embryos and embryonal mass suspensions. As in the study of Ellis et al. (1993), a low level of kanamycin selection (10μg/mL for somatic embryos and 25 μg/mL for embryonal masses) was required. With somatic embryos, secondary somatic embryogenesis was initiated to recover embryonal cell masses. Once embryonal masses growing on kanamycin selection were identified as having GUS positive sectors, they were transferred to liquid medium with a low level of kanamycin (750μg/L) to eliminate non-transformed or non-expressing tissues. A total of 12 transgenic lines were obtained with black spruce, seven from cotyledonary somatic embryo bombardment and five from embryonal cell mass suspension bombardment. The frequency of transgenic cell line recovery was low (1% for somatic embryos and 0.6% for embryonal cell masses) and labour intensive (Table 1). Two transgenic lines were obtained with tamarack following bombardment of 814 cotyledonary somatic embryos with the vector pBI426 and selection on kanamycin. Methotrexate and hygromycin were tested as selective agents for transformation with their respective resistance genes (dihodrofolate reductase and hygromycin phosphotransferase) in tamarack and the bombardment of 1064 somatic embryos did not yield any transgenic lines.

Table 1. Summary of bombardment experiments with black spruce tissues

Quantity of tissue bombarded	# of different experiments	# of tissue pieces surviving km selection	# of pieces that are GUS+ and NPT II+	# of pieces that are GUS- and NPT II+	% escape
2010 somatic embryos (67 Petri dishes)	5	711	6*	1	99.0
61.8 mL of packed cell suspension (206 Petri dishes)	9	779	4	1	99.4

*One of the lines was lost after several months of subculture.

All transgenic lines had the morphological characteristics of embryogenic tissues, with immature somatic embryos having well-formed meristematic heads with suspensor cells. The transgenic lines expressed the GUS and NPT II genes at various levels, with some black spruce lines not showing any expression of the NPT II gene. Furthermore, one transgenic line from tamarack (line #2D1) yielded an apparent tissue-specific expression of the GUS gene restricted to the early stage in somatic embryogenesis up to the germination stage and to the meristematic areas of the germinated seedlings.

Four transgenic black spruce lines were characterized further (Table 2) and complex patterns of integration of the introduced genes were detected by Southern hybridization. In three of the lines, the expected diagnostic fragments for the GUS and NPT II genes were present and several other bands hybridizing with the GUS and NPT II probes were also detected. In one line (EM-B3), only bands of unexpected sizes were detected. The number of integration sites ranged from three to eight and there was an apparent correlation with the level of NPT II enzyme detected by ELISA and the level of GUS enzyme activity detected by the MUG fluorescent assay. The level of kanamycin resistance showed by the transgenic lines did not correlate with the number of integration sites.

Transgenic trees were regenerated for both transgenic black spruce and tamarack; however, some of the transgenic lines lost their regeneration potential and, despite several attempts, transgenic trees were not recovered from those (e.g., second transgenic line #2D1 from tamarack showing constitutive expression of the introduced GUS gene.

Table 2. Characterization of black spruce transgenic lines

	Southern hybridization results[a]					Gene expression results[b]		
Line #	NPT II band	Extra NPT II bands	GUS band	Extra GUS bands	# of integration sites	NPT II enzyme level	GUS enzyme activity	Km resistance
SE-A1	+	3	+	3	4	3	2	1
SE-A4	+	-	+	4	5	2	3	2
SE-C2	+	7	+	8	>8	1	1	3
EM-B3	-	5	1	4	3	4	4	4

a) Southern hybridizations were done by digesting the DNA with restriction enzymes i) to yield a specific diagnostic fragment for both the NPT II and GUS genes and ii) to cut on one side of the introduced genes giving fragments of different sizes depending on the integration site on the transformation vector.

b) The gene expression results give a ranking of the four lines characterized, 1 being the strongest expressor and 4 being the lowest expressor. For the details of the results, see Charest et al. (1995).

4.2 TRANSIENT EXPRESSION OF CHIMERIC CONIFER GENES

Two components of black spruce genes, a cDNA from a *myb* C1 homologue and a chalcone synthase promoter, were used in chimeric gene constructs to examine their expression when reintroduced back into conifer tissue culture systems by microprojectile bombardment. The cDNA clone MBF1 was isolated from a female cone cDNA library of black spruce and was linked to a 35S promoter with a TMV translational enhancer. It was compared with the corn homologue gene C1 (Bodeau and Walbot 1992) linked to a 35S promoter (pC1) used in combination with the *myc* homologue R linked to a 35S promoter (pR) to transactivate the Bronze 2 promoter with the alcohol dehydrogenase intron linked to the luciferase gene (pBZ2IL) in tamarack embryonal masses (Figure 1A). The control vector pPCV8145 yielded high levels of transient luciferase activity. The vector combinations of pR/pC1/pBz2IL and pR/pMBF1/pBz2IL yielded higher transient luciferase activity demonstrating that both combinations of genes were effective in transactivating the Bz2 promoter in conifer tissues. Furthermore, red pigments were produced into tamarack and black spruce embryonal masses, into corn cell suspension, and into black spruce female flower tissues when both pMBF1 and pR were co-transformed by bombardment. This strongly suggests evolutionary conservation of the transactivating factors involved in the anthocyanin pathway in angiosperms and gymnosperms.

The other component tested, a chalcone synthase promoter, was isolated from a genomic bank probed with a cDNA isolated from an embryonal mass library. The promoter was linked to a GUS gene present in SLJ4D4 (Jones et al. 1992) and bombarded into embryonal masses of black spruce (Figure 1B). The promoter yielded a basal level of GUS gene activity but was activated further when co-transformed with pR/pC1 or pR/pMBF1. This finding confirmed the results with the MBF1 gene about the activation of the anthocyanin pathway in conifers because chalcone synthase is an enzyme part of this pathway.

These results indicating a conservation of gene function between gymnosperms and angiosperms suggest that knowledge on gene structure and function gain with crop plants will be applicable to gymnosperms and that, in general, gene constructs from angiosperms can be used to genetically engineer gymnosperms in a predictable manner.

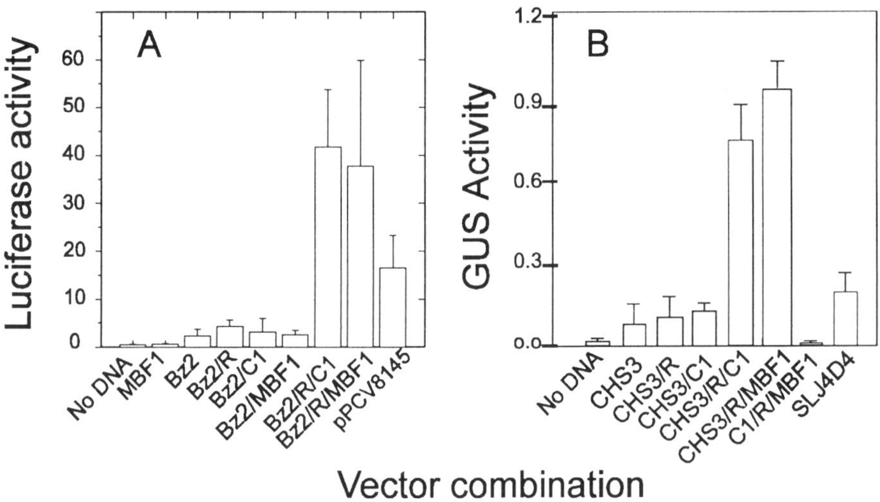

Figure 1. Expression of chimeric conifer genes following bombardment into embryogenic tissues of larch and spruce. Luciferase activity is in photon events X 1000/mg of protein extracted. GUS activity is in μole of MUG/mine/mg of protein extracted. A) Results from bombardment of tamarack embryonal cell masses and B) results from bombardment of black spruce embryonal cell masses. Results are given ± standard deviation. The results were from three repetitions and five samples. The coding in the figure correspond to the plasmid used (e.g. MBF1=pMBF1).

4.3 EVALUATION OF THE GREEN FLUORESCENT PROTEIN FROM JELLYFISH AS A MARKER FOR CONIFER TRANSFORMATION

To improve genetic transformation efficiency with the conifer species used in our laboratory, we evaluated the Green Fluorescent Protein (GFP; Chalfie et al. 1994) gene from the jellyfish *Aequorea victoria* as a screenable marker. The vector used comprised a modified GFP gene in which the coding region was mutated to remove a plant cryptic intron (mGFP4, Dr. Jim Haseloff, unpublished) linked to a 35S promoter into pBlueScript II KS (HindIII-EcoRI fragment). Black spruce pollen, embryonal masses, and somatic embryos were bombarded with this vector construct and yielded transient gene expression in recipient cells as detected by the emission of green light when the cells were excited with blue light (395nm) using fluorescence microscopy (Figure 2). The frequency of positive cells obtained with embryonal masses were equivalent to results obtained with the GUS gene in a similar construct. The wild type GFP gene (with the presumed cryptic intron) yielded a very low level of expression in the tissues tested. These results suggest that the modified GFP can be used for gene expression study in conifers and that it has the potential to be used as a screenable marker for stabley transformed tissues resulting from microprojectile bombardment.

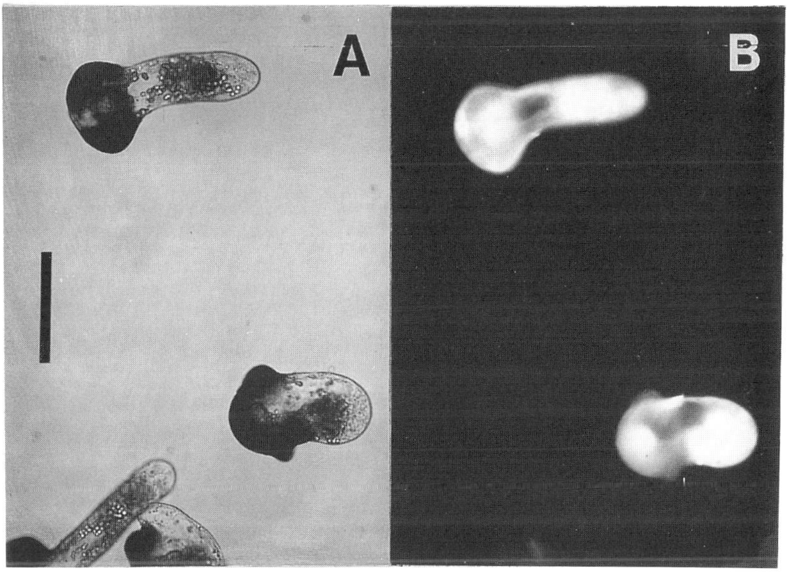

Figure 2. Expression of the Green Fluorescent Protein gene in black spruce pollen following microprojectile bombardment. A) Bombarded pollen seen under normal light and B) under fluorescent microscopy. Bar= 100μm.

5. References

Bodeau, J.P., and V. Walbot (1992) Regulated transcription of the maize Bronze-2 promoter in electroporated protoplasts requires the C1 and R gene products. Mol Gen Genet 233:379-387.

Bommineni, V.R., Chibbar, R.N., Datla, R.S.S., and Tsang, E.W.T. (1993) Transformation of white spruce (*Picea glauca*) somatic embryos by microprojectile bombardment. Plant Cell Rep 13:17-23.

Chalfie, M., Yuan, T., Euskirchen, G., Ward, W.W., and Prasher, D.C. (1994) Green fluorescent protein as a marker for gene expression. Science 263:802-805.

Charest, P.J., Calero, N., Lachance, D., Mitsumune, M., and Yoo, B.Y. (1993a) The use of microprojectile DNA delivery to bypass the long life cycle of tree species in gene expression studies. Curr. Topics Bot. Res. 1:151-163.

Charest, P.J., Calero, N., Lachance, D., Datla, R.S.S., Duchesne, L.C., and Tsang, E.W.T. (1993b) Microprojectile DNA delivery in conifer species: factors affecting assessment of transient gene expression using the β-glucuronidase reporter gene. Plant Cell Rep. 12:189-193.

Charest, P.J., Devantier, Y., and Lachance D. (1995) Stable genetic transformation *of Picea mariana* (black spruce) via microprojectile bombardement. In Vitro Cell Dev Biol (Plant) (Submitted).

Charest, P.J. and Michel, M.F. (1991) Basics of plant genetic engineering and potential applications to tree species. Petawawa National Forestry Institute, Information Report PI-X-104. 48pp.

Cheliak, W.M. and Klimaszewska, K.K. (1991) Genetic variation in somatic embryogenic response in open-pollinated families of black spruce. Theor. Appl. Genet. 82:185-190.

Datla, R.S.S., Hammerlindl, J.K., Pelcher, L.E., Crosby, W.L., and Selvaraj, G. (1991) A bifunctional fusion between β-glucuronidase and neomycin phosphotransferase: a broad-spectrum marker enzyme for plants. Gene 101:239-246.

Datla, R.S.S., Bekkaoui, F., Hammerlindl, J.K., Pilate, G., Dunstan, D.I., and Crosby, W.L. (1993) Improved high-level constitutive foreign gene expression in plants using an AMV RNA4 untranslated leader sequence. Plant Sci. 94:139-149.

Ellis, D.D., McCabe, D.E., McInnis, S., Ramachandran, R., Russel, D.R., Wallace, K.M., Martinell, B.J., Roberts, D.R., Raffa, K.F., and McCown; B.H. (1993) Stable transformation of *Picea glauca* by particle acceleration. BioTechnol. 11: 84-89.

Huang, Y., Diner, A.M., and Karnosky, D.F. *Agrobacterium rhizogenes*-mediated genetic transformation and regeneration of a conifer: *Larix decidua*. In Vitro Cell Dev. Biol. 27:201-207.

Jefferson, R.A., Kavanagh, T.A., and Bevan, M.W. (1987) GUS fusion: β-glucuronidase as a sensitive and versatile gene fusion marker in higher plants. EMBO J. 6:3901-3907.

Jones, J.D.G., Shlumukov, L., Carland, F., English, J., Scofield, S.R., Bishop, G.J., and Harrison, K. (1992) Effective vectors for transformation, expression of heterologous genes, and assaying transposon excision in transgenic plants. Transgenic Res. 1:285-297.

Kay, R., Chan, A., Daly, M., and McPherson, J. (1987) Duplication of CaMV 35S promoter sequences creates a strong enhancer for plant genes. Science 236:1299-1302.

Klimazewska, K., Devantier, Y., Lachance, D., Lelu, M.A., and Charest, P.J. (1995) *Larix laricina* (tamarack):somatic embryogenesis and genetic transformation. (in preparation)

Lelu, M.A., Klimaszewska, K., Jones, C., Ward, C., von Aderkas, P., and Charest, P.J. (1993) A laboratory guide to somatic embryogenesis in spruce and larch. Petawawa National Forestry Institute. Information report PI-X-111. 57pp..

Robertson, D., Weissinger, A. K., Ackley, R., Glover, S., and Sederoff , R. (1991) Genetic transformation of Norway spruce (*Picea abies* (L.) Karst) using somatic embryo explants by microprojectile bombardment. Plant Mol. Biol. 19:925-935.

Seguin, A., Lachance, D., and Charest, P.J. (1995) Transient gene expression and stable genetic transformation into conifer tissues by microprojectile bombardment. Plant Tiss. Cult. Man. (in press).

Topfer, R., Schell, J., and Steinbiss, H.-H. (1988) Versatile cloning vectors for transient gene expression and direct gene transfer in plant cells. Nucleic Acid Res. 16:8725.

A STRATEGY FOR THE GENETIC MANIPULATION OF ENGLISH ELM

T.M. FENNING[1], S.S. TYMENS[1], C.M. BRASIER[2], J.S. GARTLAND[3] AND K.M.A. GARTLAND[3].

[1] *Breeding and Genetics Department, HRI Wellesbourne, Warwick, U.K.,* [2] *Pathology Department, Forest Research Station, Alice Holt Lodge, Wrecclesham, Farnham, Surrey, U.K.,* [3] *School of Molecular and Life Sciences, University of Abertay-Dundee, Scotland.*

1. Abstract

Preliminary work to establish a genetic manipulation protocol for the English elm (*Ulmus procera*) led to the establishment of a microbiologically sterile shoot culture of a mature clone (SR4). High levels of shoot regeneration were obtained from leaf and internode explants, but attempts to infect them or whole shoots, with wild-type strains of *Agrobacterium tumefaciens* failed to induce tumour formation. Similarly, inoculations with disarmed strains harbouring the 35S-GUS/INTRON plasmid failed to induce *gus* expression.

However, it was found that inoculating sideways lain shoots in hormone free medium with *A. tumefaciens* 82.139 (L. Jouanin), led to the production of shooty tumours. A co-cultivation medium at pH 5.2 was found to be optimal, with 70% or more of the inoculated shoots producing viable tumours, fewer tumours being produced at pH 5.0, 5.5, 5.8 or 7.0. Numerous non-transformed shoots have been recovered from these tumours, as well as a small number of morphologically aberrant shoots transformed by the T-DNA of pTi82.139.

A binary plasmid containing the T-DNA of pTi82.139 (L. Jouanin) has been established in a number of disarmed strains, enabling the virulence of each strain to be accurately quantified on elm by tumour production, in the absence of reliable *gus* expression. A wide variation in virulence was found, including one strain which induced tumours on over 90% of inoculated shoots. Using this strain for co-cultivations in combination with 82.139, *gus* positive tumours have been produced, and it has been found to infect elm internode explants on its own. A number of GUS positive shoots have been produced by these methods.

The opportunities for using genetic manipulation to produce elm plants resistant to Dutch elm disease are discussed.

2. Introduction

2.1 THE ELM AND DUTCH ELM DISEASE

Elms (*Ulmus* spp.) are greatly valued for their landscape, amenity and timber qualities, and

M. R. Ahuja et al. (eds.), Somatic Cell Genetics and Molecular Genetics of Trees, 105–112.
© 1996 *Kluwer Academic Publishers. Printed in the Netherlands.*

for their ability to withstand high levels of environmental stress. With over 40 recognised species in the Northern hemisphere they represent one of the worlds major tree species groups (e.g. Richens, 1983; Heybroek, 1993a).

In Europe there are two main taxa: the field elm *U. carpinifolia* (or *U. minor*) and the mountain elm *U. glabra*. *U. carpinifolia* favours the warmer drier climates of central-southern Europe and the Middle east, while *U. glabra* favours the cooler climates of mountain and Northern regions, extending far into Russia. However, these taxa have adapted to a wide variety of habitats and have frequently hybridised, resulting in forms which dominate as varieties or microspecies in some areas. *U. carpinifolia* is particularly diverse, with many distinct sub-taxa identifiable, probably including the English elm *U. procera*.

Many elms also reproduce vegetatively by suckering from their root systems, and most elm species are readily propagated vegetatively via cuttings, including *U. carpinifolia* and its relatives. In some areas, the elm population may exist largely as a single clone, especially if it has been introduced and disseminated within the area by man.

Indeed, because of its utility, elm distribution in Europe has been heavily influenced by human activities since pre-historic times. Human tribes and races planted their favourite varieties as they migrated (Richens, 1983), and elms were often planted in large numbers near villages and towns to provide fodder and a ready source of building materials.

In recent centuries, elms have also been widely planted for landscape and amenity purposes in towns and parks, due to their high tolerance of salt, cold, drought, wind and many atmospheric pollutants. In parts of Russia and central Asia elms are used for shelter belts, sometimes in the form of tree hedges.

Elm timber is traditionally valued for uses which required durability, strength and water resistance. It has long been favoured by furniture makers for its characteristic 'flowing' grain. In recent years the supply of elm timber has declined due to the effects of Dutch elm disease, and its use has been largely substituted by tropical hardwoods.

2.2 DUTCH ELM DISEASE

Dutch elm disease is a vascular wilt disease of elms caused by an ascomycete fungus and spread by elm bark beetles. The disease was unknown in Europe before this century, but since the 1920's the elm populations of Europe have been devastated by two separate epidemics of the disease.

The fungus moves through the tree's vascular system, alternating between mycelial and yeast phases as each xylem vessel is breached. Whether xylem function is lost because of fungal toxin production or because of anti-fungal resins and gums which the tree exudes into the vessels lumens, or both, is unclear. Either way, vascular activity is progressively lost once a tree is infected, and death usually follows within 4-24 months depending upon the season, the size and variety of elm, and environmental conditions.

The geographic origin of the disease is unknown. The first outbreak began in northwest Europe around 1910 and was caused by a fungus of moderate pathogenic ability to elm, *Ophiostoma ulmi* Buism. Nanf. This pathogen spread rapidly eastwards across Europe and into central Asia and westwards to Britain and North America, but caused the death of only 10-40% of European elms by the 1940's (Peace, 1962).

This epidemic then declined in Europe, possibly due to the spread of deleterious cytoplasmically transmitted mycoviruses (called 'd-factors') through the fungal population

(Brasier, 1986a; Brasier, 1990a; Mitchell and Brasier, 1994). The American elm *U. americana*, however, was more susceptible to *O. ulmi*, which rapidly destroyed large numbers of elms after its arrival on the East coast in the 1920's (Gibbs, 1978).

Since the 1940's. a new species of Dutch elm disease fungus far more aggressive than *O. ulmi* has appeared in North America and Europe with disastrous results. *O. novo-ulmi* differs from *O. ulmi* in most behavioural, physiological, genetical and molecular properties (Brasier, 1986b; Brasier, 1991; Bates *et al.*, 1993; Pipe *et al.*, 1995). It has displaced *O. ulmi* as it has spread across the two continents (Brasier, 1990b).

O. novo-ulmi, can readily overcome the resistance of European elms, with the effect that in the current epidemic most mature elms are likely to be killed. In Britain alone some 25 million trees have died since 1970.

A significant proportion of juvenile elms may be unaffected, and numerous root suckers continue to regenerate from dead stems, but these remain just as susceptible as the parent trees. *O. novo-ulmi* is therefore seriously out of equilibrium with its elm host, and so long as the imbalance between this pathogen and the elm population remains unchanged, the future of the elm remains bleak (Brasier, 1983, 1986c, 1987).

The origin of *O. novo-ulmi*, like that of *O. ulmi*, is unknown. A recent survey of the Himalayas as the possible source of *O. novo-ulmi* led instead to the discovery of another hitherto undiscovered highly aggressive Dutch elm disease fungus in the area, *O. himal-ulmi* (Brasier and Mehrotra, 1995).

Although an elm breeding programme in the Netherlands, initiated in the 1930's, succeeded in producing elms with good resistance to *O. ulmi*, these proved as susceptible to *O. novo-ulmi* as the unimproved population (Heybroek, 1993b). *O. novo-ulmi*, like *O. ulmi*, is also affected by mycoviruses or 'd-factors' but these have not played a significant role in limiting the spread of the current epidemic. It is unlikely that Dutch elm disease will attenuate of its own accord in the near future, without some human influence on the system.

2.3 THE ROLE OF THE BEETLE VECTOR

The fungus is spread from diseased to healthy trees by elm bark beetles, *Scolytus* spp. (Webber and Brasier, 1984; Webber, 1989). The beetles lay their eggs in the bark of recently dead or dying trees, and in doing so may introduce spores of *O. novo-ulmi*, even if the tree died of some other cause. As the larvae feed and create galleries in which they overwinter, the fungus also proliferates and sporulates.

When the adult beetles emerge in spring, they may be contaminated by the spores of the pathogen. The beetles fly to the tops of healthy trees or saplings where they feed on the phloem in twig crotches. If sufficient spores are introduced into such a wound site (~1000 spores), then infection of the elm may occur. The pathogen then spreads downwards within the tree, which eventually dies.

After *O. novo-ulmi* arrives in an area and the elms begin to die, the local beetle population explodes because of the increase in available breeding material. This further enhances the rate of spread of the disease, until all adult trees within range of the beetles may be killed. Man also plays a role in spreading the disease into new localities by introducing beetle-infested timber (e.g. Brasier and Gibbs, 1973).

Although in northern Europe cooler springs and shorter summers may restrict beetle activity and so slow the spread of the disease (e.g. Fairhurst and Atkins, 1987), severe outbreaks may still occur in hotter years.

The beetle vectors also show a marked preference for feeding on different varieties and species of elms, which can affect the field performance of different elm taxa. For example, *U. procera* is much preferred by feeding beetles than *U. glabra* (Webber and Kirby, 1983), so the former is at greater risk of infection. Progress of the disease in *U. glabra* is also significantly reduced by the effects of a natural fungal biocontrol agent (*Phomopsis*), which prevents beetles breeding in *U. glabra* bark (Webber, 1980). However, once infected, *U. glabra* is actually more susceptible to *O. novo-ulmi* than *U. procera*.

3. Possible Solutions

Traditional elm breeding programmes have been initiated in Europe and North America, and various chemical and other control approaches have been attempted internationally in an effort to control Dutch elm disease. However, none of these have yet succeeded in reducing its impact (see Sticklen and Sherald, 1993). One problem is that of breeding for resistance to a highly aggressive introduced pathogen. Another is that of obtaining a product which is arboriculturally and environmentally suitable.

An alternative strategy to traditional breeding is to genetically manipulate elms for resistance to the disease. This approach would have the advantage that native elm species or popular varieties, suited to particular climatic conditions or horticultural requirements, could be reintroduced.

Other novel approaches to disease control currently under investigation include genetic manipulation of the mycoviruses of the pathogen as biocontrol agents (Brasier, 1986a; Webber, 1993), and a search for new biocontrol agents of the fungus and beetle in the recently discovered centre of endemism in the Himalayas (Brasier, 1994; Brasier and Mehrotra, 1995).

Regarding a strategy for the genetic manipulation of elms, previous work has already shown that both the English elm and its relative *U. carpinifolia* are suited to many of the procedures necessary for a genetic manipulation protocol, including the rapid proliferation of *in vitro* shoots from elite genotypes and high levels of shoot regeneration from explants (Mezzetti *et al.*, 1988, Fenning *et al.*, 1993).

The objective of the work presented here has been to determine the physical and culture conditions necessary to infect *U. procera* with wild-type and engineered strains of *Agrobacterium tumefaciens*, in order to produce transgenic plants.

4. Current Work

Previous work succeeded in establishing a proliferating shoot culture of *U. procera* under *in vitro* conditions, and good levels of regeneration (Fenning *et al.*, 1993).

Considerable effort is required to optimise any gene transfer system, particularly with trees (Fenning and Gartland, 1995). Factors influencing success include the nature of the regulatory element involved, the gene transfer method and, in the case of *Agrobacterium* delivery, both the plant variety and bacterial strain chosen.

In *U. procera,* β-glucuronidase (*gus*) reporter gene plasmids have been shown to be effective in biolistic transient gene expression assays (Gartland *et al.*, 1994). However, it was found that GUS activity could not be detected sufficiently early after infection with disarmed *A. tumefaciens*, to assist in the development of a transformation protocol. In order to optimise some of the plant and bacterial factors involved in virulence induction, the use of wild-type *A. tumefaciens* strains was investigated, as transformation events could be assessed by tumour formation.

The wild-type strain *A. tumefaciens* 82.139 was tried because it could give rise to 'shooty' tumours, and had been used previously for a poplar transformation protocol (Brasileiro *et al.*, 1991). This strain was found to infect *U. procera* successfully across a wide range of co-cultivation pH's (Table 1.).

TABLE 1. The percentage of English elm shoots forming tumours *in vitro* following inoculation with *A. tumefaciens* 82.139 and co-cultivation at a range of pH's.

pH of co-cultivation medium				
5.0	5.2	5.5	5.8	7.0
56% (n=41)	70% (n=40)	40% (n=40)	45% (n=40)	35% (n=40)

After co-cultivation all the shoots were maintained on hormone free DKW medium at pH 5.8, supplemented with cefotaxime and ticarcillin for 6 weeks prior to scoring.

Non-transformed shoots were readily recovered from these tumours, which could be weaned off into glasshouse conditions without showing morphological aberrations. This is an essential prerequisite for using this strain as part of a protocol, to obtain plants transformed with engineered transgenes only (Brasileiro *et al.*, 1991).
However, a small number of morphologically aberrant shoots were also recovered from these tumours. These shoots continued to proliferate and callus excessively, even in the absence of exogenously supplied phytohormones.
Plants derived from these shoots were analyzed for transformation by a nopaline test and found to be positive, indicating that they were transformed by *A. tumefaciens* 82.139. This was confirmed by Southern blot analysis (Figure 1). Full details are given in Fenning *et al.*, (1996).

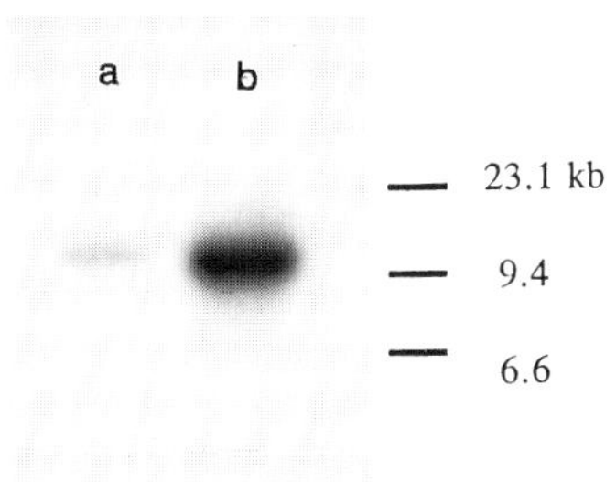

Figure 1. Bands of hybridisation obtained when DNA samples digested with *Xba*1 from two putatively transformed English elm shoots were probed for the T-DNA of pTi82.139. From the position of the bands it was determined that the T-DNA insert was 12.6 kb for shoot sample A, and 11.7 kb for shoot sample B. The positions of λ/HindIII markers are indicated on the right.

For the protocol of Brasileiro *et al.* (1991) to be used on elm, it will be necessary to use a disarmed *A. tumefaciens* strain in addition to *A. tumefaciens* 82.139. Because of the difficulty in detecting early transformation events by GUS activity, a binary plasmid harbouring most of the T-DNA of *A. tumefaciens* 82.139 (Drevet *et al.*, 1994) was used. This was placed into a number of disarmed strains so that virulence of these strains on *U. procera* could be assessed by tumour formation (Table 2.).

This is a more reliable strategy for accurately determining the virulence of disarmed *A. tumefaciens* strains, than using the wild-types, as their tumour inducing potential is affected by differences in T-DNA structure. This method enables accurate comparisons of virulence to be made between disarmed *A. tumefaciens* strains, when early transformation events are not readily detectable by GUS, or other reporter gene systems.

From these results it is clear the disarmed *A. tumefaciens* strain MP90 (Koncz and Schell, 1986) is the most virulent on *U. procera*, of those tested. Using this strain both on its own and in conjunction with *A. tumefaciens* 82.139, GUS positive elm shoots have now been obtained (Gartland *et al.*, 1994). These methods are now being used to transform *U. procera* with potential Dutch elm disease resistance genes.

Table 2. The percentage of English elm shoots forming tumours *in vitro*, following inoculation with a range of disarmed *A. tumefaciens* strains, containing a binary plasmid with a wild-type *A. tumefaciens* T-DNA region.

Disarmed Strain : (wild-type):	ALG1 (A281)	MP90 (C58)	LBA4404 (AcH5)	GV2260 (B6S3)
Tumours produced: at pH 5.2 co-cult.	5% (n=40)	97.5% (n=40)	37.5% (n=40)	42.5% (n=40)

All shoots were co-cultivated at pH 5.2 and thereafter maintained at pH 5.8 with the antibiotics cefotaxime and ticarcillin.

5. Conclusions

Dutch elm disease remains a very serious problem throughout much of the world. Genetic manipulation strategies offer a means of introducing disease resistance genes into elm genotypes suited to particular climatic conditions or arboricultural requirements.

Agrobacterium systems have been developed for the transfer and expression of foreign genes into elms, and a number of promising candidate genes and promoter sequences have been identified. These are being transformed into *U. procera* initially, and later will be used with other elm varieties.

This approach is being developed in co-ordination with other control strategies for Dutch elm disease, and with a number of research programmes on the fungus.

6. Acknowledgements

We would like to acknowledge the help and advice of Professors D.J. James of HRI East Malling, and L. Jouanin of INRA Versailles in the execution of this work.

7. References

Bates, M.R., Buck, K.W. and Brasier, C.M. (1993) Molecular relationships between *Ophiostoma ulmi* and the NAN and EAN races of *O. novo-ulmi* determined by restriction fragment length polymorphisms of nuclear DNA. *Mycological Research* **97**, 449-455.

Brasier, C.M. (1983) The future of Dutch elm disease in Europe, in D.A. Burdekin (Ed), *Forestry Commission Bulletin 60*, HMSO, London, pp 96-104.

Brasier, C.M. (1986a) The d-factor in *ceratocystis ulmi*: its biological characteristics and implications for Dutch elm disease, in K.W. Buck (Ed), *Fungal Virology*, CRC Press, Florida, pp 177-208.

Brasier, C.M. (1986b) A comparison of pathogenicity and cultural characteristics of the EAN and NAN aggressive subgroups of *Ophiostoma ulmi*. *Transactions of the British Mycological Society* **87**, 1-5.

Brasier, C.M. (1986c) The population biology of Dutch elm disease: its principal features and some implications for other host-pathogen systems, in D.S. Ingram and P.H. Williams (Eds), *Advances in Plant Pathology*, Vol. **5**, Academic Press, London, pp 55-118.

Brasier, C.M. (1987) Recent genetic changes in the *Ophiostoma ulmi* population: the threat to the future of elm, in M.S. Wolfe and C.E. Caten (Eds), *Populations of plant pathogens*, Blackwell Scientific Publications, Oxford, pp 213-226.

Brasier, C.M. (1990a) The unexpected element: mycovirus involvement in the outcome of two recent pandemics, Dutch elm disease and chestnut blight, in J.J. Burdon and S.R. Leather (Eds), *Pests, Pathogens and Plant Communities*, Blackwell Scientific Publications, Oxford, pp 289-307.

Brasier, C.M. (1990b) China and the origins of Dutch elm disease: an appraisal. *Plant Pathology* **39**, 5-16.

Brasier, C.M. (1991) *Ophiostoma novo-ulmi* sp. novi, causative agent of the current Dutch elm disease pandemics. *Mycopathologia* **115**, 151-161.

Brasier, C.M. (1994) Missing link in tree disease. *Nature* **372**, 227-228.

Brasier, C.M. and Gibbs, J.N. (1973) Origin of the Dutch elm disease epidemic in Britain. *Nature* **242**, 607-609.

Brasier, C.M. and Mehrotra, M.D. (1995) *Ophiostoma himal-ulmi* sp. nov., a new species of Dutch elm disease fungus endemic to the Himalayas. *Mycological Research* **99**, 205-215.

Brasileiro, A.C.M., Leplé, J.C., Muzzin, J., Ounnoughi, D., Michel, M.F. and Jouanin L. (1991) An alternative approach for gene transfer in trees using wild-type *Agrobacterium* strains. *Plant Mol. Biol.* **17**, 441-452.

Drevet, C., Brasileiro, A.C.M. and Jouanin, L. (1994) Oncogene arrangement in a shooty strain of *Agrobacterium tumefaciens*. *Plant Mol. Biol.* **25**, 83-90.

Fairhurst, C.P. and Atkins, P.M. (1987) Dutch elm disease: The vectors, in D. Patch (Ed), *Advances in Practical Arboriculture, Forestry Commission Bulletin* **65**, HMSO, London.

Fenning, T.M., Gartland, K.M.A. and Brasier, C.M. (1993) Micropropagation and regeneration of English elm, *Ulmus procera* Salisbury. *J. Exp. Bot.* **44**, 1211-1217.

Fenning, T.M. and Gartland, K.M.A. (1995) Transformation protocols for broadleaved trees, in K.M.A. Gartland and M.R. Davey (Eds), *Agrobacterium Protocols, Methods in Molecular Biology* vol. 44, Humana Press, Totowa, New Jersey, pp.149-165.

Fenning, T.M., Tymens, S.S., Gartland, J.S., Brasier, C.M. and Gartland, K.M.A. (1996) Transformation and regeneration of English elm using wild-type *Agrobacterium tumefaciens*. *Plant Science* (In Press).

Gartland, J.S., Gartland, K.M.A., Main, G.D., Brasier, C.M. and Fenning, T.M. (1994) Transformation studies with English elm. *Biologia Plantarum* **36**(S), 331.

Gibbs, J.N. (1978) Intercontinental epidemiology of Dutch elm disease. *Annual Review of Phytopathology* **16**, 287-307.

Heybroek, H.E. (1993a) Why bother about elm? in M.B. Sticklen and J.L. Sherald (Eds), *Dutch elm disease research. Cellular and molecular approaches*, Springer Verlag, Berlin, pp 1-8.

Heybroek, H.E. (1993b) The Dutch elm breeding programme, in M.B.Sticklen and J.L. Sherald (Eds), *Dutch elm disease research. Cellular and molecular approaches*, Springer Verlag, Berlin, pp 16-25.

Koncz, C. and Schell, J. (1986) The promoter of T_L-DNA gene 5 controls the tissue-specific expression of chimaeric genes carried by a novel type of *Agrobacterium* binary vector. *Mol. Gen. Genet.* **204**, 383-396.

Mezzetti, B., Minotta, G., Navacchi, O. and Rosati P. (1988), Propagation and rooting of *Ulmus carpinifolia in vitro*. *Acta Hort.* **227**, 396-8.

Mitchell, A.G. and Brasier, C.M. (1994) Contrasting structure of European and North American populations of *Ophiostoma ulmi*. *Mycological Research* **98**, 576-582.

Peace, F.C. (1962) *Dutch elm disease*, Forestry Commission Bulletin **33**, HMSO, London.

Pipe, N., Buck, K.W. and Brasier, C.M. (1995) Molecular relationships between *Ophiostoma ulmi* and the EAN and NAN races of *O. novo-ulmi* determined by RAPD markers. *Mycological Research* **97**, 653-658.

Richens, R.H. (1983) *Elm*, Cambridge University Press, Cambridge.

Sticklen, M.B. and Sherald J.L. (Eds) (1993) *Dutch elm disease research, cellular and molecular approaches*, Springer-Verlag, Berlin.

Webber, J.F. (1989) Relative effectiveness of *Scolytus scolytus, S. multistriatus* and *S. kirchi* as vectors of Dutch elm disease. *European Journal of Forest Pathology* **20**, 184-192.

Webber, J.F. (1993) D factors and their potential for controlling Dutch elm disease, in M.B. Sticklen and J.L. Sherald (Eds), *Dutch elm disease research. Cellular and molecular approaches*, Springer Verlag, Berlin, pp 322-323.

Webber, J.F. and Brasier, C.M. (1984) The transmission of Dutch elm disease: a study of the processes involved, in J.M. Anderson, A.D.M. Rayner and D. Walton (Eds), *Invertebrate-microbial interactions*. Cambridge University Press, Cambridge, pp. 271-306.

Webber, J.F. and Kirby, S. (1983) Host feeding preference of *Scolytus scolytus*, in D.A. Burdekin (Ed), *Forestry Commission Bulletin*, HMSO, London, pp. 47-49.

Webber, J.F. (1980) A natural biological control of Dutch elm disease. *Nature* **392**, 449-451.

FOREIGN GENE EXPRESSION IN *Pinus nigra, P. radiata* AND *P. pinea* FOLLOWING PARTICLE BOMBARDMENT

MANUEL REY[1,2,5], JAIME M. HUMARA[1,2], MARIAN LOPEZ[1,2], M. VICTORIA GONZALEZ[3], ROBERTO RODRIGUEZ[1,2], RAFFAELA TAVAZZA[4], GIORGIO ANCORA[4], RICARDO J. ORDAS[1,2]

[1] Lab. Fisiología Vegetal, Dept Biología de Organismos y Sistemas, Universidad de Oviedo, c/ Catedrático Rodrigo Uría s/n, E-33071 Oviedo, Spain; [2] Instituto Universitario de Biotecnología de Asturias - CNB (CSIC), E-33071 Oviedo, Spain; [3] Unidad Fruticultura, Centro de Investigación Aplicada y Tecnología Agroalimentaria, Apdo 13, E-33300 Villaviciosa, Spain; [4] Div. BIMO, Agriculture and Biotechnology Sector, ENEA - C.R. Casaccia, I-00060 Santa Maria di Galeria (Rome), Italy; [5] permanent address: Lab. Fisiología Vegetal, Fac. Biología y Ciencias del Mar, Universidad de Vigo, E-36200 Vigo, Spain

Abstract

Using particle bombardment we have developed efficient protocols for transient gene expression in excised cotyledons of three pine species, *Pinus nigra* Arn. ssp. *salzmanii* (Dunal) Franco (black pine), *P. radiata* D. Don (radiata pine), and *P. pinea* L. (stone pine). We have used the plasmid pBI121 to set up the bombardment parameters for transient foreign gene expression, by means of histochemical β-glucuronidase (GUS) assays 24 h after bombardment. Although *gusA* gene expression was obtained for all pine species, bombardment parameters were different for each one. In black pine, the best results of *gusA* gene expression were obtained using 1 μm gold particles covered with 0.8 μg plasmid DNA per bombardment, 4.5 MPa of helium pressure and 6 cm of bombardment distance. In radiata pine, 1.6 μm gold particles gave the best results, using 0.8 μg plasmid DNA, 7.6 MPa and 6 cm. In addition, 8-day-old cultured cotyledons gave better results than younger explants. In stone pine, the best levels of *gusA* gene expression were obsserved when 1 μm gold particles, 0.8 μg plasmid DNA, 6.2 MPa, and 6 cm were used. In addition, we have used this approach to test the effect of different promoters on *gusA* gene expression in pine cotyledons, those of the sunflower ubiquitin, the maize alcohol dehydrogenase, and the rice actin genes, as well as the doubled CaMV 35S promoter. The sunflower ubiquitin gene promoter gave the best results, confirmed by using fluorometric GUS assay. On the other hand, the use of the sunflower ubiquitin gene promoter allowed to last the expression of the *gusA* gene until 20 days of culture of radiata pine cotyledons after bombardment.

1. Introduction

Agrobacterium is the most successful method for plant genetic transformation, but its application to conifer species have been limited due to host-range problems. Although virulent strains of *Agrobacterium* have been identified which can infect a variety of conifer species (van Wordragen and Dons 1992), to date the European larch is the only conifer species in which genetic transformation using this method has been achieved (Shin et al. 1994). Bombardment of cells with high velocity microprojectiles (Klein et al. 1987), is a gene transfer system which can circumvent the limitations of the *Agrobacterium*-mediated transformation. In addition, it allows gene expression studies in organized, intact tissues, and overcomes the need of other direct transfer methods for plant regeneration systems from protoplasts (electroporation).

M. R. Ahuja et al. (eds.), Somatic Cell Genetics and Molecular Genetics of Trees, 113–117.
© 1996 *Kluwer Academic Publishers. Printed in the Netherlands.*

Transient foreign gene expression through particle bombardment has been reported in several pine species (Stomp et al. 1991, Loopstra et al. 1992, Hay et al. 1994, Aronen et al. 1995), from the species used in this study only radiata pine (Campbell et al. 1992, Walter et al. 1994) has been previously used. Protocols for particle bombardment using cotyledons have been obtained only in loblolly pine (Stomp et al. 1991) and radiata pine (González et al. 1994). We describe here the transient foreign gene expression through particle bombardment in cotyledons of three pine species, all important in the mediterranean area, as the first step towards their genetic transformation.

2. Optimization of Bombardment Parameters for Transient *gusA* Gene Expression

Cotyledons from open-pollinated seeds, collected less than one year ago in natural stands in Spain (black pine and stone pine) and Italy (radiata pine) and stored at 4°C, were used for all bombardment experiments. After surface sterilization of the seeds, cotyledons obtained in different ways according the plant regeneration protocol for each species, were used for bombardment experiments.

Embryos of radiata pine were cultured on LP medium (Aitken-Christie et al. 1988) in the presence of 22 μM BA (LP5 medium), in vertical position with the cotyledons inside the culture medium for 4 days. After that, cotyledons were excised and transferred in the same medium until bombardment. In the case of black pine, cotyledons were immediately excised from embryos after seed sterilization, and cultured on modified QL (Quoirin and Lepoivre 1977) medium in the presence of 22 μM BA and 0.05 μM NAA (QLP medium) until bombardment. On the other hand, embryos of stone pine were cultured for 24 h on ½LP medium without growth regulators in vertical position with the cotyledons outside the culture medium , and then cotyledons were excised and used immediately for bombardment.

Bombardments were performed with the Biolistic® PDS-1000/He device (Bio-Rad), using plastic Petri dishes containing the abovementioned media depending on the species. In each dish were placed about 100 cotyledons of radiata pine, about 120 of black pine, and about 70 of stone pine, respectively. Each bombardment delivered approx. 0.8 μg of plasmid DNA associated to 500 μg of gold particles, and each Petri dish was bombarded twice under a vacuum of 85 kPa. After bombardment, cotyledons were maintained in the same medium they were bombarded until GUS assays were performed. All cultures were maintained in a growth chamber at 26 ± 2°C with 16-h photoperiod under a photosynthetic photon flux of 30 $\mu mol\ m^{-2}\ s^{-1}$ provided by cool white light fluorescent tubes (TLD 58W/33, Philips, France).

Plasmid DNAs were isolated from DH5α *Escherichia coli* cells following standard procedures (Sambrook et al. 1989), and were precipitated onto gold particles (mean diameter 1.0 or 1.6 μm) by the method of Klein et al. (1987). GUS histochemical assays (Jefferson et al. 1987) were performed 24 h after bombardment, and results (data not shown) confirmed by fluorometric assays (Jefferson et al. 1987).

TABLE 1. Parameters for optimum transient *gusA* gene expression (plasmid pBI121) in cotyledons of pine species following particle bombardment.

	Pinus nigra	*Pinus radiata*	*Pinus pinea*
Helium pressure (MPa)	4.5	7.6	6.2
Bombardment distance (cm)	6	6	6
Gold particle diameter (μm)	1.0	1.6	1.0
Days in culture of cotyledons	0	8	0

Optimum conditions for transient *gusA* expression using plasmid pBI121 (Clontech) were determined for all three pine species (Table 1). This plasmid contains the *gusA* and *npt II* genes, under the control of the

CaMV 35S and nos promoters, respectively (Jefferson et al. 1987). From the bombardment parameters tested (Table 1), whereas helium pressure was different among pine species, the bombardment distance from the stopping screen to the target tissues was the same (6 cm). It is also remarkable that for radiata pine 1.6 μm gold particles were better than smaller ones, and 8-day-old cultured cotyledons gave better results than younger ones. On the other hand, cotyledons of stone pine showed the best efficiency of all three pine species tested, with a large number of blue spots per cotyledon.

3. Expression of Diffferent Promoters in Pine Cotyledons Through Particle Bombardment

Besides plasmid pBI121, several plasmids were tested for transient gene expression, using optimal parameters for particle bombardment in cotyledons of each pine species (Table 1). All pasmids contain the *gusA* gene under the control of the following promoters. Plasmid pCGUΔ1 contains the sunflower ubiquitin *UbB1* gene promoter (Binet et al. 1991). Plasmid pAIGusN contains the maize alcohol dehydrogenase *Adh1* gene promoter (Klein et al. 1988). Plasmid pAct1-D contains the rice actin *Act1* gene promoter (Zhang et al. 1990). Plasmid p35SGUSint contains the CaMV 35S promoter, and in this plasmid an intron was introduced into the coding region of the *gusA* gene giving rise to an expression vector exclusive of eukaryotic cells (Vancanneyt et al. 1990). Plasmid pBI364 contains a doubled CaMV 35S promoter (Bekkaoui et al. 1990).

TABLE 2. Effect of different promoters on the transient *gusA* gene expression in cotyledons of pine species following particle bombardment. nt = not tested; ov = overexpression, not measurable.

	Pinus nigra		*Pinus radiata*		*Pinus pinea*	
Plasmid	% GUS + cotyledons	Mean no. blue spots	% GUS + cotyledons	Mean no. blue spots	% GUS + cotyledons	Mean no. blue spots
No bomb.	0	0	0	0	0	0
TE buffer	nt	nt	0	0	nt	nt
pBI121	83	7	79	4	98	45
p35SGUSint	94	10	nt	nt	98	55
pBI364	77	5	nt	nt	95	32
pCGUΔ1	94	ov	99	14	97	ov
pAIGusN	79	5	23	2	100	50
pAct1-D	79	6	8	2	100	50

Results (Table 2) showed that the sunflower ubiquitin promoter was the best for transient *gusA* gene expression in all three pine species. In particular, it was not possible to count spots in cotyledons of black pine and stone pine due to the extremely high levels of expression using this construct. Use of the maize alcohol dehydrogenase and the rice actin promoters resulted in low levels of *gusA* expression in black pine and radiata pine, but yielded good levels in stone pine cotyledons. Although lower than the ubiquitin promoter, CaMV 35S promoter yielded good levels of expression, particularly in stone pine. The use of an intron inside the *gusA* gene did not affect expression in cotyledons of stone pine, whereas in black pine slightly increased. Both in black pine and stone pine cotyledons, the use of a doubled CaMV 35S promoter decreased expression. These results were confirmed fluorometrically (data not shown). Long-term expression experiments (Figure 1) confirmed that the sunflower ubiquitin promoter has a stronger activity than the CaMV 35S promoter in radiata pine cells.

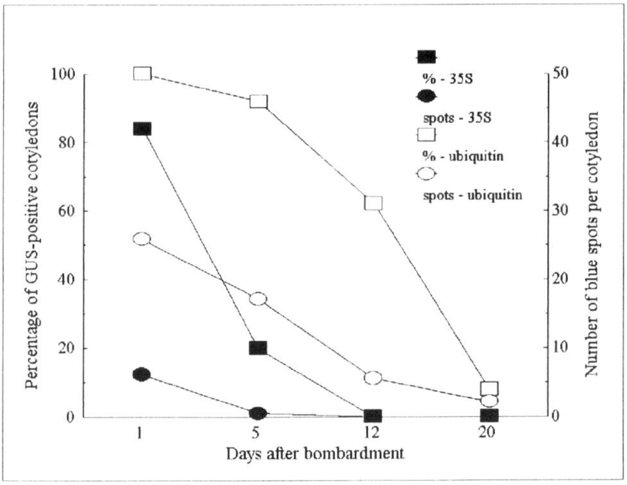

FIGURE 1. Effect of promoters on long-term *gusA* gene expression in cultured radiata pine cotyledons after particle bombardment.

In cotyledons of radiata pine, *gusA* expression after bombardment was observed in epidermal and subepidermal cells, which are those more frequently involved in shoot regeneration (Villalobos et al. 1985). Considering the high efficiency in shoot regeneration of the total area of the cotyledon, the activity of the sunflower ubiquitin promoter, and the estimated percentage of cells still expressing *gusA* 20 days after bombardment, the possibility may exist for obtaining transgenic radiata pines, by adopting the above conditions. A similar situation can be probably observed in cotyledons of black pine and stone pine (work in progress), so particle bombardment currently appears as a promising procedure for transformation of pine species in particular and conifers in general.

4. Acknowledgements

We thank E. Fusaro (SAF, Rome, Italy) for the radiata pine seeds, M. Tavazza, A. Brunetti, C. Marusic and P. Molinari for help with molecular techniques, G. Giuliano for the particle bombardment device, and L. Cuozzo for technical assistance. We acknowledge receiving plasmid samples from L.-H. Tessier (CNRS, Strasbourg, France), M. Fromm (USDA-ARS Plant Gene Expression Ctr, Albany, CA, USA), R. Wu (Cornell Univ., Ithaca, NY, USA), L. Willmitzer (Inst. Genbiol. Forsch., Berlin, Germany), and R.N. Chibbar (Plant Biotechnol. Inst., NRCC, Saskatoon, Canada). This work was supported by grants Nos. MA2B-CT91-0040 (EC-FOREST contract "Upgrading of Southern Pines"), and PB92-0997 (DGICYT, Ministerio de Educación y Ciencia, Spain). Personal support came from an EC-FLAIR postdoctoral fellowship to M. Rey, an UNESCO short-term fellowship in biotechnology to M.V. González, a FICYT (Spain) predoctoral fellowship to J.M. Humara, and an EC-FOREST predoctoral fellowship to M. López.

5. References

Aitken-Christie, J., Singh, A.P., and Davies, H. (1988) Multiplication of meristematic tissue: A new tissue culture system for radiata pine, in J.W. Hanover and D.E. Keathley (eds.), *Genetic Manipulation of Woody Plants*, Plenum Press, New York, NY, pp. 413-432.

Aronen, T., Hohtola, A., Laukkanen, H., and Häggman, H. (1995) Seasonal changes in the transient expression of a 35S CaMV-GUS gene construct introduced into Scots pine buds, *Tree Physiol.* **15**, 65-70.

Bekkaoui, F., Datla, R.S.S., Pilon, M., Tautorus, T.E., Crosby, W.L., and Dunstan, D.I. (1990) The effects of promoter on transient expression in conifer cell lines, *Theor. Appl. Genet.* **79**, 353-359.

Binet, M.-N., Lepetit, M., Weil, J.-H., and Tessier, L.-H. (1991). Analysis of a sunflower polyubiquitin promoter by transient expression, *Plant Sci.* **79**, 87-94.

Campbell, M.A., Kinlaw, C.S., and Neale, D.B. (1992) Expression of luciferase and β-glucuronidase in *Pinus radiata* suspension cells using electroporation and particle bombardment, *Can. J. For. Res.* **22**, 2014-2018.

González, M.V., Rey, M., Ordás, R.J., Ancora, G., and Tavazza, R. (1994) Transient gene expression in cultured radiata pine cotyledons, in J.A. Pardos, M.R. Ahuja and R.E. Rossello (eds.), *Biotechnology of Trees*, Invest. Agraria, INIA, Madrid, pp. 193-198.

Hay, I., Lachance, D., von Aderkas, P., and Charest, P.J. (1994) Transient chimeric gene expression in pollen of five conifer species following microparticle bombardment, *Can. J. For. Res.* **24**, 2417-2423.

Jefferson, R.A., Kavanagh, T.A., and Bevan, M.W. (1987) GUS fusions: β-glucuronidase as a sensitive and versatile gene fusion marker in higher plants, *EMBO J.* **6**, 3901-3907.

Klein, T.M., Gradziel, T., Fromm, M.E., and Sanford, J.C. (1988) Factors influencing gene delivery into *Zea mays* cells by high-velocity microprojectiles, *Bio/Technology* **6**, 559-563.

Klein, T.M., Wolf, E.D., Wu, R., and Sanford, J.C. (1987) High-velocity microprojectiles for delivering nucleic acids into living cells, *Nature* **327**, 70-73.

Loopstra, C.A., Weissinger, A.K., and Sederoff, R.R. (1992) Transient gene expression in differentiating pine wood using microprojectile bombardment, *Can. J. For. Res.* **22**, 993-996.

Quoirin, M. and LePoivre, P. (1977) Etudes de milieux adaptés aux cultures *in vitro* de *Prunus*, *Acta Hortic.* **78**, 437-442.

Sambrook, J., Fritsch, E.F., and Maniatis, T. (1989) *Molecular Cloning. A Laboratory Manual (2nd Ed.)*, Cold Spring Harbor Lab. Press, Cold Spring Harbor, New York.

Shin, D.I., Podila, G.K., Huang, Y.H., and Karnosky, D.F. (1994) Transgenic larch expressing genes for herbicide and insect resistance, *Can. J. For. Res.* **24**, 2059-2067.

Stomp, A.-M., Weissinger, A., and Sederoff, R.R. (1991) Transient expression from microprojectile-mediated DNA transfer in *Pinus taeda*, *Plant Cell Rep.* **10**, 187-190.

van Wordragen, M.F. and Dons, H.J.M. (1992) *Agrobacterium tumefaciens*-mediated transformation of recalcitrant crops, *Plant Mol. Biol. Rep.* **10**, 12-36.

Vancanneyt, G., Schmidt, R., O'Connor-Sanchez, A., Willmitzer, L., and Rocha-Sosa, M. (1990) Construction of an intron-containing marker gene: splicing of the intron in transgenic plants and its use in monitoring early events in *Agrobacterium*-mediated plant transformation, *Mol. Gen. Genet.* **220**, 245-250.

Villalobos, V.M., Yeung, E.C., and Thorpe, T.A. (1985) Origin of adventitious shoots in excised radiata pine cotyledons cultured *in vitro*, *Can. J. Bot.* **63**, 2172-2176.

Walter, C., Smith, D.R., Connet, M.B., Grace, L., and White, D.W.R. (1994) A biolistic approach for the transfer and expression of a *gusA* reporter gene in embryogenic cultures of *Pinus radiata*, *Plant Cell Rep.* **14**, 69-74.

Zhang, W., McElroy, D., and Wu, R. (1991) Analysis of rice *Act1* 5' region activity in transgenic rice plants, *Plant Cell* **3**, 1155-1165.

PLANTLET REGENERATION VIA SOMATIC EMBRYOGENESIS AND INVESTIGATIONS ON *Agrobacterium tumefaciens* MEDIATED TRANSFORMATION OF OAK (*Quercus robur*)

E. WILHELM, A. BURG, M. BERENYI, M. ENDEMANN and R. RODLER
Austrian Research Centre Seibersdorf
A-2444 Seibersdorf
Austria

Abstract

Immature zygotic embryos at different developmental stages were used as explant source to induce somatic embryogenesis in *Quercus robur* throughout several years. Material was collected from several open pollinated trees at different sites. Somatic embryogenic callus could be induced on immature zygotic embryos collected 3 weeks to 18 weeks after anthesis. The highest initiation rate of nearly 30% could be induced approximately 11 weeks after anthesis in the middle of August on 5 µM 2-4D plus 0.5 µM BAP. Embryogenic competence of the callus lines has been maintained by regular subculturing on 1 µM BAP or hormone free media for four years. Maturation media with increased agar concentration (1%) proved to enhance germination rate significantly. Conversion rate of matured somatic embryos into plantlets could be increased to 80% by a partial desiccation treatment.
Somatic embryogenic callus of oak was shown to be infectable by *Agrobacterium tumefaciens.* Additional wounding of embryogenic tissue caused significant decreases in regeneration. A cocultivation period of two days seemed to enhance the regeneration capacity of oak somatic embryogenic callus more than in four days cocultivation. Regeneration rate on selection media can be considered as relativlely high. In the optimum case 70% regeneration could be observed after two passages. More than 100 lines are presently under cultivation. After several subcultures on kanamycin media samples were tested by PCR with selected primers for the presence of the CAT (Chloramphenicol acetyltransferase) gene. About 70% of the regenerated embryogenic calli showed positive results, whereas other samples of the kanamycin resistant tissue and also the control exhibited no band of the CAT fragment. PCR-tests with regenerated somatic embryos gave similar results.

1. Introduction

Somatic embryogenesis has been considered as an efficient and cost-effective method for automated propagation, cryopreservation and production of artificial seed of plant material. During the last years increased interest in this method could be also noticed for woody plant species.
Oak species are among the most important broadleaved tree species because of their ecological and economical value in the temperate climate. *Quercus robur* is valued because of the superior timber quality. Induction of somatic embryogenesis has been reported for many oak species. But until now successful plant production proved to be difficult because the conversion frequency of somatic embryos into rooted plantlets is very low (Bueno et al. 1992, Chalupa 1990 and Gingas and Lineberger 1989). The application of physical treatments like partial desiccation have been shown to increase germination rate of somatic embryos especially in conifers (Attree et al. 1993). To overcome the bottleneck of low conversion frequency the effects of various maturation media, culture conditions and partial desiccation treatment were investigated.
Because of the high regeneration capacitiy of somatic embryogenesis, these offer a useful avenue for genetic transformation in plants. Transgenic plants may provide an excellent material to study the

M. R. Ahuja et al. (eds.), Somatic Cell Genetics and Molecular Genetics of Trees, 119–124.
© 1996 *Kluwer Academic Publishers. Printed in the Netherlands.*

expression of introduced genes involved in controlling plant growth and development. Experiments were conducted to develop a transformation system for *Quercus robur*. The effect of different parameters like wounding, inoculation and cocultivation period on the infectibility of oak somatic embryos towards *Agrobacterium tumefaciens* was studied.

2. Material and methods

Embryogenic culture initiation

Acorns at different stages of maturity were harvested in weekly intervals starting from the end of May to the end of September. Material was collected from several open pollinated trees at different sites. Desinfected acorns and later excised zygotic embryos were cultured on P24-medium (Patent No. 92902531.0 from Robert Teasdale /Forbio) using 0.8 % agar (Daishin), 3% sucrose with 5 µM 2,4-D plus 0.5 µM BAP or without hormones for three weeks and then transferred to 1 µM BAP. Cultures were kept at 24 °C (+/- 2°C) in the light (Philips TLD lamps) with a 16 hour photoperiod at 65 µ $E.m^{-2}.s^{-1}$.

Culture maintenance

The callus lines have been maintained and multiplied by secondary embryogenesis with regular subculturing (16 h photoperiod) at five to six weeks intervals on P24-medium, 0.8% agar, 3% sucrose with 0.9 µM BAP or hormone free.

Somatic embryo maturation and germination

For maturation and germination experiments one line (EAFL) was used for all experiments. This line has been maintained for four years by subculturing. Embryogenic calli (heart and torpedo stages) were transferred to maturation medium (P24) shown in table 1. ABA was filter sterilized and added to the media after autoclaving.

TABLE 1. Maturation media

No.	Agar %	Sucrose %	AC %	ABA µM
1	0.8	3	1	
2	0.8	5		
3	1.0	3		
4	1.0	5		
5	0.8	3		
6	0.8	3		2.5

Abbreviations:
AC Activated charcoal
ABA Abscisic acid

Cultures were placed either in the dark or under light conditions (16 h photoperiod) at a temperature of 24 °C (+/- 2 °C). These conditions were maintained throughout the desiccation and germination experiment.
After five weeks on maturation media single somatic embryos (size > 5 mm) were either transferred directly to germination media (1 µM BAP plus 0.1 µM IBA) or partially dehydrated for three weeks at

a relative humidity > 95% (26° C, dark) in multiwell chambers and afterwards placed on germination media. Results were evaluated for each experiment after five weeks.

Transformation

Plant material

Line EAFL was used for all experiments. Somatic embryogenic callus was subcultured on solid or liquid P24-medium supplemented with 0.9 µM BAP in the light as described above. Embryogenic friable callus with 2 - 5 mm size was transferred from a liquid to solid medium P24 (0.8% Difco Noble agar) 5 days prior to inoculation with bacteria was started.

Plasmid: the binary vectorPMOG402 (kindly provided by the Mogen company) containing the sequence of the chloraphenicol acetyltransferase (CAT)gen derived from the pACYC184 vector was used in transformations mediated by the *Agrobacterium tumefaciens* strain LBA4404. The plasmid pMOG402 is based on the binary vector pBIN19 and contains the NPTII gene driven by a NOS promotor and terminated by NOS terminator for the selection of the transgene tissues on kanamycin containing medium. The CAT sequence was used for monitoring of transfer of a non-selected marker.

Bacterial stock was cultured 24-36 h (27 °C) in 50 ml LB medium plus 50 mg/l kanamycin before used in cocultivation.

Inoculation and cocultivation: 1 ml of bacterial culture was added to 100 ml liquid P24 medium plus 0.9 µM BAP and approximately 5 ml of somatic embryogenic callus was incubated from 30 min to 24 hours by gentle shaking at room temperature.

To study the effect of wounding versus nonwounding embryogenic callus was treated with several scalpel cuts. Plant tissue was dried on sterile filter paper. Cocultivation occured in sealed petridishes on sterile filter paper soaked with 2 ml liquid P24 (0.9 µM BAP) and acetosyringone (100 µM) under 16 h photoperiod at 24°C +/-2 for 2 or 4 days.

After cocultivation the embryogenic callus was incubated by gentle shaking for 20 min in liquid 100 ml P24 with 0.9 µM BAP plus 500mg/l cefotaxime, blotted and dried on sterile filter paper and transferred to solid (0.8 % Difco Noble Agar) P24 with 0.9 µM BAP plus 50 mg/l kanamycin and 300 mg/l cefotaxime (claforan). Explants we transferred to new media every three weeks.

After four months the results of regeneration were compared by variance analysis (SAS).

DNA of oak tissue (somatic embryos, leaves and stems) was extracted and purified according to Heinze and Schmidt 1993.

PCR-test: DynaZyme ^{Tm} Taq DNA-Polymerase (F500L) was used under standard condition. Primer sequences were planed by computer programm PC/GENE for the silent CAT test sequence (CAT1: 5`ATTCAGGCGTAGCACCAGG 3`, CAT2: 5`TGAGACGTTGGATCGGCACG3`) Reaction products were resolved by agarose gel electrophoresis.

3. Results and conclusion

The highest initiation rate of somatic embryogenesis of nearly 30% could be induced approx. 11 weeks after anthesis in the middle of August on 5 µM 2-4D plus 0.5 µM BAP. Regular harvesting in weekly intervals of immature zygotic embryos revealed that during the end of June until the beginning of August the response to induce somatic embryogenic callus is highest. Thirty percent of the immature zygotic embryos started to form somatic embryos in media with hormones (Fig. 1). The initiation of somatic embryogenesis seems to be favoured during a certain time period, which is related to the development of the zygotic embryo. Similar experiments have been described from Chalupa (1987) and Gingas et al. (1989), who observed that initiation of somatic embryogenesis occurs during a certain time period in the developmental phase of the zygotic embryo. Embryogenic callus could also be induced on hormone free media, whereas only genotypes from one location responded to that treatment.

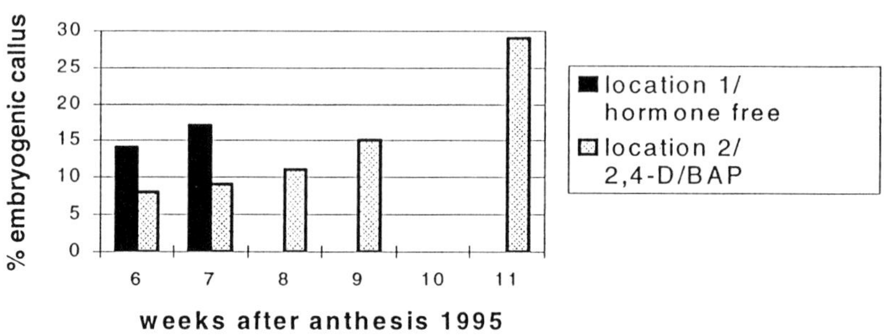

Fig. 1.:Induction of embryogenic callus in oak.

Somatic embryogenic callus lines were subcultured and maintained on media with 0.9 μM BAP or hormone free media without problems.

The development of oak somatic embryos was generally favoured on all tested maturation media during cultivation in the light. Osmotic stress induced by increased agar concentration (1%) exhibited an improved effect on maturation. This medium proved to enhance germination rate also significantly Conversion rate of matured somatic embryos into plantlets could be increased to 80% by a partial desiccation treatment during 16 h day conditions (Fig.2).

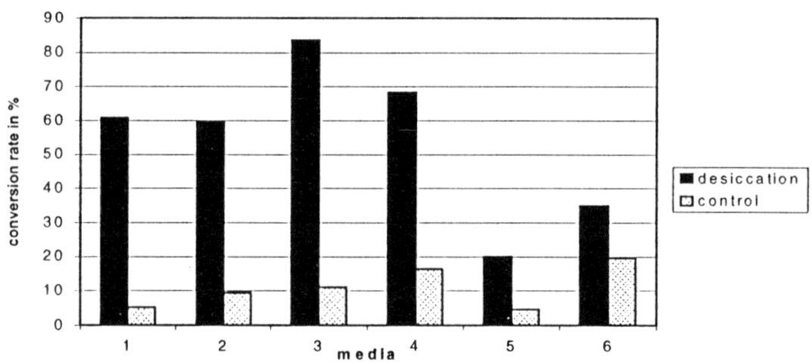

Fig. 2: Conversion of oak SE after maturation in 16 h photoperiod per day
Effect of desiccation and maturation media.

The maturation phase seems to be an important step for determining the fate and future survival of the somatic embryos. The optimization of the maturation phase during which the somatic embryos are able to accumulate storage products is of main importance for the later reprogramming of the somatic embryo. Physical stress as applied by partial drying seems to provide a switch in the programme of somatic embryogenesis to organogenesis like shoot and root forming, which is demonstrated by the low rate of secondary embryogenesis (fig. 3). Well matured somatic embryos sustain desiccation treatment, which is necessary to mobilize reserve products during the process of germination.

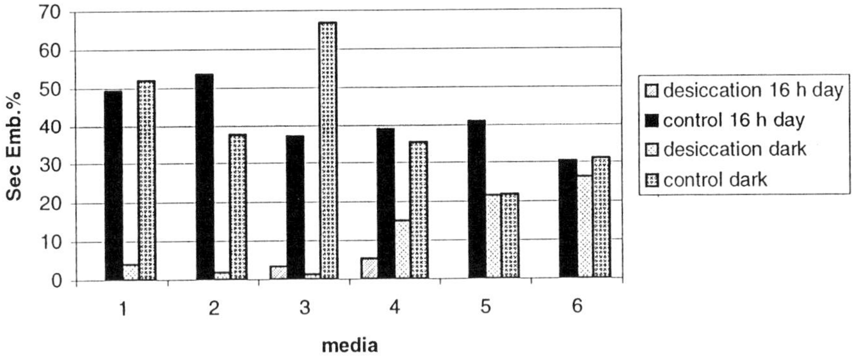

Fig. 3: Secondary embryogenesis of oak after maturation/ Effect of desiccation.

Oak somatic embryos were shown to be infectable by *A. tumefaciens*. Fourteen days after inoculation regeneration of embryogenic callus on selection medium could be observed. Km^R callus started to regenerate from brown tissue. McGranahan et al. (1988) reported that infection of walnut somatic embryos with *A. tumefaciens* was significantly enhanced by cutting the embryos during inoculation. These observations could not be confirmed in our experiments. Of the inoculation methods used, wounding of embryogenic tissue caused significant ($p<0.05$) decreases in regeneration. During inoculation treatment wounded embryos secreted phenolic compounds into the liquid medium which could be observed as brown colour. The cocultivation period of two days seems to enhance the regeneration capacity of oak somatic embryogenic callus more than in four days cocultivation (Fig. 4).

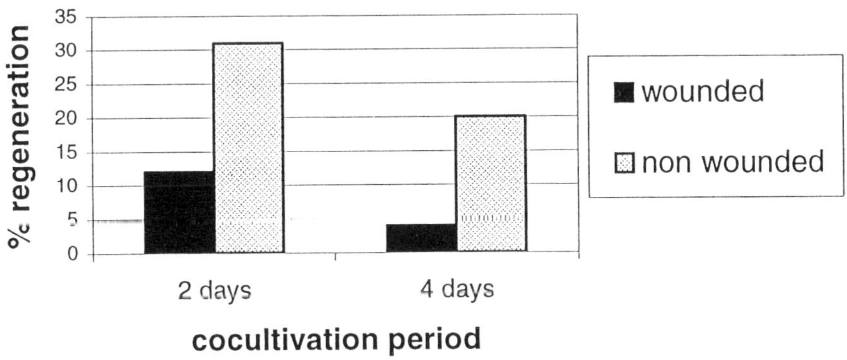

Fig. 4: Response of oak SE to inoculation method and cocultivation period.

After three subcultures on selection medium samples of regenerated callus were tested by PCR for the sleeping CAT gene. About 70% of the Km^R regenerated embryogenic calli showed the 800 bp fragment whereas 30 % and the control exhibited no band of this size. Control experiments revealed that despite of high kanamycin in the medium untransformed somatic embryos were able to regenerate, although to a very little extent (5%), which requires further investigations.

Regeneration rate on selection medium of *A. tumefaciens* treated embryogenic oak tissue can be considered as relatively high. More than 90 KmR embryogenic callus lines are presently under cultivation. The high percentage of may be due to the use of acetosyringone. Firoozabady et al. (1994) reported an 5-fold increase of transformation by inclusion of acetosyringone in the cocultivation medium. Other factors which are currently under investigation might be the preconditioning of the plant tissue. Sangwan et al. 1995 observed that pretreatment of embryogic explants with phytohormones prior to infection appeared to be a very important factor for transformation.
Plantlets have been regenerated from PCR - Positive embryogenic tissues.Such tissues are multiplied and further investigations by southern hybridisation will be necessary to confirm transformation.

Acknowledgement
Parts of the work reported here have been supported by the Austrian Ministry for Agriculture and Forestry.

Literature

Attree, S.J., and Fowke, L.C. (1993) Embryogeny of gymnosperms: advances in synthetic seed technology of conifers, *Plant Cell Tissue Organ Cult.* **35**,1-35.

Bueno,M.A., R. Astorga, and Manzanera, J.A. (1992) Plant regeneration through somatic embryogenesis in Quercus suber, *Physiol. Plant.* **85,** 30-34.

Chalupa, A. V. (1990) Plant regeneration by somatic embryogenesis from cultured immature embryos of oak (Quercus robur L.) and linden (Tillia cordata Mill.), *Plant Cell Rep.* **9**, 398-401.

Chalupa, A.V.(1987) Somatic embryogenesis and plant regeneration in Pices, Quercus, Betula, Tilia, Robinia, Fagus and Aesculus, *Communications Instituti Forestalis Cechosloveniae*, Vol.**15**, 133-148.

Firoozabady, E., Y. Moy, N. Courtney-Gutterson, and K. Robinson (1994) Regneration of transgenic rose (Rosa hybrida) plants from embryogenic tissue, *Biotechnology* **12**, 609-613.

Gingas, V.M. and R.D. Lineberger (1989) Asexual embryogenesis and plant regeneration in Quercus, *Plant Cell, Tissue and Organ Culture* **17,** 191-203.

Heinze, B., and J. Schmidt (1993) Clonal fidelitiy of norway spruce somatic embryos analysed by RAPD, *Proceedings of the IUFRO Working Party S2.04-07 Somatic Cell Genetics, Valsain, Spain, 18 -22 Oct. 1993*, 143 - 148.

McGranagan,G.H., C.A.Leslie, L. Uratsu, L.A. Martin, and A.M. Dandekar (1988) Agrobacterium-mediated transformation of walnut somatic embryos and regeneration of transgenic plants, *Biotechnology* **6**, 800-804.

Sangwan, R.S.; F. Dubois, C. Ducrocq, Y. Bourgeois, B. Vilcot, N. Pawlici, and B.S. Sangwan-Noreel (1995) The embryo as a tool for genetic engineering in higher plants. Current issues in plant molecular and ellular biology, *Proceedings of the VIIth International Congress on Plant Tissue and Cell Culture, Florence, Italy, 12-17 June 1994, Kluwer Academic Publishers*, 271-277.

HIGHLY EFFICIENT TRANSFORMATION AND REGENERATION OF TRANSGENIC ASPEN PLANTS THROUGH SHOOT-BUD FORMATION IN ROOT CULTURE, AND TRANSFORMATION OF *PINUS HALEPENSIS*

T. TZFIRA, O. YARNITZKY, A. VAINSTEIN AND A. ALTMAN.
The Kennedy-Leigh Centre for Horticultural Research and The Otto Warburg Center for Biotechnology in Agriculture, The Hebrew University of Jerusalem, Rehovot 76-100, Israel.

ABSTRACT. The natural capacity of aspen (*Populus tremula*) roots for shoot-bud regeneration was harnessed to establish a highly efficient transformation and regeneration procedure which does not require pre-selection with antibiotics. The procedure involved the use of wild type *Agrobacterium rhizogenes* and *uidA* (GUS) as the a reporter gene. High transient GUS expression was found in inoculated aspen stem segments, which then formed GUS-expressing adventitious roots. The roots proliferated in liquid culture and formed adventitious shoot buds expressing GUS activity. Kanamycin resistance, resulting from the expression of the *nptII* gene in the binary plasmid, was confirmed in transformed plants. Specific phenotypic changes in several transgenic aspen plants were monitored. The same *A. rhizogenes* strain was used for transformation of *Pinus halepensis* embryos, seedlings and adventitious shoots. Mature embryos were highly susceptible to the *Agrobacterium*, as detected by transient GUS expression in the radicle. Stable transformation was evidenced by the regeneration of GUS-expressing roots and calli from infected *P. halepensis* seedlings and adventitious shoots. PCR and Southern blot analyses confirmed the transgenic nature of aspen plants and transformed *P. halepensis* tissues.

1. Introduction

Forest trees are a valuable asset worldwide, playing an important role in preserving our global environment, due to their use for soil conservation, CO_2 assimilation, timber production and as a renewed energy source. However, forest trees are continuously threatened by desertification, pests, acid rain and deforestation. Thus, efficient tree improvement programs are required to produce fast-growing, more tolerant trees. Novel tree breeding methods, can also be achieved through recombinant DNA technology, which provides quick and reliable methods to introduce and express specific traits in transgenic plants [1]. This is in addition to classical breeding, which is limited because of the long generation times and the inability to introduce specific genes when crossing parental lines.

Populus and *Pinus* are two of the most important genera of forest trees, which are widely distributed in the northern hemisphere both as native trees and in commercial forests. The importance of *Populus* and *Pinus* has recently led to an increased interest in *in vitro* propagation [2,3], stress tolerance [4], and transformation and expression of foreign genes in transformed plants, mainly in *Populus* [5,6,7].

Agrobacterium-mediated transfer of DNA to higher plants is commonly used for successful plant transformation and for the regeneration of many herbaceous transgenic plants, e.g. tomato, potato, tobacco and other species. Regeneration of transgenic forest trees has not been as successful, although the transformation of *Populus* species using *Agrobaterium* has been described [5,8]. In contrast,

M. R. Ahuja et al. (eds.), Somatic Cell Genetics and Molecular Genetics of Trees, 125–130.
© 1996 *Kluwer Academic Publishers. Printed in the Netherlands.*

transformation of *Pinus* species is fraught with difficulties, and most studies have focused on screening for gall formation, following infection with various *A. tumefaciens* strains [9,10].

The importance of *Populus* and *Pinus* spp. for forestry on a global scale, and the amenability of *Populus* as a woody plant model for molecular research and genetic engineering, led to a series of physiological and molecular studies in our laboratory. In the following we present data on a highly efficient transformation procedure for aspen, using wild-type *A. rhizogenes* (strain LBA9402), and on the susceptibility of *Pinus halepensis* seedling explants and embryos to infection by that strain, resulting in transient and stable transformation of the two species.

2. Experimental

2.1. BACTERIAL CULTURE

The wild-type *A. rhizogenes* strain LBA9402 was grown in liquid YMB medium on kanamycin, at 28°C, on a rotary shaker. The strain harbors the pRi1855 [11] *vir* oncogenic plasmid and the binary p35SGUSINT [12] plasmid (carrying the *uidA* (GUS) and *nptII* genes).

2.2. PLANT MATERIAL AND TRANSFORMATION PROCEDURE

2.2.1. Populus. tremula

Aspen stem segments were excised from an *in vitro*-cultured aspen clone [2] and inoculated as depicted in Fig. 1A. Semisolid and liquid half-strength MS media [13] (without hormones) were used for all the transformation steps. Buds were induced from GUS-expressing adventitious roots using liquid medium according to Carmi [14]. For additional details on the transformation system see Tzfira *et al.* [15].

2.2.2. Pinus halepensis

Free-pollinated seeds were used as a source for mature embryos and seedling establishment, as described by Ya'ari [16]. Several seedling explants (e.g. apical explants, hypocotyl segments and root explants), as well as mature embryos, were inoculated by immersion in bacterial suspension, and cultured on half-strength MS medium supplemented with 10^{-5} M 6-benzylaminopurine and 10^{-6} M phenoxy-auxin-glycine conjugate. Intact seedlings were inoculated by injection, or by scratching and applying the bacterial suspension to the wound site, as illustrated in Fig. 1B. Multiple brachyblasts (axillary shoots) were induced on 2.5-year-old seedlings by pruning and spraying with cytokinins, and the resulting elongated adventitious shoots were inoculated by injecting bacterial suspension into their basal cut surface.

2.3. MONITORING TRANSFORMATION EVENTS

Transient and stable transformation in aspen explants was monitored using histochemical GUS detection [17]. Kanamycin resistance was confirmed in roots and stem segments of transformed aspen plants based on their ability to produce adventitious shoot buds or adventitious roots, respectively, in the presence of kanamycin. Transformation events in *P. halepensis* embryos, seedling explants and intact seedlings were monitored using GUS detection, as well as root and callus formation on intact seedlings and adventitious shoots.

2.4. MOLECULAR ANALYSIS

PCR was employed with aspen and *P. halepensis* tissues using primers for *uidA* amplification. DNA was extracted from cultured aspen plants and from adventitious roots of infected *P. halepensis* seedlings according to Ben-Meir and Vainstein [18]. Southern blot analysis of digested DNA was performed to detected *uidA* and *rol* genes [19]. RNA was extracted and used for northern blot analysis of *uidA* transcript in transgenic aspen plants, as described by Maniatis *et al.* [19].

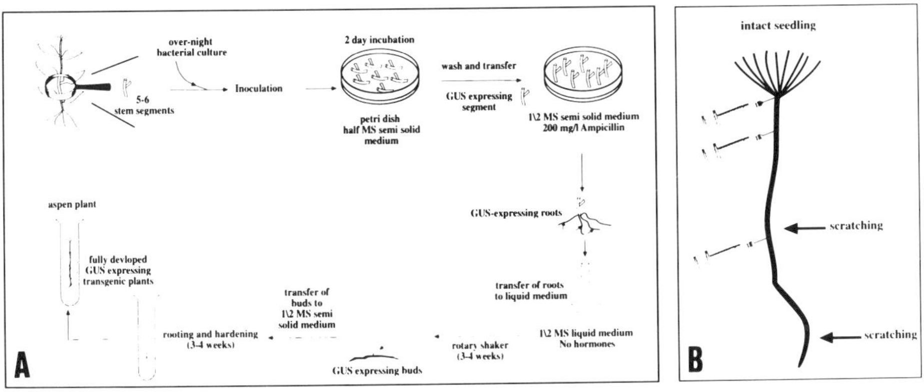

Figure 1. Schematic illustration of the transformation procedures. A. The entire transformation and regeneration cycle of *P. tremula* stem segments and the establishment of transgenic plants. B. Inoculation treatments (injection and scratching) applied to *P. halepensis* intact seedlings.

3. Results

3.1. *POPULUS TREMULA*

A 2-day co-cultivation of single-node stem explants of aspen with A. *rhizogenes*, without any pre-incubation, resulted in GUS expression at the basal cut surface of more than 87% of the inoculated stem segments. About 96% of these survived and 97% of the surviving explants produced GUS-expressing adventitious roots. Almost all of the GUS-expressing roots (74% of the total regenerated roots) yielded an average of four buds/root in regeneration medium (Fig. 2A). Adventitious shoot buds (of which more than 91% were GUS-positive) were transferred for rooting and plant establishment to semisolid half-strength MS medium, and continuously propagated *in vitro* using stem explants. Ninety-four independent clones, of which 52 were phenotypically true-to-type, were established. Southern blot analysis confirmed the presence of the *uidA* gene in HindIII-digested genomes of four out of five selected transformed clones (Fig. 3A). Southern blot analysis of EcoRI-digested DNA from the same clones confirmed the integration of the *uidA* gene in all transformed lines except line T-27 (data not shown). Northern blot analysis demonstrated the expression of the *uidA* gene in transformed lines, but not in controls or line T-27 (Fig. 3B). All transgenic lines presented here were found positive for the presence of *rolB* or *rolC* genes, including T-27 (data not shown).

Kanamycin resistance of transgenic plants was reflected in the ability of roots from transgenic lines to regenerate shoot buds in the presence of up to 30 μg/μl kanamycin, whereas roots from control plants failed to regenerate in the presence of kanamycin concentrations above 10 μg/μl. Stem segments from transgenic lines rooted and developed new shoots in the presence of up to 70 μg/μl kanamycin, whereas control stem segments failed to form roots with more than 50 μg/μl kanamycin.

Some transgenic aspen lines showed morphological abnormalities as compared to control (non-transformed) plants, e.g. loss of apical dominance, extensive root formation, increased growth rate (Fig. 2B), distorted leaf shapes and wrinkled leaves. One of the major traits of the transgenic lines was their remarkable rooting ability. These lines tended to root more quickly (Fig. 4) and to form more roots

(Fig. 2B), resulting in higher root mass than in control plants. Some of the abnormalities were lost after transferring the plants to the greenhouse. However, differences in plant height (Fig. 2C), leaf wrinkling and leaf shape were evident, even in the greenhouse.

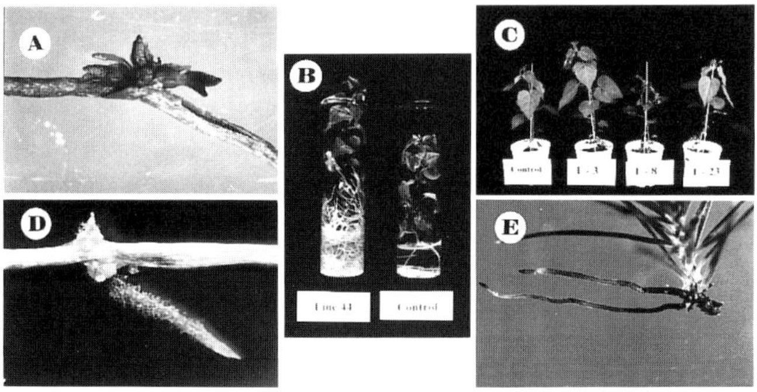

Figure 2. (A) GUS-expressing bud regenerated from transformed aspen root. (B) Transformed (left) and non-transformed (right) aspen plants: note extensive root formation and longer plant in the former. (C) Greenhouse-grown aspen plants: differences in plant height, leaf wrinkling and leaf shape. Adventitious root formation at a wound site on the injected *P. halepensis* seedling hypocotyl (D) and adventitious shoot (E).

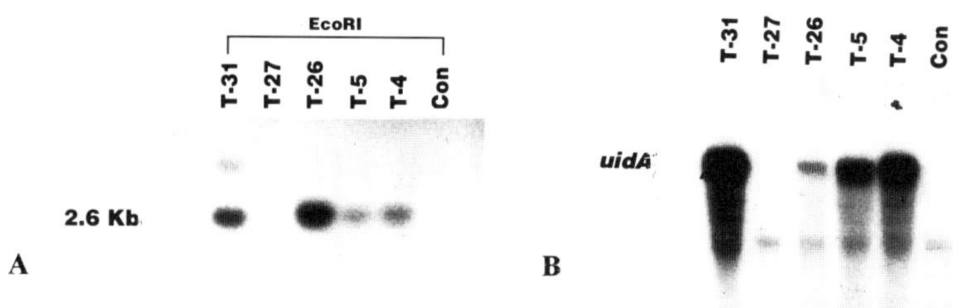

Figure 3. Molecular evidence for the transgenic nature of transformed aspen plants and *P. halepensis* adventitious roots. (A) Southern blot analysis of DNA from five different aspen plants (T-4, T-5, T-26, T-27 and T-31). Total DNA (10 μg) was digested with HindIII and hybridized with a GUS probe. A DNA fragment of the expected size (2.6 kb) is present in all transgenic lines, except line T-27, and not in control plants. (B) Northern blot analysis of RNA from the same aspen plants. Total plant RNA (40 μg) was hybridized with a GUS probe. *uidA* is expressed in four of the transgenic lines, but not in line T-27 or control plants.

3.2. *PINUS HALEPENSIS*

Mature *P. halepensis* embryos showed high susceptibility to the *A. rhizogenes,* as monitored by GUS expression, and more than 85% of the embryos exhibited considerable transient GUS expression in the radicle region. GUS-expression spots on cotyledons were also observed, but at lower rates (one to five spots/embryo), in about 24% of the embryos. Stable transformation was evidenced by the regeneration of GUS-expressing roots and calli from infected *P. halepensis* seedlings. Injection into seedling hypocotyls resulted in root (Fig. 2D) and callus formation at the wound site of 64% of the seedlings. Dipping of excised hypocotyls in bacterial suspension resulted in adventitious root formation in 71% of the hypocotyls, all of which expressed GUS activity. Two months after injection of bacterial suspension into adventitious shoots, adventitious roots (Fig. 2E) and root primordia regenerated in 74% and 40% of 2- and 5-month-old shoots, respectively. Control and auxin-treated shoots failed to form roots. PCR amplification of the *uidA* gene in different tissues (embryos, adventitious roots and calli) yielded a DNA fragment of the expected size (0.53 kb), but not in the control tissues. Southern blot analysis of PCR products using the bacterial *uidA* probe confirmed the authenticity of the amplified fragments. A *uidA* fragment smaller than 21 kb was present in EcoRI-digested DNA from adventitious roots, confirming the integration of T-DNA into the plant genome (data not shown). Further analysis revealed the presence of *rolB* and *rolC* genes in the digested genome of transformed plants, but not in non-transformed ones (data not shown).

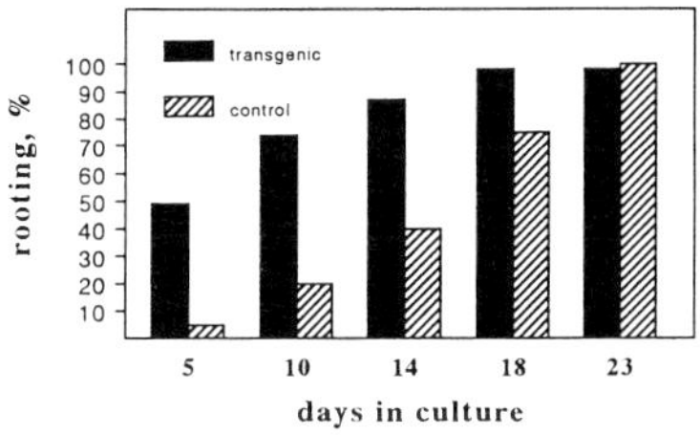

Figure 4. Rooting ability of transformed and control aspen plants. Transformed plants exhibit a higher rooting percentage throughout the culture period, as compared with non-transformed (control) plants (data represent the average of five transgenic lines).

4. Discussion

In this study we describe a unique, highly efficient transformation procedure for aspen without the need for pre-selection with antibiotics. This procedure differs from other reported methods in that it combines two advantages: (1) the ability of *A. rhizogenes* to transform stem cells, resulting in the formation of transformed roots which appear to be composed of transformed cells only, and (2) the natural ability of aspen roots to readily regenerate shoot buds [15]. Liquid root cultures are highly efficient for shoot-bud regeneration in aspen [14]. The use of transformed roots as a source of shoot buds enabled a high yield of new regenerated plantlets. Some of the transgenic plants obtained during the course of this study, as confirmed by Southern blot analysis, showed different degrees of phenotypic abnormalities. The wild-type T-DNA was most probably responsible for the abnormalities observed in some transgenic plants, as it carries gene products which interfere with the plant's hormonal balance [20]. The variability observed in these abnormalities can be explained by the random integration of the oncogenic T-DNA. The absence of some of the phenotypic abnormalities could result from gene silencing [21] or from changes in the growth environment. We should emphasize that not all transgenic plants were positive for the *uidA* and *rol* genes. Thus, the system can be used, in principle, for obtain non-oncogenic, transformed plants.

We also found that *A. rhizogenes* can be used to transform *P. halepensis*. The efficiency of the transformation events was highly dependent on both the tissue infected and the method of infection. Molecular data (PCR and Southern blot analyses) confirmed the integration of the *uidA* gene in the examined tissues, and the phenotypic changes in root formation could be related to the integration and expression of the oncogenic *rol* genes [20].

In the present study, we limited our work to an analysis of the susceptibility of *P. halepensis* to *A. rhizogenes* strain LBA9402 only. We demonstrated the susceptibility of different *P. halepensis* tissues based on GUS expression and root or callus formation. Further progress in the development of a *P. halepensis* DNA transfer system should include screening of additional *Agrobacterium* strains, the use of disarmed vectors, and the development of a successful procedure for plant regeneration from transformed pine tissues and cells. Several systems have been reviewed by Dunstan [22], and available ones include regeneration of plants from cotyledons [16], needles [3] and somatic embryos [23]. The combination of successful transformation and regeneration methods will pave the way for practical genetic engineering of pines. This is currently being studied in our laboratory.

5. Acknowledgements

This research was funded in part by the German-Israeli Agricultural Research Agreement for the benefit of the Third World (GIARA), by the Land Development Authority (Keren Kayemet Le'Israel) and by the Authority for Research and Development, The Hebrew University. The authors wish to thank Dr. D. Clapham for the bacterial strain and Ms. Ofira Tzfira for the graphic illustrations.

6. References

1. Chalupa V (1990) Biotechnology in forest tree improvement: trees of the future. Plant Aging: Basic and Applied Approaches (Rodriguez R, Sánches Tamés R, Duzran DJ eds). Plenum Press, New York, pp. 311-318
2. Nadel BL, Hazan G, David R, Huttermann A, Altman A (1992) In vitro propagation of *Populus* species: response to growth regulators and media composition. Acta. Hort. 314: 61-68
3. Yarnitzki O (1995) Clonal propagation of allepo pine (*Pinus halepensis*). M. Sc. Thesis (Hebrew with English summary)
4. Pelah D, Shoseyov O, Altman A (1995) Characterization of BspA, a major boiling-stable, water-stress-responsive protein in aspen *(Populus tremula*). Tree Physiology 15 (in press)
5. Klopfenstein NB, Shi NQ, Kernan A, McNabb HS, Hall RB, Hart ER, Thornburg RW (1991). Transgenic *Populus* hybrid expresses a wound-inducible potato proteinase inhibitor II-Cat gene fusion. Can. J. Forest. Res. 21: 1321-1328
6. McCown BH, McCade DE, Russell DR, Robison DJ, Barton KA, Raffa KF (1991) Stable transformation of *Populus* and incorporation of pest resistance by electric discharge particle acceleration. Plant Cell Rep. 9: 590-594
7. Tsai C-J, Hu WJ, Mielke M, Zhang XH, Joshi CP, Podila GK, Chiang VL (1995) Transgenic quaking aspen expressing antisense O-methyltransferase gene. Joint Meeting of the IFURO working parties S.04-07 and S.04-06. Somatic Cell Genetics and Molecular Genetics of Trees. Gent, Belgium (abstract)
8. Tsai C-J, Podila GK, Chiang VL (1994) *Agrobacterium*-mediated transformation of quaking aspen (*Populus tremuloides*) and regeneration of transgenic plants. Plant Cell Rep. 14: 94-97
9. Morris JW, Castle LA and Morris RO (1989) Efficacy of different *Agrobacterium tumefaciens* strains in transformation of pinaceous gymnosperms. Physiol. Mol. Plant Pathol. 34: 451-461
10. Stomp A-M, Loopstra C, Chilton WS, Sedroff RR and Moore LW (1990) Extended host range of *Agrobacterium tumefaciens* in the genus *Pinus*. Plant Physiol. 92: 1226-1232
11. Pomponi M, Spano L, Sabbadini MG, Costantion P (1983) Restriction endonuclease mapping of the root-inducing plasmid of *Agrobacterium rhizogenes* 1855. Plasmid 10: 119-129
12. Vancanneyt G, Schmidt R, O'Connor-Sanchez A, Willmitzer L, Rocha-Sosa M (1990) Construction of an intron-containing marker gene: splicing of the intron in transgenic plants and its use in monitoring early events in *Agrobacterium*-mediated plant transformation. Mol. Gen. Genet. 220: 245-250
13. Murashige T, Skoog F (1962) A revised medium for rapid growth and bioassays with tobacco tissue culture. Physiol. Plant. 15: 473-497
14. Carmi T (1994) Bud differentiation and growth from aspen (*Populus tremula*) root and bud explants in liquid cultures. M.Sc. Thesis (Hebrew with English summary)
15. Tzfira T, Ben-Meir H, Vainstein A, Altman A (1995) Highly efficient transformation and regeneration of aspen plants through shoot-bud formation in root culture. Plant Cell Rep. (in press)
16. Y'aari A (1995) Clonal propagation of allepo pine (*Pinus halepensis*). M.Sc. Thesis (Hebrew with English summary)
17. Jefferson RA, Wilson KJ (1991) The GUS gene fusion system. Plant Mol. Biol. Manual B14: 1-33
18. Ben-Meir H, Vainstein A (1994) Assessment of genetic relatedness in roses by DNA fingerprint analysis. Scien. Hort. 58: 115-121
19. Maniatis T, Fritsch EF, Sambrook J (1982) Molecular Cloning. A Laboratory Manual. Cold Spring Harbor, NY.
20. Klee H, Horsch R, Rogers S (1987) *Agrobacterium*-mediated plant transformation and its further applications to plant biology. Ann. Rev. Plant Physiol. 38: 467-486
21. Matzke M, Matzke AJM (1993) Genomic imprinting in plants: parental effects and *trans*-inactivation phenomena. Annu. Rev. Plant. Physiol. Plant Mol. Biol. 44: 53-76
22. Dunstan DI (1988) Prospects and progress in conifer biotechnology. Can. J. For. Res. 2: 1497-1506
23. Gupta PK and Durzan DJ (1986) Somatic polyembryogenesis from callus of mature sugar pine embryos. Bio/Tech. 4: 643-645.

EXPRESSION OF A PROTEINASE INHIBITOR AND A *BACILLUS THURINGIENSIS* δ-ENDOTOXIN IN TRANSGENIC POPLARS.

D. CORNU[1], J.C. LEPLÉ[1], M. BONADÉ-BOTTINO[2], A. ROSS[2], S. AUGUSTIN[3], A. DELPLANQUE[3], L. JOUANIN[2] and G. PILATE[1]
1: *Station d'Amélioration des Arbres Forestiers, INRA, F-45160 Ardon*
2: *Laboratoire de Biologie Cellulaire, INRA, F-78026 Versailles Cedex*
3: *Station de Zoologie Forestière, INRA, F-45160 Ardon*

Abstract

Genetic transformation has been used to improve poplar tolerance toward *Chrysomela tremulae*, a coleoptera causing severe damages in nurseries and young poplar plantations. We have shown in an *in vitro* assay that cysteine proteinase represent the major proteinase activity in the midgut of *C. tremulae*. Moreover, in this system the cysteine proteinase inhibitor OCI effectively inhibits most of the digestive proteinase activity. This proteinase inhibitor and the *Bacillus thuringiensis* δ-endotoxin CRY IIIA, also known to be active against coleoptera, were both evaluated for their toxicity against *C. tremulae*. Transgenic poplars expressing either *ocI* or *cry* IIIA gene were produced. Insects feeding on this transgenic poplars exhibit reduced larval growth, altered development and increased mortality when compared to the control.

Introduction

Poplars are important trees in Western Europe and France for wood production for pulp, matches, plywood, wood package,... In France they are widely planted in valleys mostly in "Picardie", "Bassin Parisien", the Loire Valley and the Garonne Valley,... Few clones (21) are planted and are mainly interspecific hybrids like euramerican hybrids for 65 % and interamerican hybrids for 30 % (Terrasson 1992). They cover 150 000 ha for 21 millions of cubic metre in stand, and represent 3 % of the total of the productive forests and 30 % of the hardwood production. Poplars are the first harvested broadleaf species with oak.

M. R. Ahuja et al. (eds.), Somatic Cell Genetics and Molecular Genetics of Trees, 131–136.
© 1996 *Kluwer Academic Publishers. Printed in the Netherlands.*

Unfortunately poplars are sensitive to many pathogens like canker, rust..., or pest. Among them, the poplar leaf beetles, *Chrysomela tremulae* and *Chrysomela populi* are two very important defoliating insects (Augustin *et al.*, 1993). In the nursery and in short term plantations they can completely defoliate the plants and drastically reduce the growth. During wide outbreak on young plantations, larvae and adults could attack the bark and kill the plants.

In the nursery, more than 5 chemical treatments must be applied to reduce the attack. However, even if these treatments are possible, they are expensive in the nursery, and irrealistic and environmentally dangerous in plantations. Some natural mechanisms of resistance are known like trichomes -hair cells on leaves- or phenylglucosides produced in leaves -tremulacine, tremuloidine,... (Augustin 1991). But the selection for these characters in classical improvement programs is still difficult.

In the Forest Tree Improvement Station, we have developed a new strategy to obtain insect tolerant trees : the introduction of insecticidal genes into poplar by genetic engineering to confer resistance to the poplar leaf beetle (Cornu *et al.*, 1995, Leplé *et al.*, 1995).

1. Materials and Methods

Two types of insecticidal genes have been chosen :

* A *Bacillus thuringiensis* (*B.t.*) gene *cry* IIIA (Chaufaux *et al.*, 1990) coding for a δ-endotoxin toxic against beetles.

* A cysteine proteinase inhibitor gene *ocI* from rice (Abe *et al.*, 1987).

The two types of genes correspond to two different strategies to control pest. For the *B.t.* δ-endotoxin there is recognition with receptors in the midgut of the insect and destruction of this tissue. The toxicity is specific and the death usually occurs rapidly. For the proteinase inhibitors, the aim is to block the digestive proteinases of the larvae in order to disturb the digestion (or to reduce the digestibility of the diet). The effect on the insects could be delayed and is not too specific.

These insecticidal genes have been integrated into transformation vectors, under the control of the strong constitutive P35S2 promoter which was derived from the CaMV 35S RNA promoter by doubling the enhancer sequence (Kay *et al.*, 1987). The Ω' leader sequence from the tobacco mosaic virus (Gallie 1987) was added 5' to the *ocI* sequence to enhance the translational expression. The neomycin phosphotransferase (*nptII*) gene under control of the *nos* regulation signals was used as selectable marker. The resulting plasmids were introduced into the GV3101(pMP90) *Agrobacterium tumefaciens* strain.

The transformation was performed on a *Leuce* poplar hybrid clone (*Populus tremula x P. tremuloides*) INRA 353-38 which is very sensitive to *Chrysomela*. This clone is maintained *in vitro* by micropropagation and was coinoculated with *A. tumefaciens* according to the technique developed in our laboratory (Leplé *et al.*, 1992, Leplé *et al.*, 1995). This technique allows the production of transformed plants within 4-5 months.

2. Results

2.1. EFFECT OF THE PURE PROTEINS

Before transformation, we have controlled the toxicity of the pure proteins on the insects.

When sprayed on foliage, the pure *B.t.* δ-endotoxin is as active as the bacteria.

We have extracted and characterised the proteinases from midguts of third instar larvae. The basal proteinase activities are optimal at pH 7 and pH 9. With the use of a cysteine proteinase activator (DTT = dithiothreitol) A very strong increase of the activity at pH7 was observed (approximately 7 fold).

The efficiency of 2 proteinase inhibitors (OCI and BBI, a commercial serine proteinase inhibitor) to inhibit these proteinase activities was tested. At pH 7 and pH 9, OCI is the most efficient inhibitor. However a fraction of proteinase activity was not susceptible to OCI whether DTT was present or not whereas it was partially susceptible to BBI. This demonstrates that the proteinase activity of the insect midgut is mainly due to cysteine and serine (particularly at pH 9) proteinases, sensitive to OCI and BBI respectively. Consequently, the expression in transgenic poplars of the *ocI* inhibitor gene should have a deleterious effect on *Chrysomela* survival *in planta*.

2.2. POPLAR TRANSFORMATION AND SELECTION

Transformants have been obtained with both constructs carrying either the *ocI* or the *B.t. cry* IIIA genes.

For the proteinase inhibitor OCI, according to northern blot experiments, 3 out of 12 transformants were selected as possessing the highest levels of OCI RNA expression. These 3 lines : OCI-4, OCI-33 and OCI-51 were micropropagated and acclimatised in the greenhouse for insect feeding assays.

For the *B.t.* gene, the level of RNA expression was too low to be determined by northern analysis. This is due to the bacterial origin of the *cry* IIIA gene which is transcribed at a very low level in plants. Using the RT-PCR technique we were able to show that the gene is expressed in the transformants but we were not able to quantify the transcription level. Therefore, the toxicity on *C. tremulae* of the *B.t.* transformants was tested on 15 out of the 22 transformants obtained.

2.3. FEEDING ASSAYS

It is possible to put the larvae on acclimatised plants under nets in the greenhouse. Nevertheless, there were too much perturbations on the insects during the periodical observations to obtain consistent results. Consequently for OCI transformants, we used detached leaves in plastic boxes. For each transformant, we performed 10 repetitions with 10 larvae per box. Leaves collected in the greenhouse were changed in box before drying. The boxes were maintained in a growth chamber at 22°C and under a 18 h photoperiod. For each *B.t.* transformant, we performed the feeding assays directly on *in vitro* microplants on 5 culture tubes with 10 larvae per tube and per assay.

2.3.1. *Proteinase inhibitor OCI*

The tests presented were conducted from eggs to the pupation on young leaves. For the transformant OCI-33, the OCI protein represented 2% of the total soluble proteins of young leaves. However, we have shown previously that OCI accumulated in old leaves.

TABLE 1: Percentage of mortality of *Chrysomela tremulae* feeding on OCI transgenic poplar foliage

Feeding day	3rd	7th	15th	25th
Larval stage	L 1	L 2	L 3	Pupae
Control	0,25	1,75	3,75	4,5
OCI-4	3,5	7,25	21,5	29,75
OCI-33	1	11,75	27	42,5
OCI-51	0,75	17	38	43,5

From the third day we observed (TABLE 1) significant differences between the mortalities of the control and the transformant plants. These differences increased till the pupae stage. Moreover we observed important differences between lines, with two lines OCI-33 and OCI-51 which are able to induce more than 40% of mortality on *C. tremulae*. Nevertheless, the deleterious effect was observed after quite a long delay. This allows the larvae to damage the leaves.

We did not perform measurements on larvae development and growth, but we observed that adults issuing from larvae fed on transgenic plants frequently were unable to emerge from the pupae, showed wing deformations, failed to metamorphose or died.

2.3.2. *B.t. δ-endotoxin CRY IIIA*

For the first *B.t.* feeding assays we summarised (TABLE 2) the mortality observed for three experiments performed on 6 different transformants. We can assume, even if the transcription of the *cry* IIIA bacterial gene is very low, that some lines are very toxic and able to provoke the death of almost all the larvae. Nevertheless, variability between experiments was observed for some of the transformants and their toxicity on insects needs to be confirmed.

TABLE 2 : Percentage of mortality of *Chrysomela tremulae* larvae 18 days after feeding on poplar foliage transformed for the δ-endotoxin CRY IIIA.

Exp.	Control	*Bt* 23	*Bt* 06	*Bt* 35	*Bt* 26	*Bt* 25	*Bt* 34
1	26 %	36 %	62 %	66 %	70 %	88 %	96 %
2	20 %	20 %	60 %	30 %	20 %	70 %	80 %
3	3,5 %	20 %	20 %	30 %	33 %	10 %	13 %

3. Conclusions and perspectives

We demonstrate that the expression in transgenic poplars of either a specific *B.t.* δ-endotoxin or a cysteine proteinase inhibitor presents a potential interest for poplar insect resistance and in particular for the control of *Chrysomela*. Even, if there is a delay in the mortality and if this mortality is not complete, these strategies will be efficient enough to allow to control the size and the level of the next insect population.

In the future, we intend to :

* improve the constructs and particularly the *B.t.* one with the use of a synthetic coding sequence.

* determine the effect of the transgenes on other insects, whether they are a pest or not. We have already observed after natural infestation in the greenhouse that OCI poplar transformants do not seem to be toxic toward either a tenthredae (*Stauronematus compressicornis*) or a small spider (*Eotetranychus urticae*). However, these same plants seem to be toxic for a lepidoptera (*Lymantria dispar*) in laboratory tests.

* establish a field trial to study the expression and the toxicity of the transformants in natural conditions at the scale of a short term plantation. This field trial will allow to follow the evolution of the target and non target insect populations, and to determine the potential apparition of resistant insects. For this purpose, with the agreement of the French "Commission du Génie Biomoléculaire", a field trial was set in 1995 for 8 years with 2 transformants highly expressing the proteinase inhibitor OCI.

References

Abe, K., Emori, E., Kondo, H., Suzuki, K., and Arai, S. (1987) Molecular cloning of a cysteine proteinase inhibitor of rice (oryzacystatin), *J. Biol. Chem.* **262**, 16793-16797.

Augustin, S. (1991) Relations *Melasoma* (Coleoptera Chrysomelidae) - Peupliers de la section Leuce (*Salicaceae*). Etude de quelques facteurs physico-chimiques impliqués dans l'alimentation et le développement de ces insectes, *Thèse de l'Université d'Orléans, "Biologie Animale"*, 28 Juin 1991, 179 pp.

Augustin, S., Courtin, C., Delplanque, A. (1993) Poplar clone effect on development, mortality, fecundity and longevity of *C. populi* and *C. tremulae* (Coleoptera; Chrysomelidae), *J. Appl. Ent.* **115**, 370-378.

Chaufaux, J., Marchal, M., Marchal, D., and Gilois, N. (1990) Recherche de nouvelle souches de *Bacillus thuringiensis* Berliner : Activité de la souche L.M.63 sur Coléoptères Chrysomelidae, *Rencontre Caraïbe en lutte biologique 5-7 Novembre "Les Colloques de l'INRA"* **58,** Editions INRA Paris, 317-324.

Cornu, D., Leplé, J.C., Pilate, G., Villar, M., Augustin, S., Delplanque, A., Bonadé-Bottino, M., Jouanin, L., Lê Tân, V., and Chaufaux, J. (1995) Peut-on planter des peupliers transgéniques ? Cas de la résistance aux insectes, *Séminaire sur la "Dissémination des plantes génétiquement modifiées : Bilan des recherches soutenues par le Ministère de l'Agriculture et le Ministère de l'Environnement* , INRA-Paris 11 janvier 1995, CTPS Ed., 52-59.

Gallie, D., Sleat, D., Watts, J. Turner, P,. and Wilson, T. (1987) The 5'-leader of tobacco mosaic virus RNA enhances the expression of foreign gene transcripts *in vitro* and *in vivo*, *Nucleic Acids Res.* **15**, 3257-3273.

Kay, R., Cham, A., Daly, M., McPherson, J. (1987) Duplication of CaMV 35S promoter sequences create a strong enhancer for plant genes. *Science* **236,** 1299-1302.

Leplé, J.C., Brasileiro, A.C.M., Michel, M.F., Delmotte, F. and Jouanin, L. (1992) Transgenic poplars : expression of chimeric genes using four different constructs. *Plant Cell Rep.* **11**, 137-141.

Leplé, J.C., Bonadé-Bottino, M., Augustin, S., Pilate, G., Dumanois Lê Tân, V., Delplanque, A., Cornu, D., and Jouanin, L. (1995) Toxicity to *Chrysomela tremulae* (Coleoptera: Chrysomelidae) of transgenic poplars expressing a cysteine proteinase inhibitor, *Molecular .Breeding* **1**, 319-328.

Terrasson D. (1992) Evolution de la populiculture, période de 1988 à 1992, Rapport National de la France. *Proc. 19th Session International Poplar Commission*, Zaragoza, Espagne 22-25 Sept 1992.

LIGNIFICATION GENES IN EUCALYPTUS: CHARACTERIZATION, EXPRESSION AND SUBSEQUENT USE FOR LIGNIN MANIPULATION

J. GRIMA-PETTENATI, D. GOFFNER, C. TEULIERES AND A-M. BOUDET
Centre de Biologie et Physiologie Végétales, Université Paul Sabatier, CNRS-URA 1941, 118 route de Narbonne, F-31062 Toulouse Cedex, France. fax: 33 61 55 62 10

1. Abstract

Lignin constitute a major component of the terrestrial biomass accounting for up to 25-35% of the dry weight of woody species. These complex cell wall phenolic polymers play a crucial role in plant development and defense responses.

In contrast, lignin is undesirable during the paper making process. Its necessary removal from cellulose fibers is achieved through costly and environmentally unfriendly processes.

With the goal of improving lignin quantity and / or quality in *Eucalyptus*, a species of great economic interest for the pulp and paper industry, we have, as a prerequisite, characterized genes "specifically" involved in the lignin branch pathway.

Genetic engineering (antisense) experiments are being performed in model plants while recently achieved stable transformation in eucalyptus should soon allow for the validation in this target species.

2. Introduction

Lignins are complex phenolic heteropolymers which represent the second most abundant organic compound on earth after cellulose. Lignins have played a key role in the adaptation of plants to terrestrial life, providing structural support and allowing conduction of water and solutes. They are also involved in plant defence responses to environmental stresses such as infection by microorganisms. On the other hand, lignins have a negative impact on the utilization of plant biomass for specific agricultural or industrial purposes: in the pulp and paper industry, lignins have to be removed from cellulose through costly and environmentally unfriendly processes. From both an economic and environmental point of view, it is therefore of obvious importance to modulate lignin content and/or composition in order to tailor plants more suited to agroindustrial exploitations.

One of our objectives in Toulouse is to genetically engineer commercially important forest species to fit the needs of the pulp and paper industry. As a prerequisite to this goal, it was necessary to gain a better knowledge of the lignification genes and of their regulation.

M. R. Ahuja et al. (eds.), Somatic Cell Genetics and Molecular Genetics of Trees, 137–145.
© 1996 *Kluwer Academic Publishers. Printed in the Netherlands.*

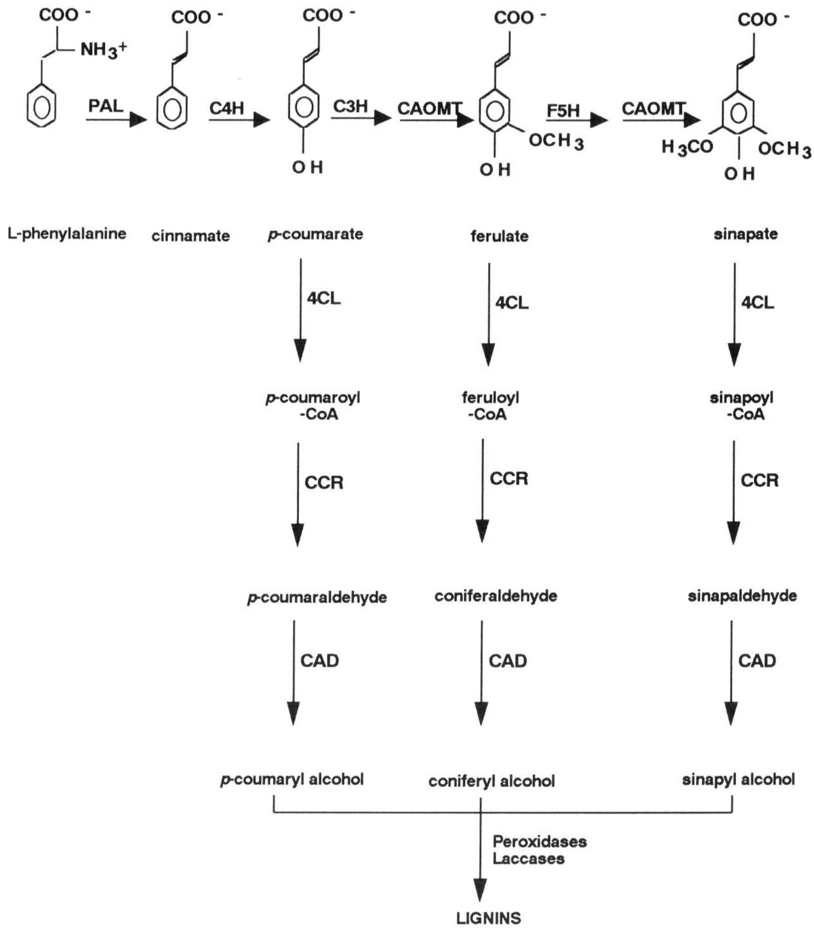

The Lignin biosynthetic pathway. *Enzymes of the common phenylpropanoid pathway include phenylalanine ammonia-lyase (PAL), cinnamate-4-hydroxylase (C4H), coumarate-3-hydroxylase (C3H), O-methyltransferase (CAOMT), ferulate-5-hydroxylase (F5H) and 4-coumarate : CoA ligase (4CL). The two reduction steps of cinnamoyl-CoAs to cinnamyl alcohols (monolignols) catalyzed by cinnamoyl-CoA reductase (CCR) and cinnamyl alcohol dehydrogenase (CAD) respectively constitute the monolignols specific pathway. Lignins result from the oxidative polymerization of the monolignols (p-coumaryl alcohol, coniferyl alcohol, sinapyl alohol) via probably the combined action of laccases and peroxidases.*

The lignin biosynthetic pathway is a branch of the more general phenylpropanoid pathway which leads to hydroxycinnamoyl-CoA esters which are also precursors for the production of various compounds such as flavonoids, tannins, phytoalexins...Once, they are channeled into the lignin branch pathway, the hydroxycinnamoyl CoA esters are converted into cinnamyl alcohols via two reductive steps which are catalyzed by cinnamoyl CoA reductase (CCR) and cinnamyl alcohol dehydrogenase (CAD) respectively.

These cinnamyl alcohols (*p*-coumaryl, coniferyl, sinapyl alcohols) or monolignols are then polymerized into lignins. The enzymes involved in this oxidative polymerization step have not been yet clearly identified although both the involvement of peroxidases and/or laccases has been suggested.

We have mainly focused our interest on the two enzymatic steps specific for lignin monomer biosynthesis: CAD and CCR. We are also investigating the putative role of laccases in the polymerization step. Our main target plant is *Eucalyptus globulus* as eucalyptus species are becoming of increasing importance for paper production throughout the world and it is extensively grown in Southern Europe (Spain, Portugal). However, stable transformation of this species has never been achieved, and is one of the challenging tasks of our group (see below). Due to this "obstacle", our transformation experiments have been performed on model plants: tobacco and/or poplar.

3. Enzymes specifically involved in the monolignols biosynthetic pathway:

3.1. CINNAMOYL CoA REDUCTASE

The enzyme CCR (E.C. 1.2.1.44) catalyzes the conversion of hydroxycinnamoyl-CoA esters to their corresponding aldehydes and, as such, is the entry point enzyme of monolignols biosynthesis.

3.1.1 Purification of CCR from Eucalyptus xylem and cloning of the cDNA

CCR has been purified to apparent homogeneity from differentiating xylem of eucalyptus (Goffner *et al.*, 1994). The purified protein had equal affinities for its three natural substrates (*p*-coumaroyl-CoA, feruloyl-CoA and sinapoyl-CoA) and was approximately three times more effective at converting feruloyl CoA than the other substrates.

Eucalytus CCR purified according to the procedure described in Goffner *et al.* (1994), was digested with endoproteinase lysine-C and the resulting peptides were separated by reverse-phase HPLC and sequenced. Degenerated oligonucleotides were designed, labelled with ^{32}P and used as probes to screen a eucalyptus xylem cDNA library. One clone hybridizing with all of the oligonucleotides was subsequently characterized and sequenced. The full length cDNA contained an open reading frame of 1005 nucleotides encoding a polypeptide of 335 amino acids (MW 36.7kDa). The identity of the isolated pEuCCR clone was demonstrated by functional expression in *E. coli*.

Searches for homology in the databases using FASTA and Blast programs revealed a significant homology with dihydroflavonol-4-reductase (DFR). This NADPH-dependent reductase of the flavonoid pathway, catalyzes the first committed step to the anthocyanins synthesis. Sixty percent homology (40% identity) was found between CCR and DFR from various plants. Whereas the homology was found over the entire sequence, the highest similitude was in the N-terminal region and particularly in the putative NADPH domain.

3.1.2 CCR-down regulation in transgenic tobacco plants

Genetically engineered modifications of lignin composition have already been obtained through down-regulation of lignin biosynthetic genes such as CAD (Halpin *et al.*, 1994, Baucher *et al.*, this volume) and caffeic acid 3-*O*-methyltransferase (COMT, Atanassova *et al.*, 1995; Van Doorsselaere *et al.*, 1996). However, none of these enzymes appeared to be control points in the lignin pathway since their down-regulation led to changes in the monomeric composition of lignins without altering significantly the lignin content. A potential candidate in tailoring hypolignified plants could be CCR.

As the first step committed to the lignin specific pathway, this enzyme is believed to play a regulatory role controlling the flux of phenylpropanoid metabolites into the lignin pathway. In order to test this hypothesis, we have constructed several CCR-antisense vectors and transferred them *via Agrobacterium*-mediated transformation into tobacco plants. The efficiency of homologous (tobacco cDNA, isolated by Zeneca Seeds) versus heterologous cDNAs (eucalyptus cDNA) sequences will be assessed as well as the efficiency of tissue specific promoters (eucalyptus CAD promoters) versus constitutive promoters (CaMV 35S with or without double enhancer). Primary transformants with severely depressed CCR activity (70 to 80% inhibition) have been identified and currently being analysed for their lignin content and composition.

3.2. CINNAMYL ALCOHOL DEHYDROGENASE

Cinnamyl alcohol dehydrogenase (CAD, E.C. 1.1.1.195) catalyzes the last step of lignin monomer biosynthesis *i.e.* the reduction of the three cinnamaldehydes to the corresponding cinnamyl alcohols which are the direct precursors of lignins.

3.2.1 Purification of CAD from Eucalyptus xylem and cloning of the corresponding cDNAs

Two biochemically, structurally and genetically distinct isoforms of CAD (CAD1 and CAD2) were purified to homogeneity from differentiating xylem of 3 to 5-year old eucalyptus branches (Goffner *et al.*,1992). CAD2 has a ten fold higher specific activity than CAD1 and utilizes all three natural substrates (*p*-coumaryl, sinapyl and coniferyl aldehydes) with a higher affinity than CAD1 which uses neither sinapaldehyde nor its corresponding alcohol as substrate. Based on biochemical data, together with our knowledge concerning the heterogeneous nature of lignin composition in Eucalyptus, CAD2 was thought to be the isoform more closely associated with constitutive lignification.

The cDNA corresponding to CAD2 was obtained by screening a *E. gunnii* cell suspension culture library (Grima-Pettenati *et al.*, 1993) with a heterologous cDNA recently isolated from tobacco (Knight *et al.*, 92). The identity of the eucalyptus cDNA was unambiguously demonstrated by functional expression of the recombinant enzyme in *E. coli*.

In order to assign a role to CAD1, molecular cloning of the gene has also been achieved and antisense CAD1 experiments are underway .

3.2.2 CAD structure/function relationship

Sequence analysis revealed that CAD2 belongs to the long chain zinc-containing alcohol dehydrogenase family. The homology of eucalyptus CAD2 with horse liver alcohol dehydrogenase (E.C. 1.1.1.1) was sufficient to allow homology-based molecular modeling of CAD2 upon the crystallographic coordinates of HLADH (Mckie *et al.*, 1993) and, hence, to identify amino acid residues of potential significance. We took advantage of the powerful tool consisting in the overproduction of catalytically active CAD2 in E. coli, harboring the same biochemical properties as the native eucalyptus CAD2, to verify the effect of site-directed mutagenesis on amino-acids potentially involved in the active site of CAD2. Site-directed mutagenesis (S212D) followed by kinetic studies on the purified wild type CAD2 and mutant S212D proteins, confirmed the putative involvment of S212 in recognizing the phosphate group of NADPH and provided information on the structural requirements for NADPH specificity (Lauvergeat *et al.*, 1995).

3.2.3 Spatiotemporal expression of the CAD promoter

We have isolated and characterized a CAD2 genomic clone from eucalyptus (Feuillet *et al.*, 1995). The promoter (2.5kb) was fused to the β-glucuronidase (GUS) reporter gene and introduced via *Agrobacterium*-mediated transformation into poplar (INRA clone 7171 B-4, *Populus tremula X Populus alba*). First histochemical staining for GUS activity were performed using a standard protocol used routinely for localizing gene expression in herbaceous species and indicated a strong expression in the vascular tissues of stems, roots, leaves and petioles (Feuillet *et al.*, 1995). Gus activity was mainly localized in xylem ray cells and in parenchyma cells surrounding lignified phloem fibers. More recently we have developed a new technique of sample preparation for GUS histochemistry specifically adapted for working with tree species containing long axial cells such as fusiform initials and phloem fibers cells) (Hawkins *et al.*, submitted). The use of this new protocol confirmed the original pattern of expression but, in addition, allowed the detection of GUS expression in the vascular cambium. These results were confirmed by both immunolocalization and *in situ* hybridization (Hawkins *et al.*, submitted).

Plants transformed with 5' serial deletions of the promoter are currently being analysed in order to identify cis-acting elements involved in vascular expression.

4. Enzymes potentially involved in the polymerization step : laccases

Historically, peroxidases have been considered as the sole enzymes implicated in the polymerization of monolignol precursors in lignin biosynthesis. However, in recent years, convincing data have been reported that suggest the involvement of laccase (*p*-diphenol:O2 oxidoreductase, EC 1.10.3.2) in lignin biosynthesis (Bao *et al.*, 1995, Sterjiades *et al.*, 1992). If laccases or other related oxidases are involved in lignification, they must meet at least two criteria : 1) they must spatially and temporally correlated with tissues /cells actively undergoing lignifcation (*i.e.* associated with cell walls of developing xylem) and 2) they must be capable of oxidizing monolignols, resulting in the formation of dehydrogenation polymers (DHPs) *in vitro*. These criteria have been met in an extremely limited number of cases.

In an attempt to identify and characterize an enzyme that meets the above mentioned requirements, ionically bound cell wall protein fractions from poplar were prepared and tested for monolignol oxidizing activity. Not only was this crude extract capable of oxidizing coniferyl alcohol, but long -term reactions led to the formation of DHPs *in vitro*.

As a prerequisite for purification and characterization of this monolignol-oxidizing enzyme, O_2-dependent phenoloxidase, activity gels (SDS-PAGE) in the presence of artificial chromogenic substates (*i.e.* DAF, ABTS) were performed. In general, this gave rise to at least 4 activity bands. However, when coniferyl alcohol was used as substrate, only one of the four polypeptides exhibited phenoloxidase activity. This simple method enabled us to target a monolignol-specific, O_2-dependent phenoloxidase potentially involved in lignification for subsequent enzyme purification. A relatively simple and efficient purification scheme consisting of ammonium sulfate precipitation, hydrophobic interaction and hydroxylapatite chromatography have been successfully used to purify an O_2-dependent phenoloxidase (specific for coniferyl alcohol) to homogeneity. The purifed protein is a monomer of approximately 90kDa. We are currently investigating its physicochemical properties in order to unambiguously identify this enzyme as either a true laccase or a related oxidase capable of coniferyl alcohol oxidation.

At the molecular level, we have screened a poplar cDNA library (provided by W. Boerjan, University of Gent) using a heterologous probe encoding *Acer* laccase (provided by L. Faye, University of Rouen and J. Dean, University of Georgia). Two distinct cDNAs were obtained. One of the clones, lac 27, is full length and encodes for a protein of 555 aa and a pI of 7,8. The other clone is partial and shows 78% homology with lac 27 in the overlapping region. Sequence analysis indicated that both poplar cDNAs exhibit the characteristic features of blue copper oxidases and show significant homologies with a previously characterized plant laccase cDNAs from *Acer* (identity : 44%, similarity : 12%) (Lafayette *et al.*, 1995). In the near future, antisense experiments will be performed in poplar, in an attempt to provide evidence to either support or refute the hypothesis of laccase involvement in lignification.

5. Development of genetic transformation procedures for *E. globulus*

As an essential tool for the manipulation of lignin by the antisense RNA strategy, efficient procedures of DNA transfer have to be developed on one of the target woody plants : *Eucalyptus globulus*. Mainly due to a very low regeneration capacity and despite efforts of several groups, genetic transformation of eucalyptus has only been reported for three eucalyptus species (Teulières *et al.*, 1994), and has never been achieved for *E. globulus*.

With the aim of designing an efficient transformation procedure for this species, two techniques were successively investigated :

- direct gene transfer through biolistics
- transformation *via Agrobacterium tumefaciens.*

The first target material used in these experiments, mature zygotic embryos, was chosen for its capacity of regeneration through organogenesis. Approximately 50% of the embryos were able to regenerate adventitious shoots after a limited callogenesis phase. The regenerated shoots could be micropropagated with a 4-fold multiplication rate. Rooting, one of the most difficult phases on *E. globulus* species (Mac Rae 1991, Willyams 1992), was achieved for 40 to 50% of the shoots. Acclimation to the greenhouse is obtained for nearly all regenerated plantlets.

Direct DNA delivery through biolistics was first set up on this material (Rochange *et al.*, 1995, Serrano *et al.*, 1996). By using either a gunpowder system of the Biorad Helium system (PDS 1000 / He) similar results were obtained : up to 130 GUS positive loci per embryo were observed at the transient expression level. However, after a kanamycin selection, calli exhibiting GUS expression were recovered with a frequency of 0,1 to 1% and transgene integration was proved by Southern blot. Among the various parameters investigated during the optimization of the procedure, three were particularly important:

- the nature of the plant material: mature zygotic embryos after a 4 to 6 day reactivation phase on hormone containing medium were the best target for bombardment ;
- the osmotic treatment prior to bombardment ;

- the DNA concentration for a controlled amount of particles.

Effectiveness of biolistics for DNA delivery and integration into *E. globulus* genome was demonstrated. However, because of the low efficiency of integration (associated to this direct transfer procedure) combined with a low number of cell regeneration sites (an average of 5 to 10 on each embryo) no transgenic shoots were recovered until now, through this procedure.

The susceptibility of *E. globulus* to *Agrobacteria* has been already demonstrated (Chriqui *et al.*, 1992). In order to improve the efficiency of the integration / regeneration phases, transformation *via Agrobacterium tumefaciens* was more recently evaluated in our lab on the same material.

A systematic survey of the experimental conditions included the nature of the bacterial strain, the wounding and the preculture of the embryos.

In the optimized procedure (Moralejo *et al.*, to be published), the embryos transiently expressed GUS reporter gene on the whole surface.

Regenerated shoots both showing GUS activity, kanamycin resistance and transgene integration (Southern blot) were recovered for the first time in *E. globulus*. Some of these shoots could be micropropagated and some are rooting. Transgenic shoot recovery was approximately 0,8% (when compared to the number of starting embryos). This result looks satisfactory when compared to data obtained on other woody plants.

The procedure is now being optimized in view of introducing lignification genes along two lines of investigation :

- increase of transformation efficiency through improvement of the regeneration system,
- adaptation of the procedure to clonal material.

In conclusion, biolistics proved successful for stable transformation of *E. globulus* cells. However due to the poor general regeneration capacity of the species, transformation *via A. tumefaciens* looks more promising. This last procedure allowed the recovery of transgenic plantlets, for the first time. An in-depth description of experimental conditions will be soon published (Moralejo *et al.*, in preparation).

6. Conclusions/Perspectives

During the last years, an increasing interest for lignification has been felt worldwide. The major reasons for this interest are the introduction of molecular biology techniques in this area of research and the exciting perspectives of lignin optimization through genetic engineering. It has already been demonstrated that genetic manipulation of lignification genes will have an important impact in tailoring plants for economical and industrial uses, but also in our understanding of the underlying mechanisms involved in what is a fundamental process for woody species.

Despite the fact that eucalyptus comprises only 15% of the paper market, it appears one of the most appropriate target for such manipulations in relation to its rapid growth rate and also because it is becoming the workhorse of the paper industry in renewable (clonal) plantations in which trees are harvested every six to eight years. Our research programs aim to develop the basis for rational and efficient optimization of lignin profiles in *E. globulus* (collaboration with ENCE, Spain) and *E. urophylla x E. grandis* (collaboration with Aracruz Celulose, Brazil).

We are confident in the future of such technologies and of their expected benefits since 1) even slight changes in wood composition would have a significant economic impact in view of the huge consumption of pulp products, 2) legfislative and consumer pressure in favour of the environment will include companies to reduce energy needs and pollution, 3) public acceptance of transgenic trees will be facilitated by the fact that they are industrial products and not food products

7. Acknowlegments

This work was supported by grants from the Commission of the European communities (ECLAIR-OPLIGE N°0021) and AIR laccases N°2CT931661). Financial support was also provided by the CNRS, UPS and the Conseil Régional Midi-Pyrénées. The authors wish to thank their coworkers for fruitful discussions and unpublished results.

8. References

Atanassova, R., Favet, N., Martz, F., Chabbert, B., Tollier, M.T., Monties, B., Fritig, B., and Legrand, M. (1995) Altered lignin composition in transgenic tobacco expressing O-methyltransferase sequences in sense and antisense orientation. *Plant J.* **8,** 465-477.

Bao, W., O'Malley, D.M., Whetten, R., and Sederoff, R.R. (1993) A laccase associated with lignification in loblolly pine xylem. *Science* **260,** 672-674.

Chriqui, D., Adam, S., Caissard, J.C., Noin, M., and Azmi, A. (1992) Shoot regeneration and Agrobaterium-mediated transformation of Eucalyptus globulus and E.gunnii, in Schönau APG (ed) Proc IUFRO Symp Intensive forestry "*The role of Eucalypts* " pp. 70-80.

Feuillet, C., Lauvergeat, V., Deswarte, C., Pilate, G., Boudet, A.-M., and Grima-Pettenati, J. (1995) Tissue and cell specific expression of a cinnamyl alcohol dehydrogenase promoter in transgenic poplar plants. *Plant. Mol. Biol.* **27,** 651-667.

Goffner, D., Joffroy, I., Grima-Pettenati, J., Halpin, C., Knight, M.E., Schuch, W., and Boudet A.M. (1992) Purification and characterization of isoforms of cinnamyl alcohol dehydrogenase (CAD) from *Eucalyptus* xylem. *Planta* **188,** 48-53.

Goffner, D., Campbell, M.M., Campargue, C., Clastre, M., Borderies, G., and Boudet A., Boudet A.-M. (1994) Purification and characterization of cinnamoyl-CoA : NADP oxidoreductase in *Eucalyptus gunnii. Plant Physiol.* **106,** 625-632.

Grima-Pettenati, J., Feuillet, C., Goffner, D., Borderies, G. and Boudet, A.-M. (1993) Molecular cloning and expression of a *Eucalyptus gunnii* cDNA clone encoding cinnamyl alcohol dehydrogenase. *Plant Mol. Biol.* **21,** 1085-1095.

Halpin, C., Knight, M.E., Foxon, G.A., Campbell, M.M., Boudet, A.M., Boon, J.J., Chabbert, B., Tollier, M.T., and Schuch, W. (1994) Manipulation of lignin quality by down-regulation of cinnamyl alcohol dehydrogenase.*Plant J.* **6,** 339-350.

Knight, M.E., Halpin C., and Schuch W., (1992) Identification and characterization of cDNA clones encoding cinnamyl alcohol dehydrogenase from tobacco. *Plant Mol. Biol.* **19,** 793-801.

Lafayette, P.R., Eriksson, K.-E.L., and Dean, J.F.D. (1995) Nucleotide sequence of a cDNA clone encoding an acidic laccase from Sycamore Maple (*Acer pseudoplatanus L.*). *Plant Physiol.* **107,** 667-668.

Lauvergeat, V., Kennedy, K., Feuillet, C., McKie, J., Gorrichon, L., Boudet, A.-M., Grima-Pettenati, J. and Douglas, K. (1995) Site directed mutagenesis of a serine residue in cinnamyl alcohol dehydrogenase affects the specificity for the coenzyme. *Biochemistry* **34**, 12426-12434.

MacKie, J.H., Jouahari, R., Douglas, K.T., Goffner, D., Feuillet, C., Grima-Pettenati, J., Boudet, A.-M., Baltas, M. and Gorrichon, L. (1993) A molecular model for cinnamyl alcohol dehydrogenase, a plant aromatic alcohol dehydrogenase involved in lignification. *Biophysica Biochemica Acta* **1202**, 61-69.

McRae, S. (1992) Agrobacterium-mediated transformation of *Eucalyptus* to improve rooting ability, in Schönau APG (ed) Proc IUFRO Symp Intensive forestry "*The role of Eucalypts* " 1 pp. 117-125.

Rochange, F., Serrano, L., Marque, C., Teulières, C., and Boudet, A.-M. (1995) DNA delivery into *Eucalyptus globulus* zygotic embryos through biolistics : optimization of the biological and physical parameters of bombardment for two different particle guns. *Plant Cell Reports* **14**, 674-678.

Serrano, L., Rochange, F., Semblat, J.-P., Marque, C., Teulières, C. and Boudet, A.-M. (1996) Genetic transformation of *Eucalyptus globulus* through biolistics : parallel development of procedures for organogenesis from zygotic embryos and stable transformation of corresponding proliferating tissue. *Journal of Experimental Botany* (in press).

Sterjiades, R., Dean, J.F.D., and Eriksson, K.-E.L. (1992) Laccase from Sycamore Maple (*Acer pseudoplatanus*) Polymerize Monolignols. *Plant Physiol.* **99**, 1162-1168.

Teulières, C., Marque, C. and Boudet, A.-M. (1994) Genetic transformation of *Eucalyptus,* in Y.P.S. Bajaj, Springer-Verlag (eds.) Biotechnology in Agriculture and Forestry, Vol. 29 "*Plant protoplasts and genetic engineering V* ", pp. 289-306.

Van Doorsselaere, J., Baucher, M., Chognot, E., Chabbert, B., Tollier, M.T., Petit-Conil, M., Leplé, J.M., Pilate, G., Cornu, D., Monties, B., Van Montagu, M., Inzé, D., Boerjan, W., and Jouanin, L. (1996) A novel lignin in poplar trees with a reduced caffeic acid / 5-hydroxyferulic acid *O*-methyltransferase activity. *Plant J.* **8**, 855-864.

Willyams, D., Whiteman, P., Cameron, J., and Chandler, S.F. (1992) Inter-and-intra-family variability for rooting capacity in micropropagated *Eucalyptus globulus* and *Eucalyptus nitens*, in Schönau APG (ed) Proc IUFRO Symp : " *Mass*

CHARACTERIZATION OF ANTISENSE CHALCONE SYNTHASE TRANSGENIC MICROCUTTINGS

L. JOUANIN[1], C. EL EUCH[1], M. PASTUGLIA[1,3], P. CAPELLI[2], P. DOUMAS[2,4], C. JAY-ALLEMAND[2]

1. Laboratoire de Biologie Cellulaire, INRA, Route de Saint Cyr F78026 Versailles cedex, FRANCE.
2. Station d'Amélioration des arbres forestiers, INRA d'Orléans, Domaine de Limère, 45160 Ardon, FRANCE
3. Present address: Laboratoire de Reconnaissance cellulaire et Amélioration des plantes, INRA-CNRS-Université, ENS, 46 allée d'Italie, 69364 Lyon cedex 07, FRANCE.
4. Present address: Laboratoire de Biochimie et Physiologie végétale, INRA, 2 place Viala, Montpellier cedex 1, FRANCE.

Abstract

Walnut somatic embryos were transformed with a vector containing the neomycin phosphotransferase (*npt*II), b-glucuronidase (*uid*A) and an antisense chalcone synthase (*chs*) genes. Thirteen transgenic lines were obtained. They were characterized at the molecular, biochemical, and physiological levels, and compared to control lines (the initial line and a transgenic line expressing only *npt*II and *uid*A). Six of the antisense lines underexpressed *chs* endogenous mRNA and CHS activity. They were flavonoid deficient. These *in vitro* plants developed a large basal callus. Rooting tests showed that decreased flavonoid content was associated with adventitious root formation.

1. Introduction

Walnuts are very useful for fruit and wood production. However, mass propagation by cuttings or micropropagation is difficult due to rooting problems. The ability to form adventitious roots varies strongly between genotypes and is rapidly lost during the ageing of woody plants. Phenolic markers (a naphtoquinone and flavonoids) have been related to the ability of microcuttings to be propagated (Jay-Allemand *et al*, 1988). Chalcone synthase (CHS) is considered as a key enzyme in flavonoid biosynthesis. In this work, the antisense strategy using a walnut *chs* gene was used to study the correlation between flavonoid content and rooting capacity.

2. Results

M. R. Ahuja et al. (eds.), Somatic Cell Genetics and Molecular Genetics of Trees, 147–151.
© 1996 *Kluwer Academic Publishers. Printed in the Netherlands.*

2.1. GENERATION OF THE TRANSGENIC LINES

A binary vector containing the β-glucuronidase (*uid*A) and the neomycin phosphotransferase (*npt*II) genes was used to obtain control lines. A 400 bp fragment of a walnut *chs* gene in inverse orientation behind the duplicated CaMV 35S promoter was cloned in this vector. These two vectors were introduced in the disarmed *Agrobacterium tumefaciens* strain C58 pMP90.

Walnut somatic embryos were obtained from immature embryos of a hybrid nut: *Juglans nigra x J. regia*. Prolifering secondary somatic embryos were cocultivated with the disarmed strains for 48 h on a basal DKW medium (Driver and Kuniyuki, 1984). Then, they were transfered to a selective medium (DKW supplemented with 500 mg/l kanamycin, 250 mg/l cefotaxime, 500mg/l ticarpen). A high kanamycin concentration was used since the non transformed embryos were resistant to kanamycin at the concentration of 200 mg/l. Transgenic secondary embryos were identified by a histochemical GUS test. The subsequent generations of transgenic embryos were maintained as homogeneous lines. Thirteen independent lines plus one control line were selected. The transgenic lines containing the *chs* fragment were named ACS and classified according to their myricitrin (a flavonoid) concentration in stems: the myricitrin content decreases when the line number increases. Morphological observations, biochemical and molecular studies, then rooting tests, were performed on microshoots produced by transferring embryonic epicotyls on a DKW medium supplemented with 1mg/l BAP and 1ug/l IBA for 3 weeks (Cornu and Jay-Allemand, 1989). These microshoots were subcultured before analysis.

2.2. MOLECULAR CHARACTERIZATION

Southern-type experiments had proved that one to three copies of the T-DNA were integrated in the ACS lines.

In walnut, CHS is encoded by a multigene family: 5 genes were observed in the hybrid line used in this study, their homology was determined to be between 67 to 95 % (El Euch, unpublished results). The fragment introduced in antisense orientation corresponds to gene IV. In some transgenic lines, northern analysis showed that expression of *chs* mRNA corresponding to gene IV and homologous genes was strongly reduced. In addition, RT-PCR analysis performed with primers able to amplify the 5 genes proved that the expression of all genes was highly reduced or absent in lines ACS 8 to 13.

2.3. BIOCHEMICAL CHARACTERIZATION

2.3.1. *Chalcone synthase activity*

CHS activity was measured in stems according to the method developed by Claudot and Drouet (1992). As shown in Table I, a decrease in enzyme activity was found in transgenic lines showing a reduced presence of the *chs* mRNA.

Lines	% CHS inhibition
ACS 1	51.3
ACS 6	55.2
ACS 9	89.6
ACS 10	100
ACS 11	90.0
ACS 13	92.1

Table I: Percentage of chalcone synthase activity inhibition in some transgenic lines. These percentages have been calculated compared to the CHS activity in the control which represents 100 % activity (0 % inhibition).

2.3.2. *Flavonoid content*

In walnut the main flavonoids are myricithin and quercitin. The soluble polyphenols were extracted according to the procedure described by Magel *et al* (1994). They were characterised and quantified by HPLC. No difference with the controls was noticed in lines ACS 1 to 3. Lines ACS 4 to 7 contained less myricitrin than control and this compound was not detected in lines ACS 8 to 13. The same results were observed with quercitin. Similar results were obtained when analyses were performed with callus or with leaf extracts.

2.4. PHYSIOLOGICAL ASPECTS

2.4.1 *Morphological aspects*. Changes were noted during proliferation and rooting. By comparison to the control lines (initial line and transgenic line expressing *npt*II and *uid*A) the antisense lines with low CHS activity develop very large calli at the base of the microshoots during the proliferation phase. Furthermore, in these lines, the IBA treatment applied during the root induction induced strong apical necrosis.

2.4.2. *Rooting ability*. The rooting ability was tested for each line. Close to 70 % of controls rooted. In ACS lines with low CHS activity, the rooting capacity reached 96 %. Moreover, different parts of the microshoots (apical and basal) were tested. These results are presented in Table 2. The rootability (probability of forming adventitious roots) in the apical part of stems of lines with low flavonoid content increased from 0.3 to 0.6.

Lines	Fragment number	Number of rooted fragments	Rootability apical segment	Rootability basal segment
Control	160	25 (16 %)	0.64	0.25
ACS 1	80	8 (10 %)	0.60	0.20
ACS 5	80	26 (33 %)	0.56	0.25
ACS 7	72	19 (26 %)	0.23	0.63
ACSB 9	76	14 (32 %)	0.58	0.58
ACS 13	48	18 (45 %)	0.57	0.60

Table 2: Rooting capacity of transformed and untransformed stem fragments. This capacity was estimated though *in vitro* fragment tests.

2.4.3. *Auxin content.* Preliminary experiments were performed to quantify indole-3-acetic acid (IAA) and IAA aspartate (IAAsp) in leaves, stems, and calli, of 5 different lines chosen for their different flavonoid content (a control line and ACS 1, 6, 9, and 11). The IAAsp concentration in the stems of ACS 9 and 11 was higher (6-8 nmol/g dry weight) than in control and ACS 1 and 6 (about 2 nmol/g DW). These results were confirmed by analyses made in leaves. However, the IAA content in stems does not appear to be significantly modified by the underexpression of CHS.

3. Discussion

Thirteen independent transgenic walnut lines containing an antisense *chs* gene were obtained. These lines were obtained *via Agrobacterium* transformation of somatic embryos by a procedure modified from Dandekar *et al* (1988). Six of the obtained lines strongly underexpressed *chs* mRNA and possess low levels of CHS activity. This high percentage of underexpressing lines could be due to the fact that CHS is highly regulated in plant cells. Therefore, a reduction of *chs* mRNA has an immediate repercussion on protein production.

The antisense *chs* gene introduced corresponds to a 400 bp fragment of the central part of one of the 5 *chs* genes. This part is well conserved among the genes. The mRNA of all genes is missing in lines with low CHS activity. Very low level or absence of flavonoids were observed in these lines. However, no repercussion on the regeneration of transformed shoots from hypocotyls were observed. New morphological phenotypes including the presence of a large calli at the base of *in vitro* microshoots and apical necrosis following an auxinic induction, were observed in lines with low CHS activity.

Previous work on CHS underexpression focused on the production of white-flowering plants (petunia: Van der Krol *et al*, 1988; gerbera: Elomaa *et al*, 1993; chrysanthemum: Courtney-Gutterson *et al*, 1994). No repercussion on other characteristics was studied. CHS is a key enzyme in plant flavonoid biosynthesis. These compounds possess multiple roles: flower color, UV, and pathogen protection... In walnut, flavonoids are highly abondant and considered to are believed to be involved in rooting ability. Our results have confirmed this hypothesis since transgenic plants lacking flavonoids have a greater rooting capacity than control plants.Changes in congugated auxin (IAAsp) and highest sensivity to IBA of transgenic walnut microshoots with low CHS activity show the relationship between flavonoids and auxin metabolism or transport. In the future, these transgenic plants will be used for physiological studies on the role of flavonoids in relation to auxin metabolism.

5. Acknowledgements

This work was developed in the frame of the European program on development of walnut trees for wood and fruit production as an alternative and extensive system to agricultural crops (AIR3 CT92-0142).

6. References

Claudot, A.C., and Drouet, A. (1992) Preparation and assay of chalcone synthase from walnut tree tissue. *Phytochemistry* **31**, 3377-3380.

Cornu, D., and Jay-Allemand, C. (1989) Micropropagation of hybrid walnut trees (*Juglans nigra x J. regia*) though culture and multiplication of embryos. *Ann. Sci. For.* **46** supl., 113-116.

Courtney-Gutterson, N., Napoli, C., Lemieux, C., Morgan, A., Firoozabady, E., and Robinson, K.E.P. (1994) Modification of flower color in florist's chysanthemum: production of a white-flowering variety though molecular genetics. *Bio/Technology* **12**, 268-271.

Driver; J.A., and Kuniyuki, A.H. (1984) *In vitro* propagation of Paradox walnut rootstock. *HortScience* **19**, 507-509.

Dandekar, A.M., Martin, L.A., and McGranahan, G. (1988) Genetic transformation and foreign gene expression in walnut tissue. *J. Amer. Soc. Hort. Sci.* **113**, 945-949.

Elomaa, P., Honkanen, J., Puska, R., Seppanen, P., Helariutta, Y., Mehto, M., Kotilainen, L., and Teeri, T.H. (1993) *Agrobacterium*-mediated transfer of antisense chalcone synthase cDNA to *Gerbera hybrida* inhibits flower pigmentation. *Bio/technology* **11**, 508-511.

Jay-Allemand, C., Cornu, D., and Macheix, J.J. (1988) Biochemical attributes associated with rejuvenation of walnut tree. *Plant Physiol. Biochem.* **26**, 139-144.

Magel, E., Jay-Allemand, C., and Ziegler, H. (1994) Formation of heartwood substances in the stemwood of *Robinia pseudoacacia* L.II. Distribution of nonstructural carbohydrates and wood extractives across the trunk. *Trees* **8**, 165-171.

Van der Krol, A.R., Lenting, P.E., Veenstra, J., Van der Meer, I.M., Koes, R.E., Gerats, A.G.M., Mol, J.N.M., and Stuijc, A.R. (1988) An antisense chalcone synthase gene in transgenic plants inhibits flower pigmentation. *Nature* **333**, 866-869.

HIGHER EXTRACTABILITY OF LIGNIN IN POPLAR (*POPULUS TREMULA x P. ALBA*) BY REDUCING CINNAMYL ALCOHOL DEHYDROGENASE ACTIVITY

M. BAUCHER[1], B. CHABBERT[2], J. VAN DOORSSELAERE[1], G. PILATE[3], D. CORNU[3], M. PETIT-CONIL[4], B. MONTIES[2], M. VAN MONTAGU[1], D. INZÉ[5], L. JOUANIN[6], and W. BOERJAN[1]

[1] Laboratorium voor Genetica, via the Department of Genetics, affiliated to the Flanders Interuniversity Institute for Biotechnology, and [5] Laboratoire Associé de l'Institut National de la Recherche Agronomique (INRA), Universiteit Gent, K.L. Ledeganckstraat 35, B-9000 Gent, Belgium; [2] Laboratoire de Chimie Biologique, INRA, F-78850 Thiverval-Grignon, France; [3] Station d'Amélioration des Arbres Forestiers, INRA, F-45160 Ardon, France; [4] Centre Technique du Papier, F-38044 Grenoble Cedex 09, France; [6] Laboratoire de Biologie Cellulaire, INRA, F-78026 Versailles Cedex, France

1. Abstract

Cinnamyl alcohol dehydrogenase (CAD) catalyses the last step in the biosynthesis of the lignin precursors, which is the conversion of cinnamaldehydes into cinnamyl alcohols. A poplar (*P. trichocarpa* x *P. deltoides*) *CAD* cDNA was introduced in sense and in antisense orientation into *P. tremula* x *P. alba* via an *Agrobacterium*- mediated transformation. The CAD activity was reduced by 70% in the xylem tissue of three antisense and two co-suppressed lines. A red coloration of the xylem was detected in all poplar lines with a residual CAD activity of less than 60%. The lignin content and the lignin composition (syringyl/guaiacyl) were similar in co-suppressed, antisense *CAD* and control lines. However, chemical properties of the cell wall were different. First, a phloroglucinol staining showed a different coloration in the xylem of the down-regulated CAD poplars, being red-brown as compared to the typical pink staining of the control xylem. Second, after NaOH treatment, lignin could be precipitated by acidification of the alkali-soluble fraction of the cell wall of the red poplars but not of that of the white poplars. A spectral characterisation of the alkali fraction from red poplars showed an increased level of vanillin and syringaldehyde. Third, pulping experiments showed less residual lignin in the pulp of CAD down-regulated poplars.

2. Introduction

Lignin is a complex phenylpropanoid heteropolymer. After cellulose, lignin is the most abundant organic polymer in the biosphere. It is composed of three monomers (or

M. R. Ahuja et al. (eds.), Somatic Cell Genetics and Molecular Genetics of Trees, 153–158.
© 1996 *Kluwer Academic Publishers. Printed in the Netherlands.*

monolignols) which differ only by their degree of methoxylation (the *p*-coumaryl, the coniferyl, and the sinapyl alcohol). Lignin molecules are classified into three groups on the basis of their major structural units. Gymnosperm lignin contains mainly coniferyl alcohol and is classified as guaiacyl (G) lignin. Dicotyledonous lignin is composed of both coniferyl and sinapyl alcohol and is classified as guaiacyl-syringyl (G-S) lignin whereas grass lignin is constituted of the three types of monomer and is called *p*-hydroxyphenyl-guaiacyl-syringyl (H-G-S) lignin. Despite its important biological role (strengthening of the cell wall, enabling solute conductance in the xylem and protecting against pathogens; Lewis and Yamamoto, 1990), lignin decreases the forage crop digestibility and is an undesirable component in paper making that has to be removed from the wood (Dean and Eriksson, 1992). Therefore, a major interest exists to decrease or to modify the lignin content in economically important plants.

Several mutants with a reduced lignin content have been identified in agricultural crops (Jorgensen, 1931; Bucholtz *et al.*, 1980). These mutants are called brown midrib because of the red-brown coloration present in the vascular tissue of the stem and the midrib. The mutation was associated with a reduction in bispecific caffeic acid/ 5-OH-ferulic acid *O*-methyltransferase (COMT) activity in the maize *bm3* mutant (Grand *et al.*, 1985a; Vignols *et al.*, 1995) and with a reduction of both COMT and CAD activities in the sorghum *bmr6* mutant (Pillonel *et al.*, 1991; Chabbert *et al.*, 1993). Moreover, Grand *et al.* (1985b) have shown that the inhibition of CAD in poplar stems by chemical inhibitors reduces the flux to lignin synthesis by 45%. Consequently, CAD and COMT have been considered as target enzymes to reduce the amount of lignin by genetic engineering.

CAD (EC 1.1.1.195) catalyses the last step in the biosynthesis of the lignin precursors (monolignols), which is the conversion of the cinnamaldehydes into their corresponding alcohols. Modification of CAD expression by antisense transformation has been reported in tobacco (Halpin *et al.*, 1994; Hibino *et al.*, 1995). By the introduction of a tobacco antisense *CAD* cDNA, Halpin *et al.* (1994) obtained several lines with a strongly reduced CAD activity (up to 7% residual activity). No difference in lignin content was detected but evidence for an alteration in lignin composition was demonstrated; more phenolics could be extracted from the cell walls by NaOH and thioglycolic acid. Peaks of coniferaldehyde and sinapaldehyde, when using pyrolysis mass spectrometry, were increased. By the same strategy, Hibino *et al.* (1995) obtained two lines with 55% and 20% reduction in CAD activity. No difference in lignin content was detected but an increase in aldehyde content was observed (by a colorimetric method). In both studies, the coloration of the stem was modified, being red in the down-regulated CAD lines. The chemical origin of this coloration has not yet been elucidated. However, synthetic lignin polymers (dehydrogenation polymers, DHPs) of a mixture of coniferaldehyde and coniferyl alcohol were red-brown as compared to the pale-brown colour of DHPs of coniferyl alcohol alone (Higuchi *et al.*, 1994; Tollier *et al.*, 1995). Other mutants in the lignin biosynthesis pathway show a similar coloration in the stem tissue, like the maize *bm3* mutant and transgenic poplars, depleted in COMT activity (Van Doorsselaere *et al.*, 1995a).

We have isolated a *CAD* cDNA from poplar (Van Doorsselaere et al., 1995b).

Here, we report on the impact of a reduction of CAD activity on the lignin extractability in transgenic poplars containing either a sense or an antisense *CAD* cDNA.

3. Results and discussion

A full-length poplar *CAD* cDNA (1.4 kb) (Van Doorsselaere *et al.*, 1995b) was inserted both in sense and in antisense orientation into the binary vector pGSJ780A between the CaMV 35S promoter (p*35S*) and the 3' terminator sequence of the nopaline synthase gene and introduced into poplar (*P. tremula* x *P. alba*, clone INRA 717-1 B4) via an *Agrobacterium*-mediated transformation (Leplé *et al.*, 1992). About 20 transformants were regenerated for both constructs and acclimated in a greenhouse. Poplars transformed with the plasmid pBI121 (containing p*35S-gus* and p*nos-nptII* constructs; Jefferson *et al.*, 1987) were used as control. All transformants were analyzed for stable integration of the T-DNA by Southern blot analysis: the T-DNA copy number varied between 1 and 5 (data not shown). In order to assess whether the transformation had resulted in a reduction or an increase in CAD expression, CAD activity measurements of xylem tissue were performed. In three antisense lines (ASCAD14, ASCAD21, and ASCAD52) and in two sense lines (CAD1 and CAD4), the CAD activity was reduced by about 70% and in one sense line (CAD24), the CAD activity was increased by 30% as compared to the control value (Table 1). In other tissues (leaves and cork), the CAD activity level, which was about two- to three-fold lower than in the xylem, was not significantly altered by the transformation (data not shown). This may be due to the presence of another CAD activity or to the different efficiency of the antisense effect in these tissues. The endogenous *CAD* transcript was detected in xylem by Northern analysis in much lower extent for the plants with a lower CAD activity (ASCAD14, ASCAD21, ASCAD52, and CAD1; CAD4 was not analyzed) (data not shown).

No morphological differences were observed between the control and the transgenic lines. However, when the bark was peeled off, the maturing xylem of the plants with a reduced CAD activity showed a red coloration. This phenotype was detected in all poplar lines with a CAD activity reduced by more than 40%. The apparition of this coloration is now well established for plants with a low CAD activity. Halpin *et al.* (1994) detected a red colour in tobacco plants with 7% residual CAD activity and Hibino *et al.* (1995) in tobacco plants with 55% residual activity. However, there is a variability in the intensity of the colour which tends to disappear in tobacco and in poplar during development. The mechanism of the disappearance of the red colour is not known.

To investigate whether a reduced CAD activity was correlated with a modified lignin amount or composition, young stem sections were subjected to phloroglucinol staining (which is considered to be indicative of aldehyde end groups, cinnamaldehydes (C6-C3) and their related benzaldehydes (C6-C1) (Monties, 1989). After staining, the colour of the lignified tissues in the control was pink with a higher intensity in the most lignified parts. In the down-regulated CAD poplars, the lignified cell walls were stained red-brown to blue, suggesting a change in the content of aldehydes in the lignin (data not shown). A difference in staining has also been observed for CAD down-regulated tobacco plants (Halpin *et al.*, 1994; Hibino *et al.*, 1995).

Lignin content was determined by the Klason gravimetric procedure (Table 1). The lignin content was similar in sense *CAD*, antisense *CAD* and control poplars. The monomeric composition of lignin was determined by thioacidolysis. This technique permits to recover the monomers involved in the ß-O-4 ether linkages, which are the major linkages in lignin (Lapierre *et al.*, 1986). The yield (S+G) and the ratio (S/G) of monomers recovered by thioacidolysis were not significantly modified among the different poplars. A high level of variation was observed between the different samples (Table 1). In contrast to what was expected, no increased amount of the expected cinnamaldehydes was identified among the thioacidolysis breakdown products.

TABLE 1. CAD activity, lignin content and lignin composition of control, sense *CAD* and antisense *CAD* poplars (June 1994)

	CAD activity	KL	G	S	S+G	S/G
ANTISENSE						
Control	0.130 ± 0.010	18.4 ± 0.1	153 ± 6	261 ± 3	414 ± 9	1.71
ASCAD3	0.138 ± 0.011	19.5 ± 0.2	133 ± 2	200 ± 2	333 ± 4	1.50
ASCAD8	0.108 ± 0.005	20.6 ± 0.1	151 ± 1	239 ± 1	390 ± 2	1.58
ASCAD9	0.106 ± 0.004	20.5 ± 0.5	168 ± 3	247 ± 4	415 ± 7	1.47
ASCAD10	0.104 ± 0.008	18.3 ± 0.1	140 ± 3	273 ± 6	413 ± 9	1.95
ASCAD14	0.047 ± 0.006	19.5 ± 0.1	136 ± 3	233 ± 7	369 ±10	1.71
ASCAD21	0.046 ± 0.003	18.6 ± 0.2	131 ± 2	173 ± 3	304 ± 5	1.32
ASCAD49	0.083 ± 0.006	17.7 ± 0.2	154 ± 1	253 ± 8	407 ± 9	1.64
ASCAD52	0.043 ± 0.003	20.3 ± 0.2	126 ± 4	196 ± 6	322 ±10	1.55
SENSE						
CAD1	0.042 ± 0.006	18.7 ± 0.1	116 ± 2	211 ± 7	327 ± 9	1.82
CAD4	0.039 ± 0.003	N.D.	N.D.	N.D.	N.D.	N.D.
CAD9	0.138 ± 0.012	N.D.	154 ± 1	271 ± 1	425 ± 2	1.76
CAD23	0.155 ± 0.017	18.8 ± 0.1	145 ± 3	256 ± 7	401 ±10	1.77
CAD24	0.175 ± 0.010	N.D.	N.D.	N.D.	N.D.	N.D.

Data are expressed as μmol g^{-1} cell wall residue (CWR) and are replicates of three measurements. Controls are transformed plants (pBI121 transgenic lines). KL, Klason lignin content (expressed as % CWR); S, syringyl unit; G, guaiacyl unit; N.D., not determined

When the reactivity of the cell wall towards alkali treatment was analyzed, wild-type and CAD down-regulated trees showed different characteristics. First, larger amounts of vanillin (up to two fold) and of syringaldehyde (up to 12-fold) (C6-C1 aldehydes) were released from the cell wall by a mild alkali treatment (Table 2). The possible relation between the accumulation of these aldehydes and the red coloration has still to be investigated. Second, lignin (about 10% of the total lignin) could be precipitated from the alkali-soluble fraction of the cell wall residue (CWR) of the red poplars but not from that of the control (data not shown). Third, chemical pulping trials were performed on young transgenic poplars in order to characterise the effect of the CAD activity on the pulp characteristics. The kappa number (which is a measure for the amount of residual lignin in the pulp) was slightly lowered (data not shown), indicating that the lignin extraction was easier in the down-regulated CAD poplars. The degree of polymerisation of cellulose and the fibre length remained unaffected. Pulp properties were thus slightly improved in the down-regulated CAD poplars resulting in an easier delignification without simultaneous degradation of the carbohydrates. It now needs to be investigated whether this positive effect will be stable over successive years of growth. Field trials with these transgenic poplars are under way.

TABLE 2. Yields of vanillin and syringaldehyde released from red and white poplar xylem by alkali extraction

	Vanillin	Syringaldehyde
WHITE POPLARS		
pBI121	1.9 ± 0.1	1.2 ± 0.2
ASCAD3	1.8 ± 0.1	1.2 ± 0.1
ASCAD8	1.7 ± 0.1	1.4 ± 0.1
ASCAD9	1.6 ± 0.1	1.2 ± 0.1
ASCAD35	1.3 ± 0.1	1.2 ± 0.1
RED POPLARS		
ASCAD14	2.2 ± 0.1	5.4 ± 0.1
ASCAD21	4.3 ± 0.1	14.7 ± 0.1
ASCAD49	1.6 ± 0.1	1.5 ± 0.1
ASCAD52	2.2 ± 0.1	7.6 ± 0.1

Data are expressed as $\mu mol \cdot g^{-1}$ CWR.

In conclusion, despite the fact that the lignin content was unaffected by down-regulating CAD, the lignin composition was altered. The increase of the lignin extractability should lead to the use of less hazardous chemicals during pulping and in an increased yield of paper that can be produced per tree.

4. Acknowledgements

This work was supported by grants of the Commission of the European Communities (ECLAIR-OPLIGE N°AGRE-0021-C(EDB)) and FAIR-PL 95424. J.V.D. is indebted to the Vlaams Instituut voor de Bevordering van het Wetenschappelijk-Technologisch Onderzoek in de Industrie for a post-doctoral fellowship. D.I. is a Research Director of the Institut National de la Recherche Agronomique (France).

5. References

Bucholtz, D.L., Cantrell, R.P., Axtell, J.D. and Lechtenberg, V.L. (1980) Lignin biochemistry of normal and brown midrib mutant sorghum. *J. Agric. Food Chem.* **28,** 1239-1241.

Chabbert, B., Tollier, M.T. and Monties, B. (1993) Expression of brown midrib mutations on grass lignin, in Japan Wood Research Society, (ed.), *Proceedings of the 38th Congress on Lignin*, Japan Wood Research Society Publishers, Kagawa, pp. 33-36.

Dean, J.F.D. and Eriksson, K.-E.L. (1992) Biotechnological modification of lignin structure and composition in forest trees. *Holzforschung* **46,** 135-147.

Grand, C., Parmentier, P., Boudet, A. and Boudet, A.M. (1985a) Comparison of lignins and of enzymes involved in lignification in normal and brown midrib (*bm*$_3$) mutant corn seedlings. *Physiol. Vég.* **23,** 905-911.

Grand, C., Sarni, F. and Boudet, A.M. (1985b) Inhibition of cinnamyl-alcohol-dehydrogenase activity and lignin synthesis in poplar (*Populus x euramericana* Dode) tissues by two organic compounds. *Planta* **163,** 232-237.

Halpin, C., Knight, M.E., Foxon, G.A., Campbell, M.M., Boudet, A.M., Boon, J.J., Chabbert, B., Tollier, M.-T. and Schuch, W. (1994) Manipulation of lignin quality by downregulation of cinnamyl alcohol dehydrogenase. *Plant J.* **6,** 339-350.

Hibino, T., Takabe, K., Kawazu, T., Shibata, D. and Higuchi, T. (1995) Increase of cinnamaldehyde groups in lignin of transgenic tobacco plants carrying an antisense gene for cinnamyl alcohol dehydrogenase. *Biosci. Biotech. Biochem.* **59,** 929-931.

Higuchi, T., Ito, T., Umezawa, T., Hibino, T. and Shibata, D. (1994) Red-brown colour of lignified tissues of transgenic plants with antisense CAD gene: wine-red lignin from coniferyl aldehyde. *J. Biotechnol.* **37,** 151-158.

Jefferson, R.A., Kavanagh, T.A. and Bevan, M.W. (1987) GUS fusions: ß-glucuronidase as a sensitive and versatile gene fusion marker in higher plants. *EMBO J.* **6,** 3901-3907.

Jorgensen, L.R. (1931) Brown midrib in maize and its linkage relations. *J. Am. Soc. Agron.* **23,** 549-557.

Lapierre, C., Rolando, C. and Monties, B. (1986) Thioacidolysis of poplar lignins: identification of monomeric syringyl products and characterization of guaiacyl syringyl-lignin fractions. *Holzforschung* **40,** 113-118.

Leplé, J.C., Brasileiro, A.C.M., Michel, M.F., Delmotte, F. and Jouanin, L. (1992) Transgenic poplars: expression of chimeric genes using four different constructs. *Plant Cell Reports* **11,** 137-141.

Lewis, N.G. and Yamamoto, E. (1990) Lignin: occurrence, biogenesis, and biodegradation. *Ann. Rev. Plant Physiol. Plant Mol. Biol.* **41,** 455-496.

Monties, B. (1989) Lignins, in P.M. Dey and J.B. Harborne (eds.), *Methods in Plant Biochemistry: Plant Phenolics*, Vol. 1, Academic Press, New York, pp. 113-157.

Pillonel, C., Mulder, M.M., Boon, J.J., Forster, B. and Binder, A. (1991) Involvement of cinnamyl-alcohol dehydrogenase in the control of lignin formation in *Sorghum bicolor* L. Moench. *Planta* **185,** 538-544.

Tollier, M.T., Chabbert, B., Lapierre, C., Monties, B., Francesch, C., Rolando, C., Jouanin, L., Pilate, G., Cornu, D., Baucher, M. and Inzé, D. (1995) Lignin composition in transgenic poplar plants with modified cinnamyl alcohol dehydrogenase activity with reference to dehydropolymer models of lignin, in R. Brouillard, M. Jay and A. Scalbert (eds.), *Polyphenols 94* (Les Colloques N° 69), INRA Editions, Paris, pp. 339-340.

Van Doorsselaere, J., Baucher, M., Chognot, E., Chabbert, B., Tollier, M.-T., Petit-Conil, M., Leplé, J.-C., Pilate, G., Cornu, D., Monties, B., Van Montagu, M., Inzé, D., Boerjan, W. and Jouanin, L. (1995b) A novel lignin in poplar trees with a reduced caffeic acid/5-hydroxyferulic acid *O*-methyltransferase activity. *Plant J.* **8,** 855-864.

Van Doorsselaere, J., Baucher, M., Feuillet, C., Boudet, A.M., Van Montagu, M. and Inzé, D. (1995b) Isolation of cinnamyl alcohol dehydrogenase cDNAs from two important economic species: alfalfa and poplar. Demonstration of a high homology of the gene within angiosperms. *Plant Physiol. Biochem.* **33,** 105-109.

Vignols, F., Rigau, J., Torres, M.A., Capellades, M. and Puigdomènech, P. (1995) The *brown midrib3* (*bm3*) mutation in maize occurs in the gene encoding caffeic acid *O*-methyltransferase. *Plant Cell* **7,** 407-416.

TRANSGENE EXPRESSION IN SPRUCE AND POPLAR: FROM THE LAB TO THE FIELD

D.D. ELLIS[1,2], J. RINTAMAKI-STRAIT[1], K. FRANCIS[1], K. KLEINER[3] K. RAFFA[3], AND B. MCCOWN[1]
Departments of Horticulture[1] and Entomology[3], University of Wisconsin, Madison, WI, USA, 53706
[2] Present Address: Forest Biotechnology Centre, BC Research Inc., 3650 Wesbrook Mall, Vancouver, B.C., V6J 2L2, CANADA

1. Abstract

As the development of transformation systems progresses in annual crop plants, one troubling theme has emerged: there is extreme variation in transgene expression, particularly between plants derived from individual transformation events even within the same petri dish. If such variation in transgene expression occurs in annual crops, what variation can be expected in long-lived perennial crops, such as trees? To answer this question, we have studied the expression over time, both in the lab and in the field, of the *uidA* gene controlled by an enhanced CaMV 35S promoter in transformants from two *Populus* hybrids and several transformed lines of *Picea glauca*. Based on MUG assays for β-glucuronidase (GUS) activity, there is a general trend: the highest and least variable GUS activity occurs in vitro and the most variable GUS activity is in field grown transgenics. Detailed analysis of GUS activity in three different developmental stages of leaves during two growing seasons indicate that *uidA* expression in young leaves of hybrid poplar NC5339 (*P. alba* x *P. grandidentata*) was highly variable and that in older fully expanded leaves, GUS activity was relatively stable during the season and from year to year. In contrast, older leaves in another hybrid poplar, NM6 (*P. nigra* x *P. maximowiczii*) had higher GUS activity relative to the younger leaves. In spruce, GUS activity in needles was only detected during needle elongation, yet activity in the stems persists throughout the season. In all transgenics in the field, GUS activity was highest early in the spring just after bud break, then activity rapidly dropped. The suppression of *uidA* expression has been noted in a limited number of transgenic lines in both spruce and poplar. After three years in the field, *uidA* expression has not been restored in any transgenic lines that had *uidA* expression suppressed. Using a wound-inducible promoter (*pinII*) to control the *uidA* gene, evidence suggests that year-to-year stability of transgene expression in poplar is not unique to CaMV 35S. It is interesting to note that the overall level of GUS activity in two consecutive seasons was similar within a species and that despite seasonal variability, predictable yearly trends in transgene expression may be possible in woody perennials.

M. R. Ahuja et al. (eds.), Somatic Cell Genetics and Molecular Genetics of Trees, 159–163.
© 1996 *Kluwer Academic Publishers. Printed in the Netherlands.*

2. Introduction

The success of plant transformation relies principally on the level, location, and temporal expression of the transgene. The goal of virtually every commercial endeavor with the expression of transgenes in plants is to express the transgene in a very precise and predictable way. One good example is the genetic engineering of male sterility in oilseed rape, where transgene expression was confined to a single cell layer, the tapetum (Mariani et al. 1990). Despite this and other examples, the precise control of transgene expression in plants has been called the exception rather than the rule (Finnegan and McElroy 1994). This is based on the huge variation that exists in the expression of identical genetic constructs in different transformed lines, even lines derived from two different transformation events on the same leaf disc.

In long-lived perennial crops, such as forest trees, not only do we have to worry about the proper spatial and temporal expression patterns, but in most cases the stability of these expression patterns, year after year, over the long life of the transgenic plant is important. This is not so with annual crops where year-to-year expression of the transgene relies on fidelity of expression patterns being transmitted sexually. Although we know of no published reports on the long-term expression of transgenes in trees, based on our experience (unpublished) and others (personal communications) we can conclude that transgenes in woody perennials are stably expressed in vitro and in the greenhouse over several years. We have transgenic cultures of both white spruce (*Picea glauca*) and poplar hybrid NC5339 (*P. alba* X *P. grandidentata*) where transgenes have continued to be expressed for over 4 years. However, changes in the level and location of transgene expression within these plants, has not been monitored.

Based on these limited studies, one might conclude that long-term expression of transgenes in trees by and large remains stable. One problem is that these studies have been conducted in controlled environments. We know little about the consistency of transgene expression after dormancy, stress, or with the changing seasons. A recent paper illustrates the complexity of transgene expression, where stresses caused by transplanting could reproducibly result in transgene silencing (Brandle et al. 1995).

3. Field Expression of Transgenes in Trees

One of our main objectives in establishing a field plot of transgenic spruce and poplar was to evaluate the long-term expression patterns of transgenes in field-planted trees. Specifically, we were interested in determining whether there was seasonal variation in transgene expression and whether the seasonal patterns of transgene expression were predictable from year-to-year.

Unless noted, all analyses were done on plants transformed with a construct containing a *uidA* gene controlled by an enhanced CaMV 35S promoter (McCown et al. 1991, Ellis et al. 1993). Expression of the *uidA* gene was quantified by MUG assays (Jefferson

1987) for β-glucuronidase (GUS) activity, the protein encoded by the *uidA* gene. Currently, the transgenic trees are in their third field season and detailed analysis of *uidA* expression, based on GUS activity, has been done for the past two seasons. In addition, we have one year of field data on transgenics from a second poplar clone, NM6 (*P. nigra* x *P. maximowiczii*).

It is interesting to note that in transgenics from the two poplar hybrids tested, the highest GUS activity was in tissues from in vitro grown material. In NC5339, GUS activity in leaves decreased by one-third as the plants were acclimated from in vitro to the greenhouse. Then, depending on the time of the year, GUS activity in field grown leaves decreases to between 10% to 75% of greenhouse levels. In spruce, GUS activity in embryogenic callus and somatic embryos, both in vitro grown was relatively high. However, as the embryos germinated into seedlings the GUS activity dropped significantly despite the fact that the seedlings are still growing in vitro.

As plants are moved to the greenhouse, differences in GUS activity between different tissues on the same plant were evident. In greenhouse grown plants of poplar hybrid NC5339, stems had a 3X higher GUS activity than any other tissue tested, and leaves of all developmental stages had higher GUS activity than roots.

In the field, transgenic NC5339 plants had detectable GUS activity throughout the year. In dormant trees, although GUS activity was very low in the stems, significant GUS activity could be detected in the buds, even when the ground was still frozen. It is unknown whether this reflects *uidA* expression during dormancy, or stability of the GUS enzyme within the bud. There was no question however that with the onset of growth, *uidA* expression was turned on as the level of GUS activity increased 2-4X in the expanding leaves during the first two weeks after bud break. Although there are fluctuations, GUS activity was maintained throughout the growing season in both poplar hybrids tested. The major difference however was the relative levels of GUS in different developmental stages of the leaves. In NC5339, fully expanded leaves (leaf plastichron index [LPI] 10) had the lowest GUS activity and LPI 2 and LPI 4 leaves generally had 2X greater GUS activity. The reverse was true with poplar hybrid NM6, where GUS activity was greatest in the fully expanded older leaves (Francis et al. 1994). Although these results are from a very limited number of transformants, they do indicate that considerable differences in the pattern of transgene expression can occur even between two poplar hybrids.

In spruce, GUS activity in the needles rapidly increased from a very low level in the dormant bud to a high level within three weeks of bud break. After this, the GUS activity declined, almost as sharply as it increased. Eight weeks after bud break, GUS activity in the needles was negligible. GUS activity in the stem was also highest in the spring, yet it gradually declined to a relatively constant level throughout the season. This pattern, of GUS activity in the needles and stems was consistent in the six transformed spruce lines analyzed, despite large differences in the overall level of GUS activity in each line.

The level and pattern of GUS activity was very consistent between the two years. Comparing the GUS activity in NC5339 on the same calendar weeks during 1994 and 1995, there were no significant differences in LPI 10 leaves of NC5339 and only a few weeks where there were significant differences in the LPI 2 and LPI 4 leaves. In spruce, the pattern of expression in the needles was offset by one week, however the pattern of expression between the two years as well as the level of GUS activity were similar. There were no significant differences in the GUS activity in spruce stems between the two years. While these comparisons of GUS activity on a weekly basis were similar, the lack of significant differences were strongly influenced by the large variation in GUS activity between similar developmental stages of tissues from different plants of the same transgenic line.

The stability of transgene expression can be dependent on the promoter or the gene used. To test the effect of promoter on the stability of transgene expression , transgenic NC5339 containing a *uidA* gene controlled by a wound-inducible potato proteinase inhibitor (*pinII*) promoter were planted out in the field in 1994. These plants were then tested for wound-inducible GUS activity throughout the growing season in 1995. As with the CaMV 35S promoter, reliable wound-inducible GUS activity was observed throughout the growing season. Further, there was a strong developmental influence on the level of GUS activity induced. Wound-inducible GUS activity was only strong in LPI 5 or older leaves. This might explain the low level of wound-inducible expression in leaves from tissue cultured shoots. Using these results, the threshold level of enzyme needed for detection of GUS activity histochemically in poplar leaves is equivalent to the amount of enzyme needed to produce 300-500pM MU/μg DNA/min.

Using the same enhanced CaMV 35S promoter, the expression of a *Bacillus thuringiensis* (B.t.) *cry1A(a)* endotoxin gene was also evaluated throughout the growing season. Using an insect bioassay, expression of B.t endotoxin was confirmed throughout the season (Kleiner et al. 1995). This work has now been replicated for a second growing season, and like *uidA,* the B.t. endotoxin gene was still expressed during the second year.

The suppression of transgene expression has been noted in several systems (Finnegan and McElroy 1994), and has been observed in both spruce and poplar. In GUS-suppressed lines from both these species, the *uidA* gene was not turned on by the field environment. In poplar, it is not clear if the two non-GUS expressing transformed lines ever expressed the *uidA* gene, although based on both PCR and Southern blot analysis these lines contain the gene (McCown et al. 1991). Analysis to confirm the presence of an intact promoter and 3' sequences has not been done. In the two non-GUS expressing spruce lines, <u>both</u> initially expressed the *uidA* gene based on histochemical detection of GUS activity. The expression of the *uidA* gene was suppressed very early in these lines and GUS activity has not been detected in the field.

Based on two years of extensive analysis, transgene expression in field-grown trees is relatively stable from year-to-year, provided the expression of the transgene has not been suppressed prior to placement in the field. Further, based on results from two promoters and two transgenes, stability of transgene expression is independent of the promoter and transgene used. Differences in the pattern, level, and inducibility of transgene expression are also present between in vitro, greenhouse, and field environments in the same transformed line. One disturbing problem is the large variation observed in transgene expression between individuals from the same transformed lines. This variation may be manageable however, as the variation exists in the level and not the pattern of transgene expression.

4. Acknowledgments

We thank The Consortium for Plant Biotechnology Research (CPBR) and Georgia Pacific for financial support and Tom Wright, of the West Madison Agricultural Research Station for assistance with the maintenance of the field plot.

5. References

Brandle, J.E., McHugh, S.G., L. James, L., Labbe, H., and Miki B.L. (1995) Instability of transgene expression in field grown tobacco carrying the *csr1-1* gene for sulfonylurea herbicide resistance. *Bio/Technology* 13:994-998.

Ellis, D.D., McCabe, D.E., McInnis, S., Ramachandran, R., Russell, D.R., Wallace, K.M., Martinell, B.J., Roberts, D.R., Raffa, K.F., and McCown, B.H. (1993) Stable transformation of *Picea* glauca by particle acceleration. *Bio/Technology* 11:84-89.

Finnegan, J. and McElroy, D. (1994) Transgene inactivation: plants fight back. *Bio/Technology* 12:883-888.

Francis, K.E., Wraith, S.J., Ellis, D.D., and McCown, B.H. (1995) Genetic transformation and analysis of transgene expression in hybrid poplar NM6 (*P. nigra* x *P. maximowiczii*). *In:* Proceedings from the International Poplar Symposium, Seattle, WA. August 20-25.

Jefferson, R. (1987) Assaying chimeric genes in plants: the GUS fusion system. *Plant. Mol. Bio. Rep.* 5(4):387-405.

Kleiner, K.W., Ellis, D.D., McCown, B.H., and Raffa, K.F. (1995) Field Evaluation of transgenic poplar expressing a *Bacillus thuringiensis* δ-endotoxin gene against forest tent caterpillar and gypsy moth. *Environ. Ent.* 24:1358-1364.

Mariani, C., DeBeukeleer, M., Truettner, J., Leemans, J., and Goldberg, R.B. (1990) Induction of male sterility in plants by a chimaeric ribonuclease gene. *Nature* 347:737-760.

McCown, B.H., McCabe, D.E., Russell, D.R., Robison, D.J., Barton, K.A., and Raffa, K.F. (1991) Stable transformation of *Populus* and incorporation of pest resistance by electric discharge particle acceleration. *Plant Cell Reports* 9:590-594.

FLORAL HOMEOTIC GENES FOR GENETIC ENGINEERING OF REPRODUCTIVE STERILITY IN POPLARS

L.A. SHEPPARD, A.M. BRUNNER, W.H. ROTTMANN, R. MEILAN, AND S.H. STRAUSS
Department of Forest Science, Oregon State University
Corvallis, Oregon 97331-7501

1. Abstract

Many floral homeotic genes are expressed early in floral development and exhibit floral-specific expression, making them excellent candidates for use in engineering reproductively sterile plants. We have cloned and partially characterized *Populus trichocarpa* cDNAs and genes homologous to the *Arabidopsis*/*Antirrhinum* genes *LEAFY* (*LFY*)/ *FLORICAULA* (*FLO*), *APETALA3* (*AP3*)/ *DEFICIENS* (*DEF*), and *AGAMOUS* (*AG*)/ *PLENA* (*PLE*). The *P. trichocarpa LFY* homolog, *PtFL*, shares 75% amino acid identity with *LFY*. *PtD* shows 81% amino acid similarity with *DEF*, and *PtAG* shows 85% homology with *AG*. *PtFL* is expressed in immature inflorescences on which floral primordia are developing. *PtD* is expressed strongly in stamen primordia from the onset of organogenesis, and is expressed at low levels in carpel primordia.

2. Introduction

1.1 RATIONALE FOR ENGINEERED STERILITY

Transgenic trees with beneficial traits such as insect, herbicide, and disease resistance are being produced in a number of laboratories. Although it remains unclear if genetic containment should be required for all transgenic trees, transgene containment at this time seems ecologically prudent and will simplify regulatory approval (reviewed in [1]). Containment of genes is more difficult to achieve in trees than in agronomic crops due to their large size and longevity. Furthermore, forest trees typically produce large quantities of wind-dispersed pollen and seeds. Reproductive sterility is the most feasible and effective method by which transgene escape from trees can be prevented.

In addition to the containment of transgenes, engineered sterility may have other beneficial effects. There is evidence that early disruption of reproductive organogenesis may result in increased allocation of resources to vegetative growth (reviewed in [2]). Also, in monoecious species male sterility will be useful for breeding purposes.

1.2 STRATEGIES FOR ENGINEERING STERILITY

We are using two strategies to engineer reproductive sterility. In one method, a reproductive-specific promoter directs the expression of a cytotoxin, resulting in ablation of

M. R. Ahuja et al. (eds.), Somatic Cell Genetics and Molecular Genetics of Trees, 165–172.
© 1996 *Kluwer Academic Publishers. Printed in the Netherlands.*

only floral cells. The second method uses antisense- and co-suppression to inhibit the expression of genes essential for fertility. Both methods require the isolation and study of floral genes, either to obtain promoters which function specifically in reproductive organs, or to obtain coding regions for suppression of endogenous genes. We are testing constructs derived from floral genes isolated in other plant species as well as isolating floral promoters and cDNAs from *Populus*.

1.3 FLORAL HOMEOTIC GENES

Floral homeotic genes appear to encode transcription factors which constitute a regulatory hierarchy controlling floral development (reviewed in [3]). Mechanisms underlying floral morphogenesis have been especially well-studied in *Arabidopsis thaliana* and *Antirrhinum majus*. Homologs of genes cloned in these two species have been isolated and characterized in several other dicots (eg.[4], [5], [6]), two monocots [7], [8] and gymnosperms [9], indicating that floral homeotic genes have fundamental roles in the reproductive development of all land plants. The expression patterns and functions of these genes seem to be reasonably well conserved from one plant species to another. Homologs of *AG*, for example, have been found in at least nine species to date; similar expression patterns and mutant phenotypes have been observed for homologs in species where they have been studied. Thus, the phenotypes of trees transformed with sterility constructs derived from floral homeotic genes can be predicted with some confidence.

Floral development proceeds in several stages: First, the transition of the vegetative meristem to an inflorescence meristem; second, the transition of the inflorescence meristem to the floral meristem; and finally, development of the floral organs. In order to achieve maximum enhancement of vegetative growth, it would be desirable to halt floral development at the earliest possible stage, the transition from a vegetative meristem to an inflorescence meristem. No genes have yet been cloned which act at this stage. However, several genes have been identified and cloned which act very early, at the transition from the inflorescence meristem to the floral meristem. *LFY*, from *Arabidopsis* [10], and its *Antirrhimum* homolog, *FLO* [11], have a major role in this transition; null-mutants of these genes cause a homeotic transformation of flowers into inflorescence shoots. Furthermore, the constitutive expression of *LFY* in transgenic *Arabidopsis*, tobacco, and aspen induces precocious flowering [12]. In the case of aspen, flowering time was reduced from several years to months.

Floral organ identity genes are expressed as the primordia of the organs they specify differentiate and emerge from the floral meristem. A model describing organ identity gene expression and function is reproduced in figure 1. Briefly, these genes are categorized into three groups, designated A, B, and C (reviewed [13]). Each group is expressed in two adjacent floral whorls and functions in establishing the identity of those whorls: A activity in whorl one specifies sepals, combined AB activities in whorl two specify petals, BC activities in whorl three specify stamens, and C activity in whorl four specifies carpels. The *Arabidopsis* gene *APETALA2* and the *Arabidopsis/Antirrhinum* gene *APETALA1/ SQUAMOSA* are A function genes, *AP3/ DEF* and *PISTILLATA/ GLOBOSA* are B function genes, and *AG/ PLE* is the only known C function gene. Organs in two adjacent whorls are homeotically transformed if one of these genes is nonfunctional. For example, in *AG* mutants, sepals and petals develop normally, but petals develop in place of stamens in the third whorl and carpels do not develop [14]. The sepal-petal-petal pattern is repeated a variable number of times, indicating that AG also functions in specifying floral meristem determinancy. Thus, we expect suppression of the poplar *AG* homolog to result in both male and female sterility. Use of the *PtAG* promoter to direct expression of a cytotoxic gene should not only result in complete sterility, but also prevent structures from developing in place of stamens and carpels.

FIGURE 1. The ABC model for organ identity gene expression and floral development (see text)

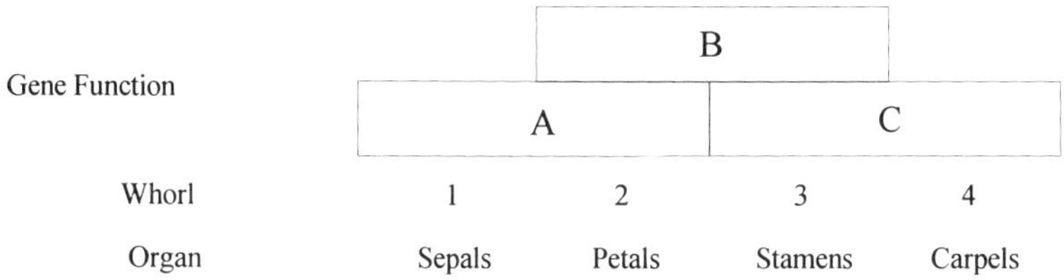

FIGURE 2. MADS-box gene structure in plants (see text)

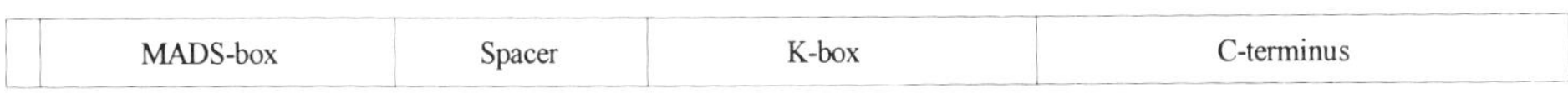

1.3.1 *MADS Box Genes*

The majority of floral homeotic genes belong to the MADS-box family of transcription factors (reviewed in [15]). The MADS-box is a conserved region of approximately 60 amino acid residues which constitutes a DNA-binding domain. The motif name is an acronym for four different MADS-box genes: yeast minichromosomal maintenance factor one (*MCM 1*), the floral homeotic genes *AG* and *DEF*, and human serum response factor (*SRF*) [16]. In plants, MADS-box genes typically contain three major domains (figure 2): the MADS-box region itself, which is located at the 5' end of the transcribed region, the K-box, which is similar to a domain known to form amphipathic α-helices in keratin, and the C-terminal region. The K-box comprises roughly 70 amino acids, and is thought to be involved in protein dimerization. Amino acid residues in the MADS-box region are highly conserved throughout the family; the K-box region is somewhat conserved, and the C-terminal domains are highly variable.

2. Methods

2.1 LIBRARY CONSTRUCTION AND SCREENING

2.1.1 *cDNA Clones*

Catkin buds were collected from a native female *P. trichocarpa* tree in late spring and frozen in liquid nitrogen as they were removed from the bud scales. Poly A^+ RNA was used to construct a cDNA library in λ Zap® (Stratagene) as suggested by the supplier. The *PtD* cDNA was obtained by probing the library under low stringency conditions with fragment containing the MADS-box region of *AGL2*. To isolate *PtAG*, the early floral cDNA library was probed with a 3' fragment of the *AG* cDNA. A partial cDNA clone of *PtFL* was isolated by PCR using degenerate primers based on the *FLO* and *LFY* sequences, and the cDNA library was screened with this fragment. In all three cases, pBluescript® (Stratagene) plasmid vectors containing the cDNAs were excised from λ ZAP® for further analysis.

2.1.2 *Genomic Clones*

Male *P. trichocarpa* genomic DNA was partially digested with *Sau3A1* and ligated into λ GEM 12 (Promega). The library was screened with a PCR fragment amplified from the 3' end of the *PtFL* cDNA to yield λ*PtFLg34*. λ*gPtAG* and λ*gPtD* were isolated by probing the genomic library with the 3' regions of the *AG* and *PtD* cDNAs, respectivly.

2.2 ANALYSIS OF GENE STRUCTURE AND EXPRESSION

2.2.1 *Southern and Northern Blots*

Southern and northern blots were performed according to standard procedures [17].

2.2.2 *Quantitative PCR*

To perform quantitative PCR, cDNA was synthesized from a variety of *P. trichocarpa* tissues using reverse transcriptase and an oligo-dT containing primer. The cDNA was quantified and used in PCR reactions containing known amounts of a homologous internal standard that had an altered size. The steady-state level of *PtFL* message was estimated by comparing the amounts of product obtained from the standards and the unknowns.

2.2.3 *DNA Sequencing*

300-500 bp fragments of the cDNA clones were subcloned into pBluescript® (Stratagene) and sequenced using fluorescent primer dye/dideoxy chain termination and an automated DNA sequencer.

2.2.4 In Situ Hybridizations

In situ hybridizations were performed as described by Kelly *et al* [18] using *PtD* sense and antisense RNAs as control and probe, respectively.

3. Results

3.1 SEQUENCE CHARACTERIZATION

Previously, we reported the isolation of *PtFL*, the *Populus* homolog of *FLO/ LFY* [19]. At the amino acid level, *PtFL, FLO,* and *LFY* are all approximately 75% identical, indicating that *PtFL* is the *Populus* homolog of *FLO* and *LFY*. Approximately 55% of the deduced *PtFL* protein sequence is highly conserved (about 90% identity), while 45% is poorly conserved (about 55% identity). Southern blot analysis demonstrated that *PtFL* is a single copy gene (data not shown). The entire *PtFL* gene as well as 3' and 5' flanking sequences have been isolated. Preliminary sequence data indicate that the intron-exon structure of *PTFL* is similar to the structure of its *Arabidopsis/ Antirrhinum* homologs, *LFY* and *FLO*.

We have also reported isolation of a *P. trichocarpa* cDNA homologous to *DEF* [20]. The *P. trichocarpa* cDNA (*PtD*) is 956 base pairs in length and has an open reading frame which corresponds to a polypeptide of 228 amino acids. Overall, 81% of the amino acid residues are conserved between *PtD* and *DEF*.(table 1) As expected, homology is strongest in the MADS-box region, where 93% of the residues are conserved. The putative phosphorylation site, RQVT, is retained. Eighty-five percent of the amino acids which make up the K-box regions are conserved, and seventy-six percent of the amino acids in the C-terminal are conserved. The gene corresponding to the *PtD* cDNA has been isolated and partially sequenced (data not shown).

TABLE 1. Percent similarity between *PtD* and *DEF* by domain based on putative amino acid sequences. Groups of similar amino acid residues were considered to be: P, A, G, S, T (neutral or weakly hydrophobic); Q, N, E, D (hydrophylic, acid amine); H, K, R (hydrophylic, basic); L, I, V, M (hydrophobic); F, Y, W (hydrophobic, aromatic); C (cross-link forming).

MADS-box	Spacer	K-box	C-terminal	Overall
93	84	85	65	81

We have also isolated cDNA and genomic clones homologous to *AG*. The deduced amino acid sequence of the cDNA shows 85% homology to *AG* (table 2). The *Arabidopsis* gene *AGL1* shares significant sequence homology with *AG*; however, *PtAG* shares substantially less homology with *AGL1*, (ca. 73%) than with *AG*, indicating that *PtAG* is the homolog of *AG*. Analysis of the *PtAG* genomic sequence is in progress.

TABLE 2. Percent similarity of amino acid residues between *PtAG* and homologs by domain. Overall % identity and the % of amino acid residues conserved between homologs are shown in the last two columns. *AGL1* is an *Arabidopsis* gene which shares significant homology with *AG*. Groups of amino acids considered to be similar in function are shown in table one.

	N-terminal	MADS box	Spacer	K-box	C-terminal	Overall	Overall Identity
AG	56.3	96.5	78.8	69.7	69.6	85.0	76.3
AGL1	31.3	89.5	66.7	77.3	36.3	73.0	63.9

3.2 GENE EXPRESSION

Quantitative PCR experiments indicate that, as expected, the strongest *PtFL* expression is in immature inflorescences which have developing floral primordia. The level in roots was below the detection limit; however, there was significant expression in vegetative buds [21].

Northern blot analysis indicates that *PtD* is expressed in both mature male and mature female floral buds, while expression in vegetative buds is extremely low or nonexistent. *In situ* hybridization shows that *PtD* mRNA is present in developing stamens and carpels from the onset of reproductive organogenesis [22].

4. Discussion

4.1 PtFL

Although *LFY* and *FLO* are expressed only in reproductive tissues, *PTFL* as well as the tobacco *LFY* homolog is expressed in vegetative meristems [23]. The level of vegetative expression observed for *PtFL* may preclude the use of the *PtFL* promoter for genetic ablation of floral tissues. However, it may be possible to identify sequences necessary for vegetative expression to obtain a reproductive-specific promoter. In addition, suppression techniques using the *PtFL* coding sequence may still be successful in preventing floral development if the function of *PtFL* is insignificant in vegetative tissues.

4.2 PtD

In *Populus* species, floral structure is highly modified compared to that of typical dicot flowers. Both petals and sepals are absent, being replaced by a perianth cup. Furthermore, *P. trichocarpa* is dioecious. In *Antirrhinum*, *DEF* is first expressed at the beginning of sepal organ differentiation [24]. It is strongly expressed in petal and stamen primordia, and its expression continues in petals and anthers throughout floral development. It is also expressed to a lesser extent in the gynoecium and very weakly in sepals. Similar expression patterns have been noted in other species [25], [26]. Likewise, we found that *PtD* is expressed strongly in stamen primordia, weakly in carpel primordia, and strongly in mature floral buds of both genders. Therefore, the *PtD* promoter may be useful for directing ablation of male and perhaps female floral organs. No expression was observed in the perianth cup.

DEF mutants exhibit a range of phenotypes, including male-sterility, which correspond to mutations at various positions within the gene [27]. *DEF* is a B-function organ identity gene (figure 1). In *DEF* mutants, stamens and petals are replaced by carpelloid and sepalloid structures, respectively, and carpel development is sometimes affected [28]. In contrast, the petunia *DEF* homolog, *pMADS1*, is generally redundant for stamen development, although it plays a minor third whorl role [29]. It will be interesting to see what floral phenotypes will be generated when *PtD* is suppressed in male and female clones of *P. trichocarpa* and relatives, or when the *PtD* promoter is used to drive a cytotoxic gene such as *DTA*. We are constructing sense and antisense *PtD* genes and will introduce them into *Populus* to suppress expression of the endogenous gene. The *PtD* promoter has been cloned and is being characterized.

4.3 PtAG

Based on studies in other species (discussed in section 1.3), we expect that suppression of *PtAG* will result in complete sterility, transforming both male and female flowers to flowers consisting only of a variable number of perianth cups. We are also studying the *PtAG* promoter and plan to use it in cytotoxin-based sterility constructs. This approach may be more successful than gene suppression in acheiving increased vegetative growth, since it may prevent any organs from developing in place of stamens and carpels.

5. References

1. Strauss, S.H., Rottmann, W.H., Brunner, A.M., Sheppard, L.A.: Genetic engineering of reproductive sterility in forest trees, *Molecular Breeding* **1** (1995), 5-26.
2. Strauss et al, *ibid.*
3. Ma, H.: The unfolding drama of flower development: recent results from genetic and molecular analyses, *Genes & Development* **8** (1994), 745-756.
4. Pneuli, L., Haraven, D., Rounsley, S.D., Yanofsky, M.F., Lifschitz, E.: Isolation of the tomato *AGAMOUS* gene *TAG1* and analysis of its homeotic role in transgenic plants, *The Plant Cell* **6** (1994), 163-173.
5. Kempin, S.A., Mandel, M.A., Yanofsky, M.F.: Conversion of perianth into reproductive organs by ectopic expression of the tobacco floral homeotic gene *NAG1*, *Plant Physiol.* **103** (1993), 1041-1046.
6. Tsuchimoto, S., van der Krol, A.R., Chua, N-H.: Ectopic expression of *pMADS3* in transgenic petunia phenocopies the petunia *blind* mutant, *The Plant Cell* **5** (1993) 843-853.
7. Lu, Z-H., Wu, M., Lon, C-S., Yeong, C-Y., Goh, C-J.: Nucleotide sequence of a flower-specific MADS box cDNA clone from orchid, *Plant Mol. Bio.* **23** (1993), 901-904.
8. Schmidt, R.J., Velt, B., Mandel, M.A., Mena, M., Hake, S., Yanofsky, M.F.: Identification and molecular

characterization of *ZAG1*, the maize homolog of the *Arabidopsis* homeotic gene *AGAMOUS*, *The Plant Cell* **5** (1993), 729-737.
9. Tandre, K., Albert, V.A., Sundås, A., Engström, P.: Conifer homologs to genes that control floral development in angiosperms, *Plant Molecular Biology* **27** (1995), 69-78.
10. Weigel, D., Alvarez, J., Smyth, D.R., Yanofsky, M.F., Meyeriwitz, E.M.: *LEAFY* controls floral meristem identity in *Arabidopsis*, *Cell* **69** (1992) 843-859.
11. Coen, E.S., Romero, J.M., Doyle, S., Elliot, R., Murphy, G., Carpenter, R.: *FLORICAULA*: a homeotic gene required for flower development in *Antirrhinum majus*, *Cell* **63** (1990), 1311-1322.
12. Weigel, D., Nilsson, O.: A developmental switch sufficient for flower initiation in diverse plants, *Nature* **12** (1995), 495-500.
13. Meyerowitz, E.M., Coen, E.S.: The war of the whorls: genetic interactions controlling flower development, *Nature* **353** (1991), 31-37.
14. Yanofsky, M.F., Ma, H., Bowman, J.L., Drews, G.N., Feldmann, K.A., Meyerowitz, E.M.: The protein encoded by the *Arabidopsis* homeotic gene agamous resembles transcription factors, *Nature* **346** (1990), 35-39.
15. Davies, B., Schwarz-Sommer, Z.: Control of floral organ identity by homeotic MADS-box transcription factors, *Results and Problems in Cell Differentiation 20*, L. Nover (ed.), Plant Promoters and Transcription Factors, Springer-Verlag, Berlin (1994).
16. Schwarz-Sommer, Z., Huijser, P., Nacken, W., Saedler, H., Sommer, H.: Genetic control of flower development by homeotic genes in *Antirrhinum majus*, *Science* **250** (1990) 931-936.
17. Sambrook,J., Fritsch, E.F., Maniatis, T.: *Molecular Cloning: A Laboratory Manual*, 2nd ed., Cold Spring Harbor Laboratory Press, N.Y., 1989.
18. Kelly, A.J., Bonnlander, M.B., Meeks-Wagner, D.R.: *NFL*, the tobacco homolog of *FLORICAULA* and *LEAFY*, is transcriptionally expressed in both vegetative and floral meristems, *The Plant Cell* **7** (1995), 225-234.
19. Rottmann, W.H., Boes, T.K., and Strauss, S.H.: Structure and expression of a *leafy* homolog from *Populus*. *J. Cell. Biochem.* Suppl 17B **23** (1993).
20. Sheppard, L.S. Rottmann, W.H., Strauss, S.H.: Floral homeotic genes for engineering sterility in poplars: cDNA cloning of a *Populus* homolog of *DEFICIENS*, International Union of Forestry Research Organizations molecular genetics workshop, Prout's Neck, Maine. May 20-23, 1994.
21. Rottmann, W.H., Sheppard, L.S., Brunner. A.M., Strauss, S.H. (manuscript in preparation)
22. Sheppard, L.S., Rottmann, W.H., Brunner, A.M., Strauss, S.H. (manuscript in preparation)
23. Kelly, A.J. *et al*, *ibid.*
24. Schwarz-Sommer, Z., Hue, I. , Huijser, P., Flor, P.J., Hansen, R., Tetens, F., Lonnig, W.-E., Saedler, H., and Sommer, H.: Characterization of the *Antirrhinum* floral homeotic MADS-box gene *deficiens*: evidence for DNA binding and autoregulation of its persistent expression throughout flower development. *EMBO* **11** (1992), 251-263.
25. Pnueli, L., Abu-Abeid, M., Zamir, D., Nacken, W., Schwarz-Sommer, Z., Lifschitz, E.: The MADS box gene family in tomato: temporal expression during floral development, conserved secondary structures and homology with homeotic genes from *Antirrhinum* and *Arabidopsis*, *The Plant Journal* **1** (1991), 255-266.
26. Hardenack, A., Ye, D., Saudler, H., Grant, S.: Comparison of MADS box gene expression in developing male and female flowers of the dioecious plant white campion, *The Plant Cell* **6** (1994), 1775-1787.
27. Schwarz-Sommer *et al*. 1992. *ibid.*
28. Trobner, W., Ramirez, L., Motte, P., Hue, I., Huijser, P., Lonnig, W.-E., Saedler, H., Sommer, H., and Schwarz-Sommer, Z.: *Globosa*: a homeotic gene which interacts with *deficiens* in the control of *Antirrhinum* floral organogenesis. *EMBO* **11** (1992), 693-4704.
29. van der Krol, A.R., Brunelle, A., Tsuchimoto, S., Chua, N.: Functional analysis of petunia floral MADS box gene *pMADS1*, *Genes & Development* **7** (1993), 1214-1228.

Section III.

Molecular markers and genome mapping

GENE DISCOVERY IN LOBLOLLY PINE THROUGH cDNA SEQUENCING

C.S. Kinlaw, T. Ho, S.M. Gerttula, E. Gladstone, D.E. Harry
Institute of Forest Genetics, PSW Station, USDA Forest Service, P.O. Box 245 Berkeley, CA 94701

and

L. Quintana, and C. Baysdorfer
Biology Department, California State University at Hayward, Hayward, CA 94542

Abstract

We have initiated efforts to identify functions of loblolly pine genes based upon partial sequences of random copy DNAs. These efforts are coordinated with the construction of genetic maps (Devey *et al.*, 1991; Neale and co-workers, unpublished data). In addition to providing possible biochemical functions encoded by individual random cDNAs, our work has allowed for the identification of classes of genes actively transcribed in tissues from actively growing seedlings or developing phloem and cambium. Also, we have gained insight into the nature of complex gene families within pine genomes.

Introduction

Two principles permit identification of gene function based upon the partial sequences of random copy DNAs (cDNAs). First, the sequences of large numbers of polypeptides and the genes which encode them are conserved among widely divergent taxa, even across bacterial, fungal, plant, and animal kingdoms (Pearson, 1991). Second, since cDNAs represent the messenger RNA populations expressed from "active" genes, unexpressed genes and huge fractions of "nongenic" DNA, which can be greater than ninety percent in large genomes such as pines, are experimentally excluded. Because sequences of large numbers of genes of known function reside in public databases, it is possible to identify probable functions for new genes based upon DNA sequence similarity. The utilization of high-throughput automated sequencing of cDNAs to identify the function of large numbers of human genes was pioneered by Craig Venter and his colleagues at the Institute of Genomic Research (Adams *et al.*, 1991). This strategy has been applied by a variety of human genome research groups world-wide, resulting in the identification of greater than fifty thousand expressed human genes, a figure which approaches the suspected number of human genes. Partial sequencing of anonymous cDNAs has been established for the identification of gene functions in other animals (e.g. McCombie *et al.*, 1992), as well as for the experimental plant, *Arabidopsis thaliana* (Newman *et al.*, 1994) and for a variety of economically important crop species including rice (Sasaki *et al.*, 1992) and corn (Keith *et al.*, 1993).

Just as the commercial availability of large numbers of chemical reagents through chemical catalogs revolutionized chemistry, so too the availability of " gene catalogs" is revolutionizing biology. The availability of gene sequence databases has greatly reduced the time it takes to isolate specific genes controlling important biological processes. In addition, it is now possible to identify and study simultaneously whole sets of genes involved in specific biological processes. Although this new biological research strategy has been most fully employed for the study of human biology, it offers great promise for answering biological questions in forest trees. Virtually every important area of biological research for tree species, ranging from tree improvement to biodiversity would be advanced if catalogs of expressed genes identified by sequence and possible biochemical function were available.

As part of a pine genome research project at the Institute of Forest Genetics, we were motivated to test the efficacy of pine cDNA sequencing by our involvement in three converging projects:

1. The placement of anonymous cDNA markers on genetic maps
2. The development of polymerase chain reaction based markers from cDNAs
3. The investigation of the structure and evolution of complex gene families.

We present here a summary of our results from an initial loblolly pine cDNA sequencing project.

M. R. Ahuja et al. (eds.), Somatic Cell Genetics and Molecular Genetics of Trees, 175–182.
© 1996 *Kluwer Academic Publishers. Printed in the Netherlands.*

Approach

Our initial approach was the sequencing of two hundred and thirty cDNAs chosen from a library prepared from mRNA of rapidly growing loblolly pine seedlings germinated from a population of open-pollinated seeds. Most of these initial cDNAs had been used previously to place molecular markers on genetic maps (Devey *et al.*, 1991). cDNAs were synthesized by random priming and non-directionally cloning into a phagemid vector as previously described (Devey *et al.*, 1991). The sequences of approximately two hundred base pairs from each end of a cDNA clone were determined according to previously reported methods (Keith *et al.*, 1993). These initial cDNAs, therefore, provided partial sequence information from two random locations of each encoded polypeptide. We later prepared cDNA libraries from the rapidly developing phloem and cambium layers underneath the bark of a single individual (clone 8-13, Weyerhaeuser company). This latter library of crude phloem was poly-T primed and directionally cloned into Uni-Zap^{TM} (STRATAGENE). Because of the manner of its construction the crude phloem library provided for the analysis of 5' (protein coding) or 3' (untranslated) sequences of encoded proteins. DNA sequences of these clones were obtained using an automated APPLIED BIOSYSTEMS, INC 373 sequencing machine at the Recombinant DNA facility of Oklahoma State University. For the purpose of gene identification, sequences were subjected to BLASTX searches (Altschul *et al.*, 1990) for protein similarity searches against the non redundant database (nr) comprised of unique GenBank and EMBO database entries maintained by the National Center for Biotechnology Information (NCBI) and to BLASTN searches (Altschul *et al.*, 1990) for nucleic acid similarity searches against the database of expressed sequence tags (dbest) also maintained by NCBI (blast@ncbi.nlm.nih.gov). To identify sequence similarity among our loblolly pine cDNAs, we performed FASTA (Pearson and Lipman, 1988) similarity searches among all clones. After analysis, sequences were submitted to public databases including dbest (see acknowledgments) and TreeGenes, the genome database for forest trees established by David Neale and Bradley Sherman (http://s27w007.pswfs.gov/).

Results and Discussion

BLASTX and BLASTN similarity searches yield alignment pairs with associated scoring statistics. In general, the higher the score values, the better the match between two sequences, and the more likely the identification of a biochemical function for an unknown gene. The score value depends both upon the overall length of the match and the degree of similarity. Figure 1 shows several examples of different alignment pairs arranged by decreasing scores. Figure 1 part A shows one example of a high scoring alignment (416) between the pine cDNA pPtIFG2892 and a *Pinus thunbergii* , Japanese black pine, gene encoding the small subunit of ribulose bisphosphate carboxylase. Figure 1 part B shows a second example of a lower scoring alignment (152) between pine cDNA pPTIFG2148 and a *Zea mays*, maize, gene encoding an ADP-ribosylation factor. Figure 1 part C shows a third example of an alignment pair with an even lower score (105) between pine cDNA pPtIFG1750 and a ferredoxin I gene from *Synechococcus. sp.* Although this sequence similarity is weak, due to the large evolutionary distance between pine and *Synechococcus sp.*, we consider it significant. In general, the degree of sequence similarity between pine cDNAs and database entries correlated inversely with the evolutionary distance of the species of origin. Figure 1 part D shows a final example of a sequence alignment with such a low score (74) that the pine cDNA and the database entry are unlikely to encode polypeptides with similar biochemical functions.

Using a BLASTX cutoff of 100 we matched 99 genes with known biochemical functions from our 230 seedling cDNAs (43%) and 19 from 100 crude phloem cDNAs (19%). The somewhat higher percentage of "hits" from seedlings may reflect the superiority of the random cloning and bidirectional sequencing approach for these clones relative to the unidirectional cloning and sequencing approach employed for phloem. The genes identified from these analyses include many genes expected to be expressed in rapidly growing tissues such as photosynthetically active seedlings and crude developing phloem. Among biochemical functions encoded by these "matched" cDNAs are polypeptides with roles in:

Figure 1. **Sequence Similarity Search Examples**

A. Blastx sequence similarity score high-probable function identified.

Pine cDNA: pPtIFG2892 (custom sequencing primer e)
Highest scoring match: RIBULOSE BISPHOSPHATE CARBOXYLASE SMALL CHAIN (rbcS)
Species: *Pinus thunbergii* (Jap. black pine)
Score: 416 P: 8.9e-53 Identities = 76/86 (88%) Positives = 80/86 (93%)

```
pPtIFG2892e    3 LPKLTEEQLVKEVDYLLRNKWVPCLEFDLDGTISRKYGSTPGSYDGRYWVMW 174
                 LP LTEEQLVKEV+YLLRNKWVPCLEFDL+G+ISRKY  +PG YDGRYWVMW
rbcS          69 LPTLTEEQLVKEVEYLLRNKWVPCLEFDLEGSISRKYNRSPGYYDGRYWVMW 120

pPtIFG2892e  175 KLPMFGCTEASQVINERYECAKAYPKAFIRVIGF 260
                 KLPMFGCTEASQVINE  ECAKAYPKAFIRVIGF
rbcS         121 KLPMFGCTEASQVINEVRECAKAYPKAFIRVIGF 154
```

B. Blastx sequence similarity score intermediate-probable function identified.

Pine cDNA: pPtIFG2148 (custom sequencing primer s)
Highest scoring match: ADP-RIBOSYLATION FACTOR (ADP)
Species: *Zea mays* (maize)
Score: 152 P: 4.4e-20 Identities = 31/33 (93%) Positives = 32/33 (96%)

```
pPtIFG2148s    6 MGLTFTKLFQRLFAKKEMRILMVGLDAAGKTSI 104
                 MGLTFTKLF RLFAKKEMRILMVGLDAAGKT+I
ADP            1 MGLTFTKLFSRLFAKKEMRILMVGLDAAGKTTI 33
```

C. Blastx sequence similarity score low-probable function identified.

Pine cDNA: pPtIFG1750 (custom sequencing primer e)
Highest scoring match: FERREDOXIN I
Database: Swissprot Accession number: P06517
Species: *Synechococcus sp*
Score: 105 P: 5.4e-29 Identities = 19/25 (76%) Positives = 21/25 (84%)

```
pPtIFG1750e    4 IDCPDDVYVLDAAEEAGIELPYSCR 78
                 ID  DD Y+LDAAEE GI+LPYSCR
ferredoxin    18 IDVADDTYILDAAEEQGIDLPYSCR 42
```

D. Blastx sequence similarity score low-probable function not identified.

Pine cDNA: pPtIFG0147 (custom sequencing primer e)
Highest scoring match: INTESTINAL MUCIN (mucin)
Database: Pir Accession number: S24169
Species: *Rattus norvegicus* (rat)
Score: 74 P: 9.6e-13 Identities = 16/41 (39%) Positives = 25/41 (60%)

```
pPtIFG0147e   78 SAGTTTCSTESTIAVTGTSSAPTPRSTVIQSWATTAATPTT 200
                 +A +T+ +TEST + + TS  PTP  T I+   T+ +T  T
mucin         59 TASSTSPTTESTSSTSTTSQTPTPTPTTIKETPTSTSTVPT 99
```

Figure 1 legend. BLASTX alignment pairs of amino acid residues shown with identical and similar (+) amino acids shown in between.

photosynthesis
glycolysis
ATP biosynthesis and hydrolysis
fatty acid biosynthesis and oxidation
protein synthesis and degradation
amino acid biosynthesis and degradation
membrane structure or transport
cellular pH maintenance
flavenoid biosynthesis
lignin polymerization
cell wall structure
DNA and RNA structure
regulation (cell cycles, environmental and hormone responses).

A number of genes appeared more than once in our seedling cDNA sample. In contrast, samples identified from the crude phloem represent unique cDNAs, probably because of the smaller sample size. From sequence comparisons, it is possible to assign possible biochemical functions encoded by a number of these "highly" abundant mRNAs (see TABLE 1). In addition, one cDNA which appeared multiple times shared no significant sequence similarity to genes encoding known biochemical functions within public databases. This gene is clearly important for seedlings as it is highly expressed, and further analysis to determine its possible function is in progress.

Significant numbers of pine cDNAs reveal complex Southern blot banding patterns suggestive of complex multi-locus gene families (Devey et al., 1991; Kinlaw and Gerttula, 1994). Analysis of the sequence and encoded biochemical function of a number of cDNAs have permitted some insight into the identity and expression of these complex gene families. As an example, we chose to analyze the sequences of light harvesting complex (lhc) cDNAs because lhc genes are the best characterized gene family in pines (Jannson and Gustaffson, 1990; Jansson and Gustafsson, 1991; Jansson, 1992; Jansson, 1994) and because six loblolly pine seedling cDNAs appear to encode lhc proteins. Conifers, like angiosperms, appear to have two light harvesting complexes, containing a total of ten distinct polypeptides (Jansson, 1992). Each polypeptide is encoded by one or more loci (Jansson, 1992), and cDNAs encoding six of the ten lhc proteins have been identified from *Pinus sylvestris*, Scots pine (Jansson, 1992). By comparing sequences of random loblolly pine cDNAs to each other and to previously identified pine lhc gene sequences,

TABLE 1. Most abundant seedling cDNAs

Probable Gene Function	Number of Clones
unidentified/anonymous cDNA 1	9
ribulose bisphosphate carboxylase, small subunit	7
light harvesting complex	6
lipid transfer protein	5
ADP-ribosylation factor	4
peroxidase	4
ubiquitin	4
unidentified/anonymous cDNA 2	4
elongation factor TufA	3
plasma membrane protein	3
Ca++ dependent protein kinase	3
carbonic anhydrase	2
actin depolymerizing factor	2
unidentified/anonymous cDNAs 3-8	2 each

we have identified four distinct classes of loblolly pine lhc cDNAs which we conclude represent four lhc gene family members corresponding to Lhca3, Lhcb1*2, Lhcb2, and Lhcb5 (using the light harvesting complex gene nomenclature of Jansson *et al.*, 1992). Homology plots using "stringent" parameters (check window size and match size of 10) between pairs of lhc sequences produced three distinct homology classes, those showing strong similarity (e.g., Figure 2A), those showing intermediate similarity (e.g., Figure 2B), and those showing weak or no similarity (e.g., Figure 2C). The summary of these comparisons is shown in Table 2. (++ indicates strong similarity, + represents intermediate similarity, and - indicates weak or no similarity). Note that there are two loci known in *Pinus sylvestris*, Scots pine, which encode the lhcb protein (Jansson, 1992). Loblolly pines showing strong similarity to the Lhcb1*2 locus show weak similarity to the Lhcb1*1 locus (see Table 2) because these two Lhcb1 loci are themselves quite similar and therefore more recently diverged than the other Lhc loci are from each other.

In addition to identifying gene families for which multiple members are expressed in the same tissue such as the lhc example above, our sequencing analysis of cDNAs from two tissue types identified gene families from which different members are expressed in different tissues. For example, only one unique sequence was identified among the five non-specific lipid transfer protein (nsltp) seedling cDNAs, whereas a very different nsltp sequence was identified from crude phloem cDNAs as shown in the amino acid alignment of Figure 3.

A final insight into the nature of pine gene families was made from a comparison of Southern banding patterns for a number of genes identified from our cDNA sequencing (see Table 3). Because Southern banding patterns provide a rough approximation of the copy number of gene sequences, we conclude that a number of genes are encoded by significantly more extended gene families in pines than are their homologs in angiosperms. This suggests that certain genes have been more extensively amplified in pines than in flowering plants.

Our success in identifying pine genes from partial cDNA sequences encourages us to continue this effort, expanding it to identify encoded functions of cDNAs from other pine tissue types, including xylem. Through the integration of our mapping and sequencing efforts (see Neale *et al.*, these proceedings) we expect to gain further insights into the evolution and function of pine gene families.

TABLE 2. Loblolly Pine Light Harvesting Complex Loci Expressed in Seedlings

Loblolly pine cDNA	Lhca1	Lhca2	Lhca3	Lhca4	Lhcb1*1	Lhcb1*2	Lhcb2	Lhcb5
pPtIFG1934m'	-	-	-	-	+	++	-	-
pPtIFG1934t	-	-	-	-	+	++	-	-
pPtIFG1949e	-	-	-	-	+	++	-	-
pPtIFG1949s'	-	-	-	-	+	++	-	-
pPtIFG2929e	-	-	-	-	+	++		-
pPtIFG2929s'	-	-	-	-	+	++	-	-
pPtIFG2006e	-	-	-	-	-	-	++	-
pPtIFG2006s'	-	-	-	-	-	-	++	-
pPtIFG2290e	-	-	-	-	-	-	-	++
pPtIFG2290s'	-	-	-		-	-	-	++
pPtIFG2361e	-	-	++	-	-	-	-	-
pPtIFG2361s'	-	-	++	-	-	-	-	-

Table 2 legend. Lhc locus sequences from *Pinus sylvestris* (Jansson, 1992) except Lhcb1*2 from *Pinus contorta* (Barret *et al.* , 1994) and Lhcb2 from *Pinus thunbergii* (Yamamoto *et al.* , 1993). Letters following clone numbers refer to custom vector sequencing primers. " ' " indicates the complement of the determined sequence. " - " indicates weak or no nucleotide sequence similarity, " +" intermediate nucleotide sequence similarity, and " + " strong nucleotide sequence similarity based upon homology plot data.

Figure 2. **Homology plot examples**

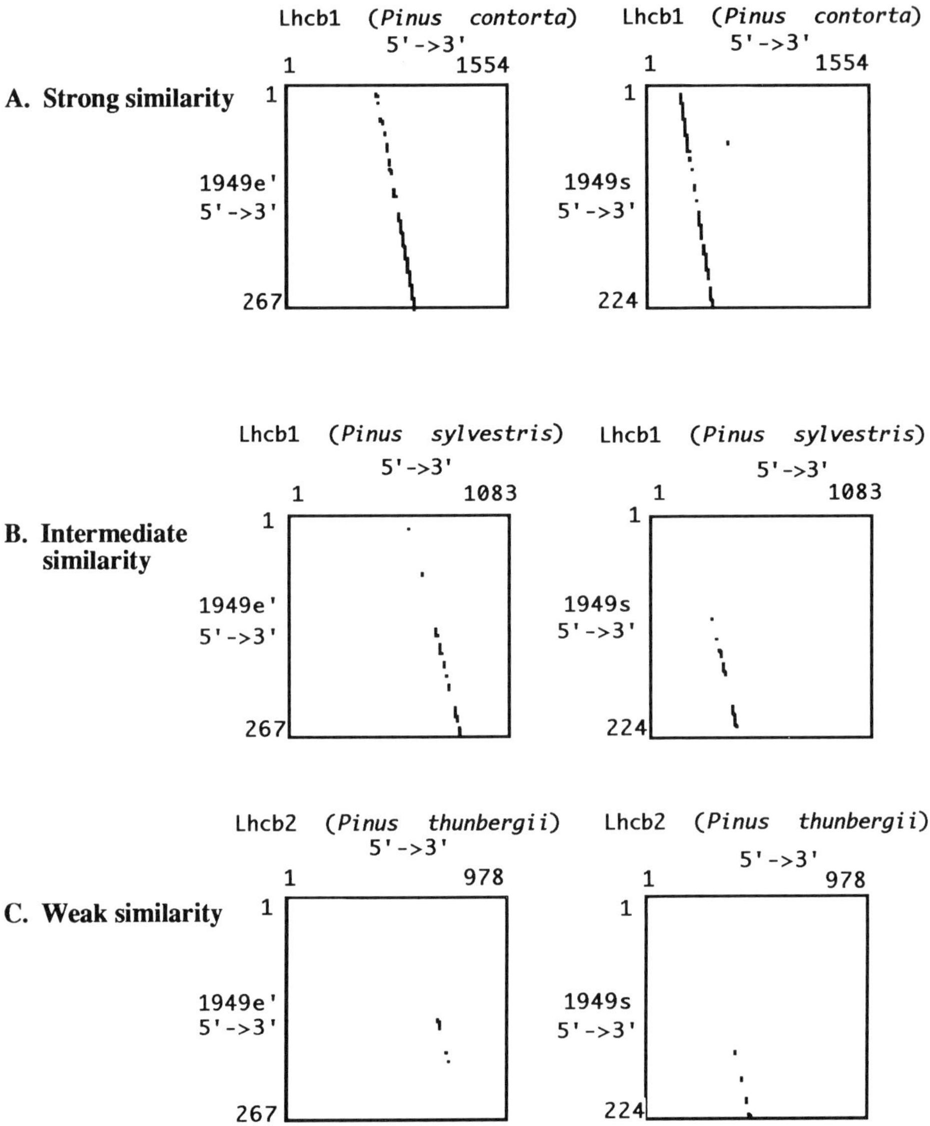

Figure 2 legend. Homology plots of loblolly pine light harvesting complex (lhc) cDNAs versus previously identified pine lhc cDNAs were constructed using DNAsis program for the Macintosh. Window check sizes and matching sizes were 10 in all cases. Representative examples are shown.

***Figure 3*. Loblolly pine lipid transfer protein gene family members**

```
wheat         aqvmlmavalvlmlaavpraavAIDCGHVDSLVRPCLSYVQGGPG-PSGQCCDGKKNLHnqa
              a   + +v+    + avp + +AI C +V S+  PC++Y+ G  + P + CC   + L  q+
Pt seedling   avfvvglvvtmmnvwgavpvegAISCNQVVSAMTPCATYLIGNAATPAATCCPSIRGLDsqv
                                       C   +     C+ Y+ G A  PA  CC  ++ L
Pt phloem     ......................QKECGSQLATLDTCIPYVQGTAKFPAKDCCDGLLKLH..
                                       CG+ ++ L  C+P+VQG  K P K CC G+ +L
pellitoria    VIVAAVLVWTSSASVAPAPAPGEETCGTVVGALMPCLPFVQGKEKEPSKGCCSGAKRLDget

wheat         rsqsdrqsacnclkgiargihnlnednarsippkcgv-nlpytislnidcsrv
              +   drq +cnclk  a+    +    a  +p  c v  l   is n+dcs+v
Pt seedling   katpdrqavcnclktqak-sygvklgkaanlpglckvtdlnvpispnvdcskvh
Pt phloem     ......................................................

pellitoria    ktgpqrvhaceciqtamktysdidgklvsevpkhcgivdsklppidvnmdcktlgvlhykgn
```

Figure 3 legend. Amino acid sequences of pine lipid transfer cDNAs from seedling and phloem are aligned with homologs from wheat and spreading pellitoria, respectively. Overlapping sequence available from the two pine cDNAs is shown in capitol letters. "..." indicate sequence is not available. Identical and similar (+) amino acids are indicated in between pairs of sequences. "-" indicates positions of insertions/deletions. Alignment of wheat and pine seedling lipid transfer protein from Kinlaw et al. (1994). Pine phloem ltp cDNA aligned to BLASTX match, ltp from spreading pellitoria (Costa *et al.* ,1994).

TABLE 3 Genes encoded by gene families which are more complex in pine

Protein encoded	Pine Southern pattern	Other species Southern pattern	 Species
alcohol dehydrogenase	≥ 10	2	corn
lipid transfer protein	≥ 10	4	corn
ribosomal protein B2	≥ 10	1-5	carrot
glutamine synthetase	≥ 10	6	corn
carbonic anhydrase	4-10	2	tobacco
protochlorophyllide reductase	4-10	1	pea
acid phosphatase	4-10	1	tomato
3 ketoacyl CoA thiolase	4-10	1	cucumber

Acknowledgments

This project was funded by the USDA Forest Service. We gratefully acknowledge Ernest Retzel and his colleagues including Kristi Swope of the Biological Computing Facility of the Medical School of the University of Minnesota for the batch submission of our sequences to dbest. We also express our gratitude to Weyerhaeuser Company for providing loblolly pine seed, and logs.

References

Adams, M.D., Kelley, J.M., Gocayne, J.D., Dubnick, M., Polymeropoulos, M.H., Xiao, H., Merril, C.R., Wu, A., Olde, B., Moreno R.F., Kerlavage, A.R., McCombie, W.R., and Venter, J.C. (1991) Complementary DNA sequencing: expressed sequence tags and human genome project. *Science* **252**, 1651.

Altschul, S.F., Gish, W., Miller, W., Myers, E.W., Lipman, D. (1990) Basic local alignment search tool. *J. Mol. Biol.* **215**, 403-410.

Barrett, J.W., Beech, R.N., Dancik, B.P., and Strobeck, C. (1994) A genomic clone of a type I cab gene encoding a light harvesting chlorophyll a/b binding protein of photosystem II identified from lodgepole pine. *Genome* **37**, 166-172.

Costa, M.A., Colombo, P., Izzo, V., Kennedy, H., Venturella, S., Cocchiara, R., Mistrello, G., Falagiani, P. and Geraci, D. (1994) cDNA cloning, expression and primary structure of Par jI, a major allergen of Parietaria judaica pollen. *FEBS Lett.* **341**, 182-186.

Devey, M.E., Jermstad, K.D., Tauer, C.G., and Neale, D.B. (1991) Inheritance of RFLP loci in a loblolly pine three-generation pedigree. *Theor. Appl. Genet.* **83**, 238-242.

Jansson, S. (1992) The Chlorophyll a/b-Binding Proteins: Studies on the Lhca and Lhcb genes of Scots pine. *Ph.D. thesis*, Department of Plant Physiology University of Umea.

Jansson, S. (1994) The light-harvesting chlorophyll a/b-binding proteins. *Biochimm. Biophys. Acta* **1184**, 1-19.

Jansson, S. and Gustafsson, P. (1990) Type I and type II genes for the chlorophyll a/b-binding protein in the gymnosperm Pinus syllvestris (Scots pine) : cDNA cloning and sequence analysis. *Plant Mol. Biol.* **14**, 287-296.

Jansson, S. and Gustafsson, P. (1991) Evolutionary conservation of the chlorophyll a/b-binding proteins: cDNAs encoding type I, II, and III LHC I polypeptides from the gymnosperm Scots pine. *Mol. Gen. Genet.* **229**, 67-76.

Jansson, S., Pichersky, E., Bassi, R., Green, B., Ikeuchi, M., Melis, A., Simpson, D., Spangfort, M., Staehelin, L.A., and Thornber, J.P. (1992) A nomenclature for the genes encoding the chlorophylla/b-binding proteins of higher plants. *Plant Mol. Biol. Reporter* **10**, 242-253.

Keith, C.S., Hoang, D.O., Barrett, B.M., Feigelman, B., Nelson, M.C., Thai, H. and Baysdorfer, C. (1993) Partial sequence anlaysis of 130 randomly selected maize cDNA clones. *Plant Physiol.* **101**, 329-332.

Kinlaw, C.S. and Gerttula, S.M. (1994) Complex pine gene families, in *Proceedings 22nd Southern Forest Tree Improvement Conference* , Southern Forest Tree Improvement Committee, Lantz and Moorhead, pp 275-283.

McCombie, W.R., Adams, M.D., Kelley, J.M., FitzGerald, M.B., Utterback, T.R., Khan, M., Dubnick, M., Kerlavage, A.R., Venter, J.C., and Fields, C. (1992) *Caenorhabditis elegans* expressed sequence tags identify gene families and potential disease gene homologues. *Nature Genet.* **1,** 124-131.

Newman, T., de Bruijn, F.J., Green, P., Keegstra, K., Kende, H., McIntosh, L., Ohlrogge, J., Raikhel, N., Somerville, S., Thomashow, M., Retzel, E., and Somerville, C. (1994) Genes galore: A summary of methods for accessing results from large-scale partial sequencing of anonymous *Arabidopsis* cDNA clones. *Plant Physiol.* **106**, 1241-1255.

Pearson, W.R. (1991) Identifying distantly related protein sequences, *Curr. Op. Struct. Biol.* **103**, 321-326.

Pearson, W.R. and Lipman, D.J. (1988) Improved tools for biological sequence comparison. *Proc. Nat. Acad. Sci., USA* **85**, 2444-2448.

Sasaki, T. (1992) The current strategy of cDNA cataloguing. *Rice Genome Newsletter* **1**, 5.

Yamamoto, N, Kojima, K. and Matsuoka, M. (1993) The presence of two types of gene that encode the chlorophyll a/b-binding protein (LHCPII) and their light-independent expression in pine (*Pinus thunbergii*), *Plant Cell Physiol..* **34**, 457-463.

MOLECULAR APPROACHES TO STUDY BUD DORMANCY IN *POPULUS*

A. ROHDE, M. VAN MONTAGU, and W. BOERJAN
Laboratorium voor Genetica, via the Department of Genetics, affiliated to the Flanders Interuniversity Institute for Biotechnology, Universiteit Gent, K.L. Ledeganckstraat 35, B-9000 Gent, Belgium

1. Abstract

Most physiological research on tree growth and dormancy has focused on the environmental control of the growth periodicity. Much less has been contributed to the understanding of the endogenous factors responsible for the integration of environmental signals into the organogenesis of the bud. This is partially due to a lack of early indicators that allow the monitoring of the effects of subtle changes in the environment on the growth activity of the bud. To reveal these differences in single buds we have analyzed their cell cycle activity individually. *Populus* has been transformed with chimeric ß-glucuronidase (*gus*) genes under the control of cell cycle-specific promoters (*Arabidopsis cdc2* and *cyc1*). After having confirmed a similar expression pattern as in *Arabidopsis*, we have analyzed the expression levels of both chimeric genes in individual buds of the same tree, and under different conditions known to promote or retard growth in the buds. The expression levels of both chimeric genes were found to act as early and sensitive indicators of the growth activity in the buds. Second, common cytological and physiological characteristics of bud and seed dormancy might suggest that they are based on similar processes at the molecular level. Promoter-*gus* fusions, with the promoters conferring expression in the seed, were assayed for *gus* expression in the apical buds of *Arabidopsis* under growth-inhibiting conditions. P*abi3-gus*, so far believed to be seed specific, was demonstrated to be expressed in arrested *Arabidopsis* apices and *abi3*-homologous transcripts were detected in poplar buds.

2. Introduction

Periodic rather than continuous growth is almost universal among trees. Periods of rapid shoot growth (flushing) usually alternate with periods of rest throughout the seasons. In addition, not all buds perform uniformly during flushing. Axillary buds may burst to contribute to the crown formation or may stay arrested (Borchert, 1991). Bud dormancy serves the persistence of the tree during the different periods of non-growth. Dormancy is defined as "the temporary suspension of visible growth of any plant structure containing a meristem" (Lang, 1987). Three major types of dormancy have been described by Lang (1987). In ecodormancy, the growth inactivity is imposed purely by the environment and growth is resumed when conditions become favourable. In

M. R. Ahuja et al. (eds.), Somatic Cell Genetics and Molecular Genetics of Trees, 183–188.
© 1996 *Kluwer Academic Publishers. Printed in the Netherlands.*

paradormancy, the cause for growth suspension in a dormant structure resides in the plant itself, but outside the dormant tissue. Paradormancy establishes, for example, in the axillary buds during apical dominance and is often described as correlative inhibition. Endodormancy is caused by endogenous factors within the dormant tissue and no growth can be achieved even under the most environmentally favourable conditions. Poplars undergo all three kinds of dormancy in the course of one growth cycle in a temperate climate (Critchfield, 1960; Nelson and Dickson, 1981; Richards and Larson, 1981).

That dormancy remained largely a "black box" phenomenon can in part be attributed to a lack of assays to monitor components in early-affected processes. Ideally, a candidate indicator would allow the detection of the effects of subtle environmental changes on the growth activity of the bud, even before phenological or morphological changes can be observed (for reviews, see Champagnat, 1983; Saure, 1985; Crabbé, 1987; Borchert, 1991; Dennis, 1994; Lang, 1994). Recently, candidate molecular indicators to reveal the meristematic activity in single organs or cells became available. In *Arabidopsis*, several genes, involved in the regulation of the cell cycle have been cloned (*e.g.*, Ferreira *et al.*, 1991; Hemerly *et al.*, 1992). The analysis of promoter-ß-glucuronidase (*gus*) fusions indicated that the expression of the *cdc2a* and *cyc1At* genes correlated with active cell division and that *cdc2a* is additionally expressed in cells with a high competence for division (Hemerly *et al.*, 1993; Ferreira *et al.*, 1994). Both chimeric genes are thus good candidates to describe differences in the growth activity of single buds, when introduced into poplar. The knowledge about temporal and spatial regulation of cell divisions in the apical and, in particular, in the axillary buds of a tree would allow to better describe the processes of growth and dormancy. Therefore, we emphasize here the evaluation of the growth potential in single buds by the analysis of their cell cycle activity.

Next to this approach, we examine the possibility that a common molecular process underlies bud and seed dormancy. Already in 1956, Wareing hypothesized that bud dormancy bears essential similarities to seed dormancy. Later on, various common characteristics of bud and seed dormancy have been reported at physiological, cytological, and biochemical levels (Black, 1991; Powell, 1987; Sagisaka, 1991). The most striking feature strengthening that hypothesis is the chilling requirement for breaking dormancy, which is the same for seeds and buds in trees (Saure, 1985). This suggests that bud and seed dormancy might share similar events at the molecular level too.

3. Results and discussion

3.1. CELL CYCLE REGULATION IN BUDS OF *POPULUS*

3.1.1. Pcdc2a-gus *and* Pcyc1At-gus *show a similar expression pattern in poplar as in* Arabidopsis

Populus tremula x P. alba has been transformed with chimeric promoter-*gus* constructs, with the promoters being derived from the *Arabidopsis* cell cycle genes *cdc2a* and *cyc1At*. A thorough analysis of GUS activity at the histochemical and fluorimetrical levels has shown that P*cdc2a-gus* expression in poplar strongly correlated with all tissues where active cell divisions occur (young tissues) and with cells and tissues where meristems with the competence for cell division are preserved (buds, vascular cambium, root tips, pericycle). P*cyc1At-gus* expression in poplar was restricted to tissues undergoing active cell divisions such as the root and shoot apical meristems.

A similar spatial expression pattern for both constructs was observed in poplar and *Arabidopsis*. Thus, also in poplar similar conclusions can be drawn: the expression of the P*cyc1At-gus* gene relates to cell division and the expression of P*cdc2a-gus* reflects cell division and additionally competence for cell division.

3.1.2. *The expression of* Pcdc2a-gus *and* Pcyc1At-gus *reflects the activity of the cell cycle during bud development and dormancy*

P*cdc2a-gus* expression levels did vary quantitatively in the buds at different positions of the stem and throughout the season. An example of the GUS activities in the different buds along the shoot at maximal growth is given in Figure 1. The highest expression of P*cdc2a-gus* was always detected in the apex. This correlates with the fact that the apex is the strongest growth centre during active flushing. Axillary buds retained a relatively lower level of P*cdc2a-gus* expression, as long as they were kept paradormant or dormant. Generally, the GUS activity in the axillary buds reached 60-80% of the GUS activity in the apex.

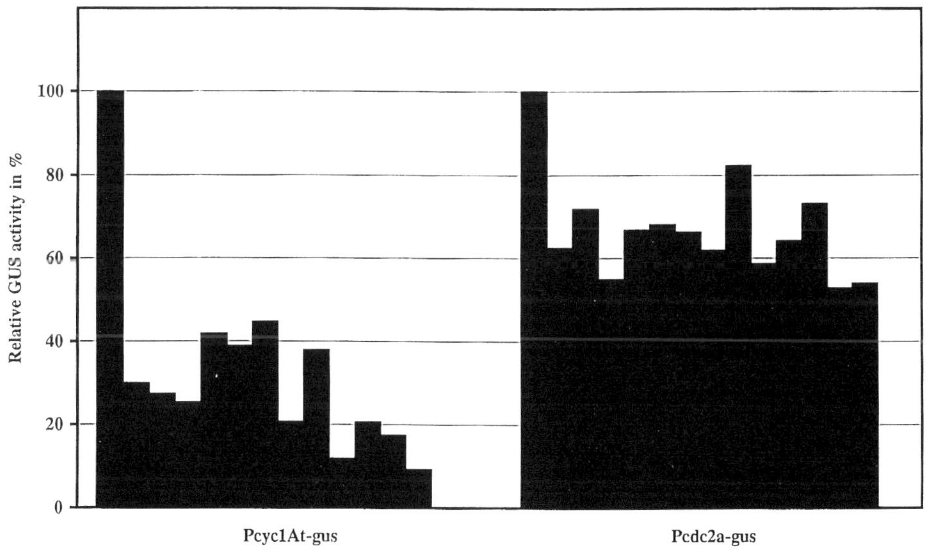

Figure 1. Relative GUS activity in the buds of young actively growing greenhouse plantlets. The typical expression of P*cyc1At-gus* and P*cdc2a-gus* is shown with one representative plant. In each panel the GUS activities are given relative to the expression in the apex, starting from the apex (100%, left) and proceeding basipetally with the axillary buds.

For P*cyc1At-gus* the highest expression was found in the apex during active flushing. In the axillary buds the expression level decreased to about 20-40% of the GUS activity of the apex throughout the season. This suggested that the expression of P*cyc1At-gus* diminished to a lower level after the creation of a distinct number of cells. Only in case buds start to develop into new shoots the levels increased again.

3.1.3. *Decapitation reveals different degrees of dormancy in the axillary buds*

The strength of dormancy differs for each individual axillary bud, although the apparent picture at the different physiological stages is all over "non-growth". Changing the growth activity of one or more buds on purpose, as it was done by shoot decapitation, yielded prompt changes in the promoter activities of P*cdc2a-gus* and P*cyc1At-gus*.

The decapitated poplars initially showed an equal increase in P*cdc2a-gus* expression in almost all buds (except for the lowest ones), indicating that they were previously prevented from outgrowth by the apical bud. The removal of the apex released the axillary buds from paradormancy. The activation of P*cyc1At-gus* expression was temporally delayed and concentrated more to the upper buds, where active cell division is provoked for the formation of new shoots. In low-pruned plants (above the 10th axillary bud), the increase in GUS activity in the axillary buds was observed later than that in high-pruned plants (above the 5th axillary bud). This suggests that the strength of dormancy, not attributable to correlative inhibition, gradually increases downwards the shoot. The expression of P*cdc2a-gus* or P*cyc1At-gus* did not change in the lowest two buds over the whole experimental period. Hence, their cell cycle activity was previously not determined by a factor originating from the apex. They were prevented from growth by another mechanism than apical dominance. We speculate that these buds either differ markedly in depth of paradormancy or that they are endodormant, meaning that the reason for dormancy is residing in the tissue itself.

3.1.4 *Light and temperature differentially influence the activity of the* Pcyc1At-gus *and the* Pcdc2a-gus *promoters in buds*

During the process of dormancy entry, the two chimeric cell cycle genes reacted markedly differently to four combinations of daylength and temperature (photoperiod of 8 h or 16 h, temperature at 15°C or 22°C). The daylength had a pronounced effect on the P*cyc1At-gus* expression. Growth continues (P*cyc1At-gus* high) under long-day conditions at 15°C and growth ceases (P*cyc1At-gus* low) under short-day condition at 22°C, which fits the growth behaviour of *Populus*. P*cdc2a-gus* expression was time mainly influenced by temperature. This demonstrates the value of the molecular indicators: the dormancy-limiting factor, light, is correlated with the expression of the cell division gene (P*cyc1At-gus*) and not to the competence indicator (P*cdc2a-gus*).

3.2. ARE BUD DORMANCY AND SEED DORMANCY BASED ON A SIMILAR MOLECULAR BASIS?

Several genes related to seed dormancy have been isolated (McCarty *et al.*, 1991; Morris *et al.*, 1991; Giraudat *et al.*, 1992; Goldmark *et al.*, 1992). The abscisic acid-insensitive (*abi3*) gene of *Arabidopsis* was of particular interest, since it encodes a transcription factor, which is crucial for seed dormancy and abscisic acid perception (Giraudat *et al.*, 1992). *abi3* mutants show among other effects a reduced seed dormancy. Assuming that the *abi3* gene establishes a link between abscisic acid and dormancy induction, it appeared interesting to test its involvement in bud dormancy. We studied the presence of this and other seed dormancy-related genes in arrested *Arabidopsis* apices and in poplar buds during autumnal dormancy induction.

The expression of *abi3* has so far been believed to be restricted to the seed (Giraudat *et al.*, 1992; Parcy *et al.*, 1994). When studying the expression of a P*abi3-gus* fusion in *Arabidopsis*, GUS activity was detected in the *Arabidopsis* apex under growth-arresting conditions, *i.e.*, after 9 days of germination in darkness. The in parallel assayed P*cyc1At-gus* construct indicated that no active cell divisions occur anymore after 9 days of germination in darkness. Hence, the apex was forced into a resting state. Under these conditions, 17% of the apices from the P*abi3-gus* plants showed GUS activity, whereas GUS activity could not be detected in any of the apices of light-grown P*abi3-gus* plants. *abi3*-homologous transcripts were detected in poplar buds.

This evidence argues for at least partially common molecular mechanisms of seed and bud dormancy. These indications will be strengthened by the expression analysis of other "seed-specific" genes.

4. Acknowledgements

Seeds of P*abi3-gus*-transformed *Arabidopsis* plants were a generous gift of Dr. Jerôme Giraudat and Dr. François Parcy. This work was supported in part by a grant from the European Union (STD3 CT94-0278). A.R. is a Research Assistant of the National Fund for Scientific Research (Belgium).

5. References

Black, M. (1991) Involvement of ABA in the physiology of developing and mature seeds, in W.J. Davies, and H.G. Jones, H.G. (eds.), *Abscisic Acid - Physiology and Biochemistry*, Bios Scientific Publishers, Oxford, pp. 99-124.

Borchert, R. (1991) Growth periodicity and dormancy, in A.S. Raghavendra (ed.), *Physiology of Trees*, John Wiley & Sons, New York, pp. 221-245.

Champagnat, P. (1983) Quelques réflexions sur la dormance des bourgeons des végétaux ligneux. *Physiol. vég.* **21**, 607-618.

Crabbé, J. (1987) *Aspects particuliers de la morphogenèse caulinaire des végétaux ligneux et introduction à leur étude quantitative*. Bruxelles, I.R.S.I.A.

Critchfield, W.B. (1960) Leaf dimorphism in *Populus trichocarpa*. Am. J. Bot. 47, 699-711.

Dennis, F.G. Jr (1994) Dormancy — what we know (and don't know). *HortScience* **29**, 1249-1255.

Ferreira, P.C.G., Hemerly, A.S., Villarroel, R., Van Montagu, M., and Inzé, D. (1991) The *Arabidopsis* functional homolog of the $p34^{cdc2}$ protein kinase. *Plant Cell* **3**, 531-540.

Ferreira, P.C.G., Hemerly, A.S., de Almeida Engler, J., Van Montagu, M., Engler, G., and Inzé, D. (1994) Developmental expression of the Arabidopsis cyclin gene *cyc1At*. *Plant Cell* **6**, 1763-1774.

Giraudat, J., Hauge, B.M., Valon, C., Smalle, J., Parcy, F., and Goodman, H.M. (1992) Isolation of the Arabidopsis *ABI3* gene by positional cloning. *Plant Cell* **4**, 1251-1261.

Goldmark, P.J., Curry, J., Morris, C.F., and Walker-Simmons, M.K. (1992) Cloning and expression of an embryo-specific mRNA up-regulated in hydrated dormant seeds. *Plant Mol. Biol.* **19**, 433-441.

Hemerly, A., Bergounioux, C., Van Montagu, M., Inzé, D., and Ferreira, P. (1992) Genes regulating the plant cell cycle: isolation of a mitotic-like cyclin from *Arabidopsis thaliana*. *Proc. Natl. Acad. Sci. USA* **89**, 3295-3299.

Hemerly, A.S., Ferreira, P.C.G., de Almeida Engler, J., Van Montagu, M., Engler, G., and Inzé, D. (1993) *cdc2a* expression in *Arabidopsis thaliana* is linked with competence for cell division. *Plant Cell* **5**, 1711-1723.

Lang, G.A. (1987) Dormancy: a new universal terminology. *HortScience* **22**, 817-820.

Lang, G.A. (1994) Dormancy — the missing links: molecular studies and integration of regulatory plant and environmental interactions. *HortScience* **29**, 1255-1263.

McCarty, D.R., Carson, C.B., Stinard, P.S., and Robertson, D.S. (1989) Molecular analysis of *viviparous-1*: an abscisic acid-insensitive mutant of maize. *Plant Cell* **1**, 523-532.

Morris, C.F., Anderberg, R.J., Goldmark, P.J., and Walker-Simmons, M.K. (1991) Molecular cloning and expression of abscisic acid-responsive genes in embryos of dormant wheat seeds. *Plant Physiol.* **95**, 814-821.

Nelson, E.A., and Dickson, R.E. (1981) Accumulation of food reserves in cottonwood stems during dormancy induction. *Can. J. For. Res.* **11**, 145-154.

Parcy, F., Valon, C., Raynal, M., Gaubier-Comella, P., Delseny, M., and Giraudat, J. (1994) Regulation of gene expression programs during Arabidopsis seed development: roles of the *ABI3* locus and of endogenous abscisic acid. *Plant Cell* **6**, 1567-1582.

Powell, L.E. (1987) Hormonal aspects of bud and seed dormancy in temperate-zone woody plants. *HortScience* **22**, 845-850.

Richards, J.H., and Larson, P.R. (1981) Morphology and development of *Populus deltoides* branches in different environment. *Bot. Gaz.* **142**, 382-393.

Sagisaka, S. (1991) Occurrence of unidentified bodies that resemble "plastid initials" in poplar cells after breaking of dormancy in midwinter. *Trees* **5**, 143-148.

Saure, M.C. (1985) Dormancy release in deciduous fruit trees. *Hortic. Rev.* **7**, 239-300.

Wareing, P.F. (1956) Photoperiodism in woody plants. *Ann. Rev. Plant Physiol.* **7**, 191-214.

MOLECULAR ANALYSIS OF PINE FERREDOXIN-DEPENDENT GLUTAMATE SYNTHASE

ANGEL GARCIA-GUTIERREZ, FERNANDO GALLARDO, FRANCISCO R. CANTON, REMEDIOS CRESPILLO AND FRANCISCO M. CANOVAS
Laboratorio de Bioquímica y Biología Molecular, Facultad de Ciencias, Universidad de Málaga, Campus de Teatinos, E-29071 Málaga, Spain.

Abstract.

Fd-GOGAT is a chloroplast-located enzyme involved in ammonia assimilation and amino acid biosynthesis in plants. In this work, we have characterized ferredoxin-dependent glutamate synthase (Fd-GOGAT) from pine. As in angiosperms, pine Fd-GOGAT is a monomeric enzyme of about 165 kD. Polyclonal antibodies raised against the purified pine protein were used for the screening of a pine cDNA library. The clone with longest cDNA was 2.1 kbp and included an open reading frame encoding for 550 amino acid residues of the C-terminal region of pine Fd-GOGAT. The comparative analysis of the derived amino acid sequence revealed that Fd-GOGAT is a well-conserved enzyme in plants. In contrast to angiosperms, conifers are able to develop chloroplasts in absence of light during early stages of seedling development. Fd-GOGAT expression in pine was unaffected by light and Fd-GOGAT is expressed in several dark-grown conifer species suggesting that the enzyme is involved in the biosynthesis of glutamate for chloroplast biogenesis and seedling development in darkness. Southern blot analysis suggest that pine genome contains more that one gene encoding the protein.

1. Introduction.

Glutamine synthetase (GS), and glutamate synthase (GOGAT) are key enzymes involved in ammonia assimilation and amino acid biosynthesis in plants (Lea *et al*, 1990). GS [EC 6.3.1.2] catalyses the synthesis of glutamine from glutamate and ammonia and exists in plants as cytosolic and chloroplastic isoenzymes. The chloroplastic GS isoform is involved in the assimilation of ammonia from nitrate reduction and photorespiration (McNally and Hirel, 1983; Wallsgrove *et al*, 1987), while cytosolic GS mainly assimilates ammonium released in protein catabolism and other metabolic reactions (Edwards *et al.* 1990). The amido group of glutamine is transferred to 2-oxoglutarate by the action of GOGAT [EC 1.4.1.14 and 1.4.7.1] to produce glutamate. The reactions catalysed by GS and GOGAT are of special biological importance since both glutamine and glutamate are the amino donors for the biosynthesis of all major nitrogen compounds in higher plants, chlorophyll, amino

M. R. Ahuja et al. (eds.), Somatic Cell Genetics and Molecular Genetics of Trees, 189–195.
© 1996 *Kluwer Academic Publishers. Printed in the Netherlands.*

acids, nucleic acid bases and polyamines. The GOGAT reaction requires reducing power and the enzyme also exists in plants as two molecular forms which differ in the nature of the electron donor, reduced ferredoxin (Fd-GOGAT) or NADH (NADH-GOGAT). (Suzuki and Gadal, 1984). In green tissues, Fd-GOGAT is the predominant form and it is a chloroplastic enzyme (Botella *et al*, 1988). The protein is also present in plastids of non-photosynthetic tissues such as roots (Suzuki *et al.*, 1985) and fruits (Gallardo *et al.* 1993). In angiosperms, Fd-GOGAT is a monomeric protein, with a molecular mass of 160-180 kD (Matoh *et al*, 1979; Márquez *et al*, 1988) and it is expressed in a light-dependent manner during greening of etiolated seedlings as occurs with other photosynthetic proteins (Sakakibara *et al*, 1992). The photoreceptor phytochrome has been suggested to be involved in the light-dependent Fd-GOGAT expression (Becker *et al*, 1993). In contrast to angiosperms, conifers are able to synthesize chlorophyll and photosynthetic proteins such as ribulose bisphosphate carboxylase oxygenase (Rubisco) and light harvesting chlorophyll a/b protein (LHCP) even when growing in absence of light (Mukai *et al*, 1991; Cánovas *et al*, 1993). Therefore, the regulation of pathways associated to the chloroplast, as the GS/GOGAT reactions, could be different than in angiosperms.

2. Results and Discussion.

2.1. MOLECULAR CHARACTERIZATION OF PINE FD-GOGAT

In order to get molecular probes to study Fd-GOGAT expression during early development of pine, we purified the protein from pine cotyledons by ammonium sulphate precipitation and ion-exchange and hydrophobic chromatography (García-Gutiérrez *et al*, 1995). As in angiosperms, pine Fd-GOGAT is a monomeric enzyme of about 165 kD. Denaturing preparations of the purified protein were injected into rabbits to raise a polyclonal antiserum. The antibodies were further purified against the protein by immunoafinity methods. The purified antiserum was highly specific and therefore tested for obtaining sequences encoding pine Fd-GOGAT. About 300,000 recombinant phages from a pine cDNA library constructed in the expression vector λgt11 were screened with the GOGAT antibodies. A single plate showing a strong signal (λGOP4) was identified and purified by two consecutive plate spreading and picking out. The cDNA was subcloned into pGEM-3Z and sequenced. The insert, of 156 bp in size, encoded an open reading frame of 51 amino acid residues homologous to a short stretch in the C-terminal region of the maize Fd-GOGAT (88%) (Sakakibara *et al*, 1991) and tobacco Fd-GOGAT (90%) (Zehnacker *et al*, 1992). Further screening of the same library using the insert of λGOP4 as a probe allowed us the isolation of several cDNA clones. All the new clones corresponded to the same Fd-GOGAT mRNA. The longer one, λGOP44, had an insert of 2.1 kbp which was fully sequenced, showing a whole 3' noncoding region including a poli-A tail, and a open reading frame encoding for 550 amino acids residues of the C-terminal region of pine Fd-GOGAT polypeptide (Figure 1). This region includes three conserved cysteines putatively involved in an iron-suphur centre and a putative FMN-binding domain which are also present in other Fd-GOGAT sequences (Sakakibara *et al*, 1991; Zehnacker *et al*, 1992). The derived amino acid sequence was compared to Fd-GOGAT sequences isolated from angiosperms. The comparative analysis revealed that the pine sequence was highly homologous to monocot and dicot Fd-GOGAT sequences (83-85% identity; García-Gutiérrez *et al.* 1995). These data indicate that Fd-GOGAT primary structure is well conserved, in spite of the evolutionary distance between angiosperms and conifers.

```
  1  PGEGGQLPGKKVSTYIARLRNSKPGVPLISPPPHHDIYSIEDLAQLIFDL   50
 51  HQVNPMAKVSVKLVAEAGIGTVASGVAKGNADIIQISGHDGGTGASPISS  100
101  IKHAGGPWELGLTETHQTLIENGLRERVVLRVDGDFKSGVDVMLAAAMGA  150
151  DEYGFGSVAMIATGCIMARICHTNNCPVGVASQREELRARFPGVPGDLVN  200
201  FFLYVAEEVRGILAQLGYAKLDDVIGRTDLLRPRHISLMKTQQLDLSYLL  250
251  SSVGRPKWSSTKIRQQNVHTNGPVLDDILLADPQISDAIEKEREVKKTVQ  300
301  IYNVDRSVCGRIAGVIAKKYGDTGYAGRLEILFTGSAGQSFGCFLTPGMN  350
351  IRLVGEANDYVAKGMAGGEIVITPSDKTGFVPEDATIVGNTCLYGATGGR  400
401  LFVRGKAGERFAVRNSLAEAIVEGTGDHCCEYMTGGCVVVLGKVGRNVAA  450
451  GMTGGLAYLLDEDDTLIPKVNKEIVKIQRVVAPVGQLQLKQLIEAHVDKT  500
501  GSTKGTAVLKEWEKYLPLFWQLVPPSEEDSPEANPEFGDRTVERITVKSA  550
```

Figure 1. C-terminal end amino acid sequence for *Pinus sylvestris* Fd-GOGAT. The sequence was deduced from a partial cDNA clone. A putative FMN-binding domain, which is well conserved in all glutamate synthases, is underlined. The locations of three conserved cysteines putatively involved in an iron-sulphur centre are marked by arrows.

2.2. DEVELOPMENTAL EXPRESSION OF FD-GOGAT IN CONIFER SEEDLINGS.

In pine, the seed storage proteins are mainly constituted by glutelines and globulines (Allona *et al*, 1992). The seed-seedlings transition involves drastic changes in polypeptide patterns. The content of storage proteins, which are quantitatively important in embryos, decrease during germination, while photosynthetic proteins such as LHCP and large and small Rubisco subunits are barely detectable in embryos and accumulate in cotyledons during seedling development in light or darkness (Cánovas *et al.*, 1991; García-Gutiérrez *et al.*, 1993). These data reflect that storage protein breakdown provides nitrogen for the biosynthesis of chlorophyll and photosynthetic proteins which will permit autotrophic plant growth. Therefore, in pine seedlings, the functionality to the GS/GOGAT pathway is crucial in nitrogen recycling during early plant development in light and dark conditions.

The antiserum raised against the pine protein and the isolated cDNA were used as molecular probes to study Fd-GOGAT expression during early development of pine. Fd-GOGAT expression appears to be organ-specific, the corresponding polypeptide was only present in shoots and mature cotyledons and was barely detectable in embryos or incipient cotyledons. Nitrogen nutrition source did not exert significant effect on the expression in seedlings. However, the levels of Fd-GOGAT enzyme activity and polypeptides increased in cotyledons during development of light- and dark-grown pine seedlings (García-Gutiérrez *et al.*, 1995). The accumulation of Fd-GOGAT during pine development was concomitant with the biosynthesis of chlorophyll and the accumulation of GS and photosynthetic proteins in the tissue (Cánovas *et al.*, 1991; Mukai *et al.*, 1991; Cantón *et al.*, 1993; García-Gutiérrez *et al.*, 1993).

Fd-GOGAT expression at mRNA level was studied by northern blot analysis. The Fd-GOGAT message was detected in cotyledons of light and dark-grown seedlings as well as the messages of *cab* and Rubisco genes (García-Gutiérrez *et al.*, 1995). These data indicate that Fd-GOGAT and other photosynthetic proteins are expressed in a light-independent manner during the initial stages of pine development.

Similar results were observed in other conifers. Figure 2 shows a western blot analysis of crude extracts prepared from different pine species, larch and pinsapo fir. The detection of Fd-GOGAT polypeptide in all extracts prepared from cotyledons of light-grown and dark-grown conifer seedlings suggest that light-independent expression of Fd-GOGAT seems to be common to all conifer species.

Seedling development in darkness is supported by the utilization of storage proteins and lipid reserves, which provided amino acids, carbon skeletons, reducing power and energy. The activity of GS and GOGAT enzymes will supply the amino acids glutamine and glutamate required for chloroplast biogenesis and seedling development in darkness.

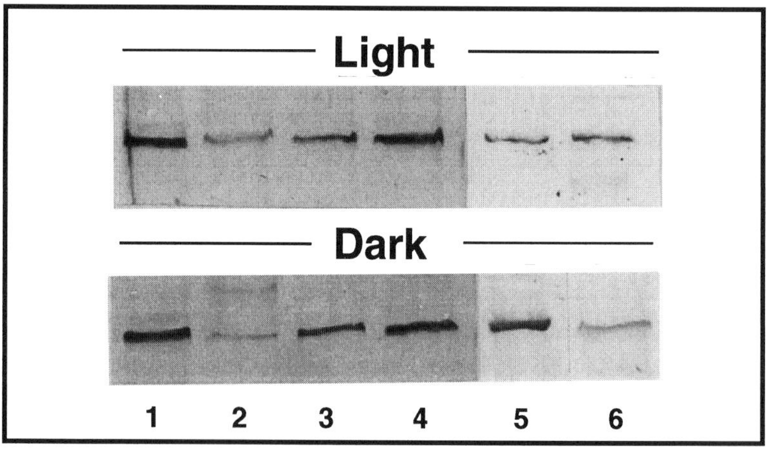

Figure 2. Fd-GOGAT accumulation in light and dark grown seedlings of different conifer species. Crude extracts from seedlings were immunoanalysed by Western blot with pine Fd-GOGAT specific antibodies. Lanes: 1, *Pinus pinaster*; 2, *Pinus pinea*; 3, *Pinus sylvestris*; 4, *Pinus halepensis*; 5, *Abies pinsapo*; 6, *Larix decidua*. 60 mg of proteins was loaded per lane.

2.3. FD-GOGAT SEEMS TO BE ENCODED BY MORE THAN ONE GENE IN PINE GENOME.

Current research in our laboratory is focused on the isolation and characterization of genomic sequences encoding pine Fd-GOGAT. Southern blot analysis of *Pinus sylvestris* DNA revealed the presence of several bands which hybridized with Fd-GOGAT cDNA (Figure 3). Genomic libraries have been constructed in λEMBL3 and screened with the cDNA as a probe. The characterization of two clones having different restriction pattern is underway. Thus, in accordance with the multiple bands detected in the southern blot, these data suggest that Fd-GOGAT is encoded by more than one gene in pine genome. No data about structure of Fd-GOGAT encoding genes are available so far in higher plants. Therefore, the molecular characterization of pine genomic clones will permit to define the genetic structure of Fd-GOGAT genes and get insights on the control of gene transcription. The functional properties of promoter sequences will be studied by promoter-GUS gene fusion analysis in transgenic plants. This molecular approach will allow us the identification of regulatory sequences involved in observed expression in conifer seedlings. Very recently, Kojima *et al.* (1994) have reported that pine cab-6 promoter confers light-independent and tissue-specific expression of a gene construct in transgenic tobacco. The functional analysis of promoters from other light-independent genes will be necessary to stablish if they share common regulatory elements involved in their characteristic expression patterns.

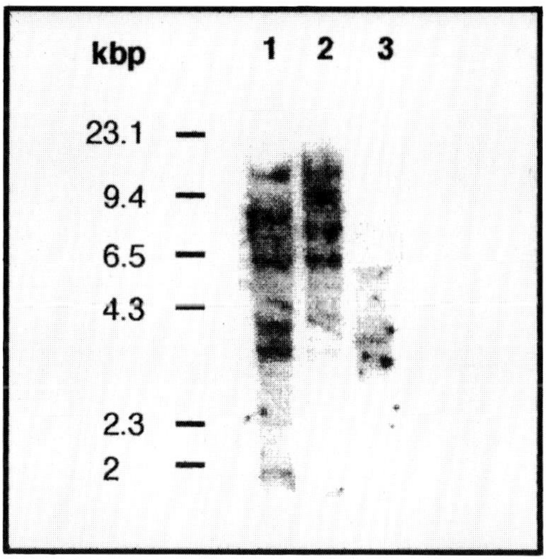

Figure 3. Southern blot analysis of *Pinus sylvestris* genomic DNA. Genomic DNA (10 mg) was digested with *Eco*RI (lane 1), *Bam*HI (lane 2) and *Hin*dIII (lane 3) and probed with the pGOP44 cDNA clone. The size of the *Hin*dIII digested lambda DNA markers are indicated on the left in kbp .

3. Acknowledgments.

This research was supported by a Grant (PB92-0423) from the Dirección General de Investigación Científica y Técnica (Spain) to F.M.C.

4. References

Allona, I., Casado, R. and Aragoncillo, C. (1992) Seed storage proteins from *Pinus pinaster* Ait.: Homology of major componentes with 11S proteins from angiosperms, *Plant Sci.* **87**,9-18.

Becker, T.W., Nef-Campa, C., Zehnacker, C. and Hirel, B. (1993) Implication of the phytochrome in light regulation of tomato gene(s) encoding ferredoxin-dependent glutamate synthase, *Plant Physiol.Biochem.* **31**,725-729.

Botella, J.R., Verbelen, J.P. and Valpuesta, V. (1988) Immunocytolocalization of Ferredoxin-GOGAT in the cells of green leaves and cotyledons of *Lycopersicon esculentum. Plant Physiol.* **87**,255-257.

Cánovas, F.M., Cantón, F.R., Gallardo, F., García-Gutiérrez, A. and de Vicente, A. (1991) Accumulation of glutamine synthetase during early development of maritime pine (*Pinus pinaster*) seedlings, *Planta* **185**,372-378.

Cánovas, F. M., McLarney, B. and Silverthorne, J. (1993) Light-independent synthesis of LHCIIb polypeptides and assembly of the major pigmented complexes during the initial stages of *Pinus palustris* seedling development, *Photosyn. Res* . **38**,89-97.

Cantón, F.R., García-Gutiérrez, A., Gallardo, F., de Vicente, A. and Cánovas, F.M. (1993) Molecular characterization of a cDNA clone encoding glutamine synthetase from a gymnosperm, *Pinus sylvestris*, *Plant Mol. Biol.* **22**,819-828.

Edwards, J.W., Walker, E. L. and Coruzzi, G. M. (1990) Cell-specific expression in transgenic plants reveals nonoverlaping roles for chloroplastic and cytosolic glutamine synthetase, *Proc. Natl. Acad. Sci. USA* **87**,3459-3463.

Gallardo, F., Cantón, F. R., García-Gutiérrez, A. and Cánovas, F. M. (1993) Changes in photorespiratory enzymes and glutamate synthases in ripening tomatoes, *Plant Physiol. Biochem.* **31**,189-196.

García-Gutiérrez, A., Cantón, F. R., Gallardo, F. and Cánovas, F. M. (1993) Immunochemical analysis of chloroplast polypeptides from maritime pine, *Phytochemistry* **34**,337-341.

García-Gutiérrez, A., Cantón, F. R., Gallardo, F., Sánchez-Jiménez, F. and Cánovas, F. M. (1995) Expression of ferredoxin-dependent glutamate synthase in dark-grown pine seedlings, *Plant Mol. Biol.* **27**,115-128.

Kojima, K., Sasaki, S. and Yamamoto, N. (1994) Light-independent and tissue-specific expression of a reporter gene mediated by the pine cab-6 promoter in trasngenic tobacco, *Plant J.* **6**,591-596.

Lea, P. J., Robinson, S. A. and Stewart, G. R. (1990) The enzymology and metabolism of glutamine, glutamate, and asparagine, in B. J. Miflin and P. J. Lea (eds), *The Biochemistry of Plants* (Intermediary Nitrogen Metabolism). Academic Press, New York, vol 16, pp.121-159.

Lederer, F., Cortial, S., Becam, A. M., Haumont, P. Y. and Pérez, L. (1985) Complete amino acid sequence of flavocytochrome b_2 from bakers yeast, *Eur. J. Biochem.* **152**,419-428.

Matoh, T., Suzuki, T. and Ida, S .(1979) Corn leaf glutamate synthase: Purification and properties of the enzyme, *Pant Cell Physiol.,* **20**,1329-1340.

Márquez, A. J., Avila, C., Forde, B. G. and Wallsgrove, R. M. (1988) Ferredoxin-glutamate synthase from barley leaves: Rapid purification and partial characterization, *Plant Physiol. Biochem.* **26**,645-651.

McNally, S.F. and Hirel, B. (1983) Glutamine synthetase isoforms in higher plants, *Physiol. Veg.* **21**,761-774.

Mukai, Y., Yamamoto, N. and Koshiba, T. (1991) Light-independent and Tissue-specific accumulation of light-harvesting chlorophyll a/b binding protein and ribulose bisphosphate carboxylase in dark-grown pine seedling, *Plant Cell Physiol.* **32**,1303-1306.

Nalbantoglu, B., Hirasawa, M., Moomaw, C., Nguyen, H., Knaff, D. B. adn Allen, R. (1994) Cloning and sequencing of the gene encoding spinach ferredoxin-dependent glutamate synthase, *Biochem. Biophys. Acta* **1183**,557-561.

Wallsgrove, R. M., Turner, J. C., Hall, N. P., Kendall, A. C. and Bright, S. W. J. (1987) Barley mutants lacking chloroplast glutamine synthetase. Biochemical and genetic analysis, *Plant Physiol.* **83**,155-158.

Sakakibara, H., Watanabe, M., Hase, T. and Sugiyama, T. (1991) Molecular cloning and characterization of complementary DNA encoding for ferredoxin-dependent glutamate synthase in maize leaf, *J. Biol. Chem.* **266**,2028-2035.

Sakakibara, H., Kawabata, S., Takahashi, H., Hase, T. and Sugiyama T (1992) Molecular cloning of the family of glutamine synthetase genes from maize: Expression of genes for glutamine synthetase and ferredoxin-dependent glutamate synthase in photosynthetic and non-photosynthetic tissues, *Plant Cell Physiol.* **33**, 49-58.

Suzuki, A. and Gadal. P. (1984a) Glutamate synthase: Physicochemical and functional properties of different forms in higher plants and other organisms, *Physiol. Veg.* **22**,471-486.

Yamamoto, N., Mukai, Y., Matsuoka, M., Kano-Murakai, Y., Tanaka, Y., Ohashi, Y., Ozeki, Y. and Odani, K. (1991) Light-independent expression of *cab* and *rbcS* genes in dark-grown pine seedlings, *Plant Physiol.* **95**,379-383.

Zehnacker, C., Becker, T. W., Suzuki, A., Carrayol, E., Caboche, M. and Hirel, B. (1992) Purification and properties of tobacco ferredoxin-dependent glutamate synthase, and isolation of corresponding cDNA clones. Light-inducibility and organ-especificity of gene transcription, *Planta* **187**,266-274.

DIFFERENTIALLY REGULATED GENE SETS IN DOUGLAS FIR SEEDS AND SOMATIC EMBRYOS.

SANTOSH MISRA, MALINEE CHATTHAI, TIMOTHY J. TRANBARGER, BENJAMIN S. FORWARD, AND KARIA H. KAUKINEN
Department of Biochemistry and Microbiology
University of Victoria, Victoria, B.C. V8W 3P6.

ABSTRACT. Differential screening of cDNA libraries of mid-maturation stage zygotic embryo 4-6 day old seedlings yielded a set of cDNAs preferentially expressed during embryogenesis and germination, respectively. Sequence analysis of embryo cDNAs identified these clones as enc isoforms of 2S storage proteins- albumins, and an MT-like protein. The expression of these ger different stages of zygotic and somatic embryogenesis was examined by Northern blot analysis different patterns of expression were observed for various isoforms of albumins. The MT-like gen expressed in embryos and needles and its expression was regulated by ABA, osmoticum as well as i ions. Partial DNA sequence analysis and predicted amino acid sequences of germination - asso cDNAs show we have isolated clones with homology to a range of plant proteins: including a chape 60-β, low molecular weight heat shock proteins (LMW HSPs), a luminal binding protein (BiP), a t chlorophyll a/b binding protein (CAB), a thiol-protease precursor and a NADPH-P450 reductase. I seeds all the clones detected transcripts which increased by 2 days of germination.

1. Introduction

Seed development and germination are distinct metabolic events in which the status of storage pr contrast markedly and hence provide a model system to investigate the regulation of gene expre Desiccation, an essential part of normal seed development, may activate genes for germination an off the synthesis of embryogenesis related proteins. Several studies have revealed the presence of m in mature seed tissues, referred to as stored mRNA, which may be immediately available on rehyd for translation of the proteins required for germination (Bewley and Marcus, 1990). Imbibition of n dormant seeds leads to the appearance of new and temporally distinct sets of mRNAs and gene pr that are absent in mature seeds (reviewed by Lane, 1991). Several of these gene products are involv the transformation of storage compounds into energy and anabolic substrates for germination (Bewle Marcus, 1990). Examples include enzymes of the glyoxylate cycle and enzymes involved in st product mobilization and nitrogen assimilation (reviewed in Bewley and Marcus, 1990). The storage tissue in conifer seeds is the haploid, maternally derived megagametophyte. In contr angiosperm storage tissues that originate and develop after fertilization, the megagametophy Douglas-fir and other conifers is fully developed before fertilization. Storage protein accumulation i megagametophye and the embryo starts during early and mid embryogenesis, repectively, and termi during the desiccation phase of seed maturation (Owens *et al.*, 1993; Misra, 1994). During germin proteins are mobilized from the megagametophyte to provide free amino acids to the embryo. elucidate the regulatory mechanisms involved in the expression of genes during embryoge germination and early seedling development of conifer seeds, we have initiated studies on the clonin characterization of developmentally regulated gene sets. We report here the cloning and analysis of of cDNAs from a Douglas-fir embryogenesis and germination-specific libraries. Northern blot an showed that these mRNAs are differentially expressed during zygotic embryogenesis, germinatio early seedling development. The temporal expression pattern of genes in somatic embryos paralle observed in zygotic embryos indicating that somatic embryos may provide a useful system to fac studies of gene regulation in conifers.

M. R. Ahuja et al. (eds.), Somatic Cell Genetics and Molecular Genetics of Trees, 197–204.
© 1996 *Kluwer Academic Publishers. Printed in the Netherlands.*

2. Material and Methods

2.1. PLANT MATERIAL

Immature female cones of Douglas-fir (*Pseudotsuga menziesii* [Mirb] Franco) were obtained from a clonal orchard of the Pacific Forest Product Ltd. The collections were made biweekly beginning May 17, 1993 until August 23, 1993. Somatic embryos of Douglas fir were supplied by Dr. Pramod K. Gupta of the Weyerhaeuser Company, Tacoma, Washington. Four developmental stages were examined: embryonal mass, precotyledonary, cotyledonary and mature somatic embryos. In addition, somatic embryos matured on standard media or modified standard media (lacking PEG or ABA, or containing high sucrose) were examined (Gupta and Grob, 1995). Mature Douglas-fir seeds were provided from BC ministry of Forests. Seeds were imbibed, stratified and germinated as described in Tranbarger and Misra (1995).

2.2. CONSTRUCTION AND DIFFERENTIAL SCREENING OF DOUGLAS-FIR cDNA LIBRARY

An embryogenesis cDNA library was constructed using poly $(A)^+$ mRNA isolated from embryonic axes at the cotyledonary stage (43 DAF) as described in Leal and Misra (1993). Briefly, total RNA was isolated and pooled for Poly $(A)^+$ mRNA isolation. Poly $(A)^+$ mRNA isolated with oligo (dT) cellulose spun columns (Pharmacia, Uppsala, Sweden) was used to synthesize the double-stranded cDNA library (Pharmacia, Uppsala, Sweden). Double-stranded cDNA was recovered and ligated into the λ ZAP II arms (Stratagene, La Jolla, California, U.S.A.) and packaged *in vitro* into phage particles using the Gigapack system (Stratagene, La Jolla, California, U.S.A.).

To differentially screen the library, mRNAs from mature seeds and 45 DAF immature embryonic axes were used to synthesize single-stranded cDNAs. A total of $1x10^4$ phage clones were transferred to nitrocellulose filters. The positive cDNA clones which were present only on the filters that hybridized with the single-stranded cDNAs labeled with ^{32}P synthesized from mRNA isolated from 45 DAF immature embryonic axes were picked and excised from λ ZAP II as a recombinant pBluescript SK(-) plasmid (Stratagene, La Jolla, California).

For a germination cDNA library, total RNA was pooled from 4-6-day-old Douglas-fir seedlings for Poly$(A)^+$ mRNA isolation. Five μg Poly$(A)^+$ mRNA isolated with oligo (dT) cellulose spun columns (Pharmacia, Uppsala, Sweden) was used to synthesize the double stranded cDNA library (Pharmacia, Uppsala, Sweden). A total of 1.2 μg of double stranded cDNA was recovered, a portion of which was then ligated into the λ ZAP II arms (Stratagene, La Jolla, California, U.S.A.) and packaged in vitro into phage particles using the Gigapack system (Stratagene, La Jolla, California, U.S.A.). To differentially screen the library, mRNAs from mature seeds and 6-day-old seedlings were used to synthesize single stranded cDNAs. A total of $1x10^4$ phage clones were transferred to nitrocellulose filters. The positive cDNA clones which were present only on the filters that hybridized with the single stranded cDNAs labeled with ^{32}P synthesized from mRNA isolated from 6-day-old seedlings were picked and excised from λ ZAP II as a recombinant pBluescript SK(-) plasmid (Stratagene, La Jolla, California, U.S.A.).

2.3. NORTHERN BLOTTING AND HYBRIDIZATION

The RNA was electrophoretically separated on agarose gels (20 μg total RNA per lane) containing formaldehyde, stained with ethidium bromide to assure equal loading, and transferred to a Zeta-Probe GT membrane (BIO-RAD, Mississauga, Ontario, Canada). The cDNA inserts were labeled by random priming with ^{32}P and hybridized (BIO-RAD, Mississauga, Ontario, Canada).

2.4. DNA SEQUENCE ANALYSIS

The cDNA insert sequences were determined by the Taq Dye Primer Cycle Sequencing and the Taq DyeDeoxy Terminator Cycle Sequencing Kits using the Applied Biosystems Model 373A DNA Sequencing System (Applied Biosystems, Inc., Foster City, California, U.S.A.), and by the dideoxynucleotide chain termination method using Sequenase version 2.0 (United States Biochemical, Cleveland, Ohio, U.S.A.). Primers used for sequencing were synthesized using the Applied Biosystems PCR-Mate DNA Synthesizer (Applied Biosystems, Inc., Foster City, California, U.S.A.). DNA similarity searches were performed using the BLAST network service at NCBI.

3. Results

3.1. ISOLATION OF EMBRYO-SPECIFIC cDNAs

The embryonic cDNA library was screened differentially with cDNA probes synthesized using mRNA isolated from dry seeds and developing embryos and several cDNAs expressed during embryogenesis were isolated (Table 1). Sequence analysis and predicted amino acid sequences showed that we have isolated cDNAs encoding major storage proteins i.e. 11S legumin and 2S albumin. Sequence characterization of a full length cDNA encoding an 11S legumin like storage protein precursor (pseudotsugin) was previousdly described (Leal and Misra, 1993). In addition to storage proteins, a cDNA encoding a metallothionein-like (MT) protein was also isolated (Table 1). Comparison of the deduced amino acid sequence of four clones (MDF3.5, MDF4.4, MDF4.7, and MDF5.1) show significant homology to some angiosperm 2S storage proteins such as napin, allergen Sin aI, and alpha-amylase/trypsin inhibitor. Generally the 2S proteins are coded for by multigene families, which is supported by the presence of several isoforms in Douglas fir (Table 2).

TABLE 1. Embryogenesis-associated cDNA clones

cDNA clone	Homology	Transcript size (Kb)	Reference
MDF 2.1	MT-like	0.5	Ledger and Gardner, 1994
MDF 3.5	2S storage protein	0.7	Krebbers *et al.*, 1988
MDF 4.4	2S storage protein	0.7	Krebbers *et al.*, 1988
MDF 4.7	2S storage protein	0.7	Krebbers *et al.*, 1988
MDF 5.1	2S storage protein	0.7	Krebbers *et al.*, 1988
MDF-1	11S legumin	1.8 Kb	Leal and Misra, 1993

TABLE 2. Comparison of sequences of cDNA clones encoding 2S storage protein isoforms (MDF 3.5, MDF 4.4, MDF 4.7 and MDF 5.1) at the amino acid and nucleotide (in parenthesis) levels.

	MDF 4.4	MDF 4.7	MDF 5.1
MDF 3.5	92.56% [86.49%]	62.57% [54.96%]	61.33% [60.36%]
MDF 4.4		62.34% [57.73%]	62.00% [60.14%]
MDF 4.7			69.33% [60.72%]

Percentages of nucleotide and amino acid sequence identity were determined using Sequence-Application program. The data of nucleotide sequence comparison correspond to the result obtained by Southern blot analysis of plasmid DNAs digested with EcoRI using selected labelled insert cDNAs. Clone MDF 3.5 was detected with the insert MDF 4.4. The insert MDF 4.4 and the insert MDF 4.7 did not hybridize to each other. The insert cDNA of MDF 5.1 was positively hybridized with both MDF 4.4 and MDF 4.7 cDNA probes.

Northern analysis with all four clones showed that they are expressed at high levels at the mid-stages of zygotic embryogenesis and decline as the seeds mature. There were differences in time and quantity of expression of various cDNA clones. All isoforms were expressed in somatic embryos. The temporal pattern of their expression paralled that of zygotic embryos. One of the cDNA clone designated MDF 2.1 carried a unique insert of 417 bp of which open frame encoded 68 amino acids and showed significant homology to metallothionein-like proteins of some angiosperms, with the exception of fewer conserved cysteine residues. Southern analysis suggest that the Douglas fir MT-like gene belongs to a gene family of atleast 8-10 members. Northern analysis show that the MT-like gene were expressed at high levels at the mid-stages of both zygotic and somatic embryo development and decreased during maturation of embryos. The transcripts were also detected in 10 week old needles but not in roots and pollen.

3.2. ISOLATION OF GERMINATION ASSOCIATED cDNAs

To identify Douglas-fir genes that are developmentally regulated during germination and post-germination, a Douglas-fir cDNA library was constructed and differentially screened using cDNA probes synthesized with mRNA from mature seeds and 6-day-old seedlings. After a primary and secondary screening, 10 putative positive clones were found to hybridize preferentially with the ^{32}P labeled cDNA probes derived from 6-day-old seedlings and to a lesser degree with mature seed cDNA probes. Some clones hybridized with transcripts that were present six days after germination (Table 3). Based on the differential screening signal, clone DF14 was picked as a representative of a gene expressed at similar levels in both mature seeds and 6-day-old seedlings. Northern blot analysis failed to reveal hybridization of DF-9, DF-4, and DF-10 to Douglas fir mRNA (Table 3).

TABLE 3. Estimated sizes of germination associated cDNA clones and their corresponding transcripts. Insert and transcript sizes in kilobases (kb) were estimated by electrophoretic mobility through 1% agarose gels and 1% agarose formaldehyde gels, relative to the migration of DNA and RNA molecular weight standards respectively.

cDNA Clone	Insert size (kb)	Transcript size (kb)
DF3	1.3	2.6
DF3-3	2.0	2.1
DF4-5	0.7	0.9
DF5-1	2.6	2.9
DF6-3	0.9	1.2
DF6-7b	0.9	1.4
DF14	1.1	3.0
DF9	1.8	NH
DF6-7a	0.7	0.9
DF4	3.7	NH
DF10	1.6	NH

NH = No hybridization (Northern)

To determine the expression pattern of the genes corresponding to the individual clones, we constructed a time course of total RNA isolated from mature seeds, stratified seeds, 2-, 4-, 6-, 8-, and 14-day-old seedlings, and 10-week-old needles. In Northern blot analysis, cDNA probes DF3, 3-3, 4-5, 5-1, 6-3 and 6-7, hybridized with distinct transcripts which increased in abundance by 2 days and were relatively low in abundance in mature seeds.

The temporal pattern of increased expression of germination-associated clones was strikingly different to the pattern of decreasing transcript levels detected by pseudotsugin (DF-1), a major storage protein in Douglas-fir (Leal and Misra, 1993).

3.2.1. *Identification of germination clones through DNA sequence and similarity search analysis*
The germination associated clones and DF14 were subjected to DNA sequence and similarity search analysis to determine their possible identities. Based on partial nucleotide sequence data (NCBI blastn) and predicted amino acid sequences (NCBI blastx), all clones except DF-4, and DF-10 were found to share identities with known plant genes. Their respective identities are shown in Table 4. The insert size and size of the transcripts as detected by Northern hybridization is shown in Table 3.

TABLE 4. Germination associated cDNA clones have homology to previously reported sequences.

cDNA clone	Homology	% nt identity	% Predicted amino acid identity	References
DF3	cpn60β	79	94	Zabeleta *et al.*, 1993
DF3-3	cysteine protease (EC 3.4.22)	ND	71	Koizumi *et al.*, 1993
DF4-5	LMW HSP	ND	72	Tseng *et al*, 1992
DF5-1	BiP	78	87	Denecke *et al.*, 1991
DF6-3	type II CAB	92	97	Kojima *et al.*, 1992
DF6-7b	Malate Dehydrogenase NADPH (EC 1.1.1.37)	ND	56	-
DF14	Cytochrome P450 Reductase	76	78	Shet *et al.*, 1993
DF6-7a	LMW HSP	ND	71	-
DF9	Succinate dehydrogenase flavoprotein	ND	30	-
DF4	?	?	?	-
DF10	?	?	?	-

ND = No data
? = No homology/identity

Clone DF6-3 contained the complete coding region for a mature type II chlorophyll a/b binding protein (CAB) of photosystem II, along with a partial sequence which encodes the transit peptide needed for proper intracellular sorting to the chloroplast (Tranbarger and Misra, 1995). Clone DF3 shared homology with an internal sequence of the *Arabidopsis thaliana* 60-kDa chaperonin beta (cpn60ß) cDNA. DF3-3 showed the highest identity to a drought-induced cysteine protease (RD21) from *Arabidopsis thaliana*.

Clone DF4-5 was found to share identity to the carboxyl-terminal portion of a class I low molecular weight heat shock protein (LMW HSP) from *Oryza sativa* seedlings on both the nucleotide (138 bp) and amino acid levels (Table 4). The predicted amino acid sequence included the conserved Pro and Gly-Val-Leu sequence common to the LMW HSPs of plants and other eukaryotes.

A 215 bp nucleotide sequence of DF5-1 aligned with an internal coding region of the BiP cDNA (BLP4) characterized from tobacco and is highly conserved on the amino acid level (Table 4).

Clone DF14 shared the highest identity with a NADPH-cytochrome P450 reductase (CPR) (EC 1.6.2.4) from *Vigna radiata* on both the nucleotide and the predicted amino acid level (Table 4). The predicted amino acid sequence from a 267 bp open reading frame was homologous with an internal sequence of the CPR containing a conserved region proposed as one of the NADPH binding sites .

The identification of clone DF 6-7, DF-4 and DF10 has not been determined.

4. Discussion

We report the isolation of cDNAs for 2S storage protein gene family, an MT-like protein, molecular chaperones (e.g. cpn 60β and BiP), a LMW HSP, and a cysteine protease from conifers. The expression of this set of cDNAs and a CAB clone is developmentally regulated. We have also isolated a cDNA for a NADPH-cytochrome P450 reductase which appears to be constitutively expressed in mature conifer seeds and seedlings.

Seed storage proteins are prominent reserves in developing seeds. They are synthesized and accumulated in large quantities during plant embryogenesis. In Douglas fir 11S legumins and 2S albumins are predominant storage proteins (Leal and Misra, 1993; Misra and Green, 1994). Molecular cloning of cDNAs encoding 11S and 2S storage proteins will allow for investigation of their temporal and spatial regulation in Douglas-fir somatic and zygotic embryos.

A major metabolic event during seed germination is the production of enzymes responsible for the mobilization and utilization of storage reserves. The proteinases and peptidases that are responsible for the storage protein breakdown have been characterized at a biochemical level in several Pinus sp. (reviewed by Misra, 1994). However, there are no reports detailing the molecular characterization and regulation of these enzymes. We have found a cDNA clone that shares identity with a cysteine protease from *A. thaliana* (Koizumi *et al*., 1993). The transcripts of this protease (DF 3-3) were detected in mature seeds and their level increased significantly during germination, suggesting a role in the mobilization of storage proteins during germination and early seedling development.

During germination and early seedling growth, there is a high degree of germination-specific protein synthesis, sorting and targeting to the various organelles of the cell. Newly synthesized proteins are folded into their native forms which for some proteins have been shown to be facilitated by proteins referred to as molecular chaperones (Hemmingsen *et al*., 1988). We have found cDNAs that encode several molecular chaperones, e.g. cpn 60β subunit (DF 3), LMW HSP (DF 4-5), and BiP (DF 5-1), a member of the HSP 70 gene family. In Douglas-fir the expression of their corresponding mRNAs increased significantly during germination or post-germination. The HSP 70 family members have also been characterized by their ability to bind polypeptides and their key role in the folding and stabilization of some proteins. BiP is a HSP 70 member that functions in the ER, is specifically involved in the secretory pathway and is not induced by heat shock. In angiosperms, BiP mRNA levels are high in tissues undergoing a large amount of cell divisions (Denecke *et al*., 1991). For example, in tobacco, BiP transcript levels were high in 4-day-old seedlings which were characterized by meristematic cells actively secreting high levels of cell wall proteins (Denecke *et al*., 1991). We observed similar expression patterns of BiP mRNAs in Douglas-fir where highest levels were reached in 4 day old seedlings. BiP mRNA levels also increased during the stratification of Douglas-fir seeds. We propose that BiP may also be involved in the folding and secretion of proteins synthesized during stratification (Taylor *et al*., 1993) and at the onset of germination.

Based on partial nucleotide sequence analysis, the clone DF 4-5 is related to the most predominant HSPs found in a range of angiosperms referred to as LMW HSPs (Mansfield and Key, 1987). Recent reports have implicated mammalian (Jacob *et al*., 1993) as well as plant LMW HSPs (Vierling, 1995) as molecular chaperones *in vitro*. In Douglas-fir, the LMW HSP transcripts (DF 4-5) were first detected in 2 day old seedlings and their high level expression continued throughout the post-germination phase (Kaukinen *et al*., in press). Transcripts corresponding to the DF 4-5 clone were not detected in developing zygotic embryos of Douglas-fir (Misra, unpublished data). The clone DF 14 described here shares homology with CPR, an integral membrane flavoprotein which is an essential component of microsomal cytochrome 450 mono-oxygenase system of eukaryotes (Lu *et al*., 1969).

In summary, in addition to CAB, which has been previously isolated and sequenced from conifers, we have isolated cDNAs that share identity with 2S storage proteins, an MT-like protein, a cpn60ß, a cysteine protease, a LMW HSP and a BiP. The developmental expression of this set of cDNAs suggests their gene products may play critical roles during the process of embryogenesis and germination. Future work will address the hormonal, environmental, and developmental regulation of these genes in relation to their roles during germination and post-germination. Characterization of the genomic sequences will lead to dissecting the elements that control their inducible and/or developmental regulation.

5. Acknowledgements

This research was supported by an NSERC operating grant (number OG P0042346) and a Forest Renewal B.C. Research Grant to Santosh Misra. M. C. is supported by a Royal Thai Government Scholarship. B.S.F. is a recipient of an NSERC post-graduate fellowship and a University of Victoria fellowship. K.H.K. is a recipients of an NSERC post-graduate fellowship. Thanks also to the Ministry of Forest Tree Seed Center, Surrey, B.C. for the Douglas-fir seeds used for these experiments, and to Dr. Pramod K. Gupta, Weyerhaeuser Company, for providing the somatic embryo samples.

6. References

Bewley, J.D. and Marcus, A. (1990) Gene expression in seed development and germination, *Prog. Nuc. Acids Res. Mol. Biol.* **38**, 165-193.

Denecke, J., Goldman, M.H.S., Demolder, J., Seurinck, J. and Botterman, J. (1991) The tobacco luminal binding protein is encoded by a multigene family, *Plant Cell* **3**, 1025-1035.

Gupta, P.K. and Grob, J.A. (1995) Somatic embryogenesis in conifers, in S. Jain, P. Gupta and R. Newton (eds.), *Somatic Embryogenesis in Woody Plants*, Kluwer Academic Publishers, Dordrecht, Vol 1, pp. 81-89.

Hemmingsen, S.M., Woolford, C., van der Vies, S.M., Tilly, K. and Dennis, D.T. (1988) Homologous plant and bacterial proteins chaperone oligomeric protein assembly, *Nature* **333**, 330-334.

Jakob, U., Gaestel, M., Engel, K. and Buchner, J. (1993) Small heat shock proteins are molecular chaperones, *J. Biol. Chem.* **268**, 1517-1520.

Kaukinen, K., Tranbarger, T.J. and Misra, S. (in press) Post-germination-induced and hormonally-dependent expression of low molecular weight heat shock protein genes in Douglas fir, *Plant Mol. Biol.*

Kojima, K., Yamamoto, N. and Sasaki, S. (1992) Structure of the pine (*Pinus thunbergii*) chlorophyll a/b-binding protein gene expressed in the absence of light, *Plant Mol. Biol.* **19**, 405-410.

Koizumi, M., Yamaguchi-Shinozaki, K., Tsuji, H. and Shinozaki, K. (1993) Structure and expression of two genes that encode distinct drought-inducible cysteine proteinases in *Arabidopsis thaliana*, *Gene* **129**, 175-182.

Krebbers, E., L. Herdies, De Clerco, A., Seurink, J., Leemans, J., Van Damme, J., Segura, M., Gheysen, G., Van Montagu, M. and Vandekerckhove, J. (1988) Determination of the processing sites of an *Arabidopsis* 2S albumin and characterization of the complete gene family, *Plant Physiol.* **87**, 859-866.

Lane, B.G. (1991) Cellular desiccation and hydration: developmentally regulated proteins, and the maturation and germination of seed embryos, *FASEB J.* **5**, 2893-2901.

Leal, I. and Misra, S. (1993) Molecular cloning and characterization of a legumin-like storage protein cDNA of Douglas-fir seeds, *Plant Mol. Biol.* **21**, 709-715.

Ledger, S.E. and Gardner, R.C. (1994) Cloning and characterization of five cDNAs for genes differentially expressed during fruit development of kiwifruit (*Actinidia delicosa* var. deliciosa), *Plant Mol. Biol.* **25**, 877-886.

Lu, A.Y.H., Junk, K.W. and Coon, M.J. (1969) Resolution of the cytochrome P-450-containing ω-hydroxylation system of liver microsomes into three components, *J. Biol. Chem.* **244**, 3714-3721.

Mansfield, M.A. and Key, J.L. (1987) Synthesis of the low molecular weight heat shock proteins in plants, *Plant Physiol.* **84**, 1007-1017.

Misra, S. (1994) Conifer zygotic embryogenesis, somatic embryogenesis, and seed germination: biochemical and molecular advances, *Seed Sci. Res.* **4**, 357-384.

Misra, S. and Green, M. (1994) Legumin-like storage polypeptides of conifer seeds and their antigenic cross-reactivity with 11S globulins from angiosperms, *J. Exp. Bot.* **45**, 269-274.

Owens, J.N., Morris, S.J. and Misra, S. (1993) The ultrastructural, histochemical and biochemical development of the post-fertilization megagametophyte and the zygotic embryo of *Pseudotsuga menziesii* (Mirb.) Franco, *Can. J. For. Res.* **23**, 816-827.

Shet, M.S., Sathasivan, K., Arlotto, M.A. and Mehdy, M.C. (1993) Purification, characterization, and cDNA cloning of an NADPH-cytochrome P450 reductase from mung bean, *Proc. Natl. Acad. Sci. USA* **90**, 2890-2894.

Tseng, T.S., Yeh, K.W., Yeh, C.H., Cheng, F.C., Chen, Y.M. and Lin, C.Y. (1992) Two rice (*Oryza sativa*) full length cDNA clones encoding low molecular weight heat shock proteins, *Plant Mol. Biol* **18**, 963-965

Taylor, M.A., Davies, H.V., Smith, S.B., Abruzzese, A. and Gosling, P.G. (1993) Cold-induced changes in gene expression during dormancy breakage in seeds of Douglas-fir (*Pseudotsuga menziesii*), *J. Plant Physiol.* **142**, 120-123.

Tranbarger, T.J. and Misra, S. (in press) Molecular characterization of a set of cDNAs differentially expressed during germination and early seedling growth, *Physiol. Plant.*

Vierling, E. (in press) Structure and in vitro molecular chaperone activity of cytosolic small heat shock proteins from pea, *J. Biol. Chem.*

Zabeleta, E., Oropeza, A., Jiménez, B., Salerno, G., Crespi, M. and Herrera-Estrella, L. (1992) Isolation and characterization of genes encoding chaperonin 60β from *Arabidopsis thaliana*, *Gene* **111**, 175-181.

THE USE OF BULKED SEGREGANT ANALYSIS TO IDENTIFY AFLP™ MOLECULAR MARKERS CLOSELY LINKED TO *MELAMPSORA LARICI-POPULINA* RESISTANCE IN *POPULUS*

M.-T. CERVERA[1], J. GUSMÃO[1], M. STEENACKERS[2], V. STORME[1], A. VANDEN BROECK[2], M. VAN MONTAGU[1,] AND W. BOERJAN[1]

[1] *Laboratorium voor Genetica, via the Department of Genetics, affiliated to the Flanders Interuniversity Institute for Biotechnology, Universiteit Gent, K.L. Ledeganckstraat 35, B-9000 Gent, Belgium;* [2] *Instituut voor Bosbouw en Wildbeheer, Gaverstraat 4, B-9500 Geraardsbergen, Belgium*

1. Abstract

We have identified three AFLP markers tightly linked to the locus conferring resistance to the leaf rust *Melampsora larici-populina* in *Populus* by Bulked Segregant Analysis. The study was carried out using a hybrid progeny derived from an inter-specific, controlled cross between a resistant *Populus deltoides* female and a susceptible *P. nigra* male. The segregation ratio of resistant to susceptible plants suggested that a single, dominant locus (*Mer*) defined the resistance to *Melampsora*.

2. Introduction

Melampsora larici-populina is one of the most damaging fungi for poplar in Central- and North-Europe. Infection by *M. larici-populina* causes premature defoliation and can reduce growth by more than 20%. Trees defoliated early in the growing season are more susceptible to other pathogens and to environmental stress. Infections over successive years can result in a complete loss of the plantation. Therefore, tree breeders have generated and selected inter-specific hybrids that are resistant to different races of *M. larici-populina*.

Since association between genetic markers and traits of interest was first reported by Sax (1923), much attention has been given to the potential applications of markers to improve plant breeding. This approach opens a multitude of new ways in tree improvement. It provides the basis for accelerated breeding through increased intensity of selection of larger progenies, allows early selection of traits, and permits more efficient selection of parents for subsequent breeding programs. Different molecular techniques, such as Restriction Fragment Length Polymorphism (RFLP) or polymerase chain reaction (PCR)-based techniques, have been used with the aim of detecting molecular markers associated with simply inherited or complex traits (quantitative trait loci, QTLs) of poplar, especially those related to disease resistance and commercially important traits (Bradshaw and Stettler, 1995a, 1995b). The efficiency of the indirect

M. R. Ahuja et al. (eds.), Somatic Cell Genetics and Molecular Genetics of Trees, 205–210.
© 1996 *Kluwer Academic Publishers. Printed in the Netherlands.*

selection will depend on the easiness of marker detection and on the degree of linkage to the locus of interest. Several high-density marker technologies have recently been developed. The AFLP™ technique developed by Vos *et al.* (1995) can be considered the most powerful and is based on the selective PCR amplification of restriction fragments from a total digest of genomic DNA. The AFLP technique combines the reliability of the RFLP technique, avoiding Southern blotting, with the specific detection of restriction fragments by PCR amplification (Vos *et al.*, 1995). The AFLP technique detects polymorphisms such as the presence or absence of restriction enzyme sites, sequence polymorphisms adjacent to these sites, insertions, deletions, and rearrangements. Michelmore *et al.* (1991) described a screening method, the Bulked Segregant Analysis (BSA) based on bulking DNAs from individual plants to rapidly screen for molecular markers, tightly linked to a locus of interest. We report the identification of three molecular markers tightly linked to the locus conferring resistance to *M. larici-populina* in *Populus* using AFLP and BSA. This strategy combines one of the most important advantages of the AFLP technology, the high number of loci that can be analyzed per experiment, with the rapid screening approach of BSA.

3. Materials and Methods

3.1. PLANT MATERIAL

The three-generation *Populus* pedigree was founded in 1948 by crossing a female *P. deltoides* V5, resistant to races E1, E2, and E3 of *M. larici-populina* and a susceptible male *P. deltoides* V1. The obtained F1 hybrid family S 9 consisted of 500 clones. A controlled cross between one clone of the family S 9, the female S 9-2 resistant to races E1, E2, and E3 of *M. larici-populina* and the susceptible male *P. nigra* Ghoy was carried out in 1987, to produce the hybrid family 87001 which contained 262 clones.

3.2. *MELAMPSORA LARICI-POPULINA* RESISTANCE TEST

Resistance tests were carried out in the experimental nursery and by artificial infections in the laboratory. For the nursery tests, 10 cuttings of all 262 seedlings were planted in the nursery in June. Plants were observed for their susceptibility to *M. larici-populina* in October. The score system used comprised 11 values ranging between 0 and 5 including half-values: 0 (tree is entirely free of rust), 0.5 (rust spores on 1-10% of the leaves, no leaves are shed), 1 (rust spores on 11-20% of the leaves, no leaves are shed), 1.5 (rust spores on 21-30% of the leaves, no leaves are shed), 2 (rust spores on 31-40% of the leaves, no leaves are shed), 2.5 (rust spores on 41-50% of the leaves, no leaves are shed), 3 (rust spores on 51-60% of the leaves, 1-20% of the leaves are shed in the lower crown-half), 3.5 (rust spores on 61-70% of the leaves, 21-40% of the leaves are shed in the lower crown-half), 4 (rust spores on 71-80% of the leaves, 41-60% of the leaves are shed), 4.5 (rust spores on 81-90% of the leaves, 61-80% of the leaves are shed), and 5 (rust spores on 91-100% of the leaves, 81-100% of the leaves are shed). The score reflects an average for the 10 cuttings per seedling.

For the laboratory tests, three cuttings of all seedlings were grown in the greenhouse. Leaf discs were cut and floated upside-down on de-mineralized water. The artificial infections were performed inoculating the leaf discs by pulverization with a suspension of uredospores (5,000 spores/ml). Three rust races E1, E2, and E3 were used for inoculation. Leaf discs floating on de-mineralized water were incubated at 22°C under continuous illumination during 14 days. All the leaf discs were scored for the appearance (susceptibility) or absence (resistance) of uredospores. The susceptibility of the seedlings to the three races of *M. larici-populina* was scored separately. Each seedling/race interaction was studied on two leaf discs.

3.3. AFLP ANALYSIS

Genomic DNA was extracted from young frozen leaves using the procedure described by Dellaporta *et al.* (1983). AFLP analysis was performed according to Vos *et al.* (1995) with some modifications. The DNA was digested using two restriction enzymes *Mse*I (frequent-cutter) and *Eco*RI (rare-cutter). Two different adaptors, designed to avoid the reconstruction of these restriction sites, one for the *Eco*RI sticky ends and one for the *Mse*I sticky ends, were ligated to the DNA. The adaptor for *Eco*RI contains a 5'-biotinylated top strand, which is covalently linked by ligation to the DNA fragments containing the *Eco*RI site. The *Mse*I adaptor is not biotinylated. All fragments containing *Eco*RI ends were subtracted from the reaction mixture using streptavidin beads (Dynal) reducing the complexity of the DNA which eases subsequent amplification reactions. This selection of biotinylated fragments was performed using 10 μl (100 μg) of beads per reaction. The beads were washed with 100 μl of STEX buffer, containing 100 mM NaCl, 10 mM Tris-HCl, 1 mM EDTA, 0.1% Triton X-100, pH 8.0, and resuspended in 140 μl of STEX buffer. The beads were added to the DNA to a final volume of 200 μl. To optimize the binding of biotinylated DNA to the beads, the mix was incubated for 30 min at room temperature with gentle agitation. The beads were collected with a magnet and, after the removal of the supernatant, they were washed three times with 200 μl of STEX buffer. Finally, the beads were resuspended in 200 μl of T01.E (10 mM Tris HCl, pH 8.0, 0.1 mM EDTA) and kept at 4°C. A preamplification step, prior to the selective radioactive PCR, was done. The preamplification consists of a PCR reaction using primers which are complementary to the adaptors *Eco*RI and *Mse*I with an additional selective 3' nucleotide. In this way, only 1/16 of the DNA fragments will be amplified. The PCR reactions were carried out either in a Techne PCH-3 or in a Perkin Elmer 9600 according to Vos *et al.* (1995) using 28 cycles, each cycle consisting of 30 sec at 94°C, 1 min at 60°C, and 1 min at 72°C. For the selective radioactive amplification, only the *Eco*RI primer was labelled; both the *Eco*RI primer and the *Mse*I primer contain the same sequences as those used in the preamplification but with three selective nucleotides at the 3' end. The PCR reaction was carried out according to Vos *et al.* (1995) using the following cycling parameters: 1 cycle of 30 sec at 94°C, 30 sec at 65°C, 1 min at 72°C followed by 12 cycles in which the annealing temperature decreases 0.7°C per cycle, followed by 23 cycles of 1 min at 94°C, 30 sec at 56°C, and 1 min at 72°C. The start at very high annealing temperature allows optimal primer selectivity. By decreasing gradually the annealing temperature, the efficiency of primer binding increases. At the end of the selective radioactive PCR, the samples were separated in polyacrylamide gels.

3.4. AFLP MARKER ANALYSIS AND LINKAGE ANALYSIS

AFLP markers tightly linked to the resistance locus were identified by BSA (Michelmore *et al.*, 1991). The bulks were made by pooling equal amounts of 17 preamplified DNAs obtained from resistant plants and 19 from susceptible plants from the family 87001. The putative positive primer combinations were analyzed in each individual DNA from the family 87001 to determine the linkage between AFLP markers and the resistance locus.

Linkage estimations were based on the maximum likelihood method using the appropriate formulas of Mather (1938). This recombination fraction was transformed in centiMorgans (cM) according to Kosambi (1944).

4. Results

4.1. GENETICS OF RESISTANCE TO *MELAMPSORA LARICI-POPULINA* IN *POPULUS*

P. deltoides S 9-2 and its progenitor *P. deltoides* V5 were resistant to E1, E2 and E3, three different races of *M. larici-populina* (V. Steenackers, personal communication). The susceptibility of the progeny obtained from the cross between the resistant female S 9-2 and the susceptible male *P. nigra* Ghoy to the three races of *M. larici-populina* was tested separately giving the same results. In the nursery, about 50% of the seedlings was fully resistant to all three races of *M. larici-populina* (133 R:129 S; $\chi^2_{1:1}=0.06$, $P>0.25$, 1 df) whereas the other 50% was susceptible to various degrees. All clones that were scored as being resistant in the nursery (score 0) were also scored as resistant in the laboratory. Similarly, all clones that were susceptible in the nursery (scores 0.5 to 5) were also scored as susceptible in the laboratory. The distribution of the resistance in the progeny is consistent with the combination of two different factors: a monogenic (vertical) resistance and a multigenic (horizontal) resistance suggested by the normal distribution shown by the susceptible progeny. The data suggest that the vertical resistance is determined by a single dominant locus and that both females *P. deltoides* V5 and S 9-2 were heterozygous for the resistance trait. We designated this locus *Melampsora* resistance (*Mer*).

4.2. IDENTIFICATION OF MARKERS LINKED TO RESISTANCE IN *POPULUS*

Primer combinations (144) were used to test parents and bulks (see Materials and Methods) and produced ≈11,500 selectively amplified DNA fragments ranging in size from 70 to 800 nucleotides. On average, 60-100 bands were observed after selective amplification with each primer combination. On average, more than 50% of these AFLP bands were polymorphic between both parentals and usually 12 to 25 of these polymorphic bands segregated in a 1:1 ratio (e.g. markers for resistance to *Melampsora* in family 87001). Primers E39/G01 (*Eco*RI+AGA/*Mse*I+AAG), E40/G37 (*Eco*RI+AGC/*Mse*I+TGT), and E44/G09 (*Eco*RI+ATC/*Mse*I+ACT) each revealed one AFLP marker with a size of ≈270, 320, and 100 nucleotides, respectively. These markers were present in the resistant parent and the resistant bulk but absent in the susceptible parent and the susceptible bulk. A total of 123 DNA samples, obtained from individual clones of the family 87001 was analyzed to determine the linkage between those AFLP markers and the *Mer* allele: only one recombinant tree was detected using markers E39.G01.270 and E44.G09.100, whereas the same recombinant and two more were detected with marker E40.G37.320, indicating that all three markers were tightly linked. Linkage between the *Mer* locus and E39.G01.270, E40.G37.320, and E44.G09.100 was 0.81 $\pm$ 0.81, 2.44 $\pm$ 1.39, and 0.81 $\pm$ 0.81 cM, respectively.

5. Discussion

The data obtained suggest that the resistance to three different races of *M. larici-populina* in *P. deltoides* V5 and S 9-2 is determined by a single dominant locus. Three different genetic markers tightly linked to the *Mer* locus were identified by using BSA combined with AFLP to analyze a hybrid family belonging to a three-generation *Populus* pedigree.

Recently, several high-volume marker technologies have been developed. Among them AFLP is considered one of the most powerful technologies. This method was developed at Keygene (Wageningen, The Netherlands; Vos *et al.*, 1995). One of the most important advantages of the AFLP technique is the high number of loci that can be analyzed per experiment. In our conditions, using *Eco*RI/*Mse*I and primers containing three additional selective 3' nucleotides, between 60-100 bands can be screened per PCR reaction for the *Populus* genome (2,400-2,600 cM; Bradshaw *et al.*, 1994).

Another advantage of the AFLP technique compared to other PCR marker technologies, is its reproducibility. RAPD analysis, based on the use of random primers, is known to suffer from a lack of reproducibility due to mismatch annealing (Neale and Harry, 1994). The primers used to obtain AFLP markers combine two key characteristics: they are complementary to the adaptor oligonucleotide, thus allowing a highly specific primer annealing and they are selective; changing the selective 3' nucleotides allows to amplify a different set of DNA fragments from the same population of preamplified fragments (Vos *et al.*, 1995). The combination of BSA, as a strategy to rapidly screen for markers tightly linked to monogenic traits and AFLP analysis, a high-density marker technology, allows a very efficient identification of markers adjacent to the locus of interest. The tightly linked AFLP markers will be particularly useful for *Populus* breeding by accelerating the disease resistance screening. This AFLP marker system satisfies the requirements for gene introgression and facilitates further genetic characterization of resistance to different races of *M. larici-populina*. We are constructing a high density AFLP molecular map of *P. deltoides* S9-2 and *P. nigra* Ghoy which might serve in cloning this locus by chromosome landing when YAC or BAC libraries are available (Tanksley *et al.*, 1995; Thomas *et al.*, 1995).

6. Acknowledgments

The authors thank M. Zabeau for a fruitful collaboration with Keygene N.V., V. Steenackers and J. Van Slycken for a long-standing collaboration with the IBW-Geraardsbergen, C. Malpica for his valuable comments during the project, T. Gerats and G. Cnops for critical reading of the manuscript, A. Van Gysel for the use of AFLP facilities, and M. Fernández for his help when collecting the material. This work was supported by grants from the Flemish Government (BNO/BB/6/1994) and the European Union AIR program (AIR1-CT92-0349). M.-T.C. is funded by an individual fellowship from the Human Capital and Mobility program of the European Union (41AS8694).

7. References

Bradshaw, H.D. Jr, Villar, M., Watson, B.D., Otto, K.G., Stewart, S., and Settler, R.F. (1994). Molecular genetics of growth and development in *Populus*. III. A genetic linkage map of a hybrid poplar composed of RFLP, STS, and RAPD markers. *Theor. Appl. Genet.* **89**, 167-178.

Bradshaw, H.D. Jr. and Stettler, R.F. (1995a) Molecular genetics of growth and development in *Populus*. IV. Mapping QTLs with large effects on growth, and phenology traits in a forest tree. *Genetics* **139**, 963-973.

Bradshaw, H.D. Jr. and Stettler, R.F. (1995b) Field trials to identify quantitative trait loci underlying growth, morphology, and stress response in *Populus*. Abstract presented at the International Poplar Symposium, Seattle (WA, USA), August 20-25, 1995.

Dellaporta, S.L., Wood, J. and Hicks, J.B. (1983) A plant DNA minipreparation: Version II. *Plant Mol. Biol. Rep.* **1**, 19-21.

Kosambi, D.D. (1944) The estimation of map distances from recombination values. *Ann. Eugen.* **12**, 172-175.

Mather, K. (1938) *The Measurement of Linkage in Heredity* (Methuen's Monographs on Biological Subjects). Methuen, London.

Michelmore, R.W., Paran, I. and Kesseli, R.V. (1991) Identification of markers linked to disease-resistance genes by bulked segregant analysis: A rapid method to detect markers in specific genomic regions by using segregating populations. *Proc. Natl. Acad. Sci. USA* **88**, 9828-9832.

Neale, D.B. and Harry, D.E. (1994) Genetic mapping in forest trees: RFLPs, RAPDs and beyond. *AgBiotech. News Inform.* **6**, 107N-114N.

Sax, K. (1923) The association of size differences with seed-coat pattern and pigmentation in *Phaseolus vulgaris*. *Genetics* **8**, 552-560.

Tanksley, S.D., Ganal, M.W. and Martin, G.B. (1995) Chromosome landing: a paradigm for map-based gene cloning in plants with large genomes. *Trends Genet.* **11**, 63-68.

Thomas, C.M., Vos, P., Zabeau, M., Jones, D.A., Norcott, K.A., Chadwick, B.P. and Jones, D.J.G. (1995) Identification of amplified restriction fragment polymorphism (AFLP) marker tightly linked to the tomato *Cf-9* gene for resistance to *Cladosporium fulvum*. *Plant J.* **8**, 785-794.

Vos, P., Hogers, R., Bleeker, M., Reijans, M., van de Lee, T., Hornes, M., Frijters, A., Pot, J., Peleman, J., Kuiper, M. and Zabeau, M. (1995) AFLP: a new technique for DNA fingerprinting. *Nucleic Acids Res.* **23**, 4407-4414.

QUANTITATIVE TRAITS AND GENETIC MARKERS: ANALYSIS OF A FACTORIAL MATING DESIGN IN LARCH

A. ARCADE, P. FAIVRE RAMPANT, B. LE GUERROUE, L.E. PAQUES, D. PRAT
Station d'Amélioration des Arbres Forestiers
INRA, 45160 ARDON, FRANCE

Abstract

A factorial mating design involving European and Japanese larches was carried out in order to produce interspecific hybrids. Hybrid families have been measured for 11 years for quantitative traits including growth characters, morphological characters and wood quality. Parent trees were genetically assessed using RAPD markers. Ninety-eight markers showing intraspecific polymorphism in one species at least were used in this study. Firstly, family frequencies for a marker were deduced from parental frequencies for this marker, and statistical relationship between marker frequency and performance of the family was investigated. Significant relationships were determined for 19 markers and quantitative characters including height, branch angle, stem straightness and wood specific gravity. This suggests that these markers are linked to QTL involved in these traits. Secondly, correlations between hybrid family performances and genetic distance of respective parents were investigated. A positive correlation between genetic distance and growth characters (height, volume, B.H. girth) and a negative correlation between genetic distance and branch angle were observed, and demonstrated the effect of heterozygosity level on these characters. Thus, genetic distance between parents is a first selection criteria if hybrid growth is a requested characteristic. More precisions (complete parent genotype, genetic map, QTLs distribution) are required for a better understanding of QTLs effects and expression and a possible application in marker-assisted selection.

1. Introduction

Genetic improvement of forest tree species takes a long time since the evaluation of plant material for quantitative traits is not possible before several years. Now that access to the genome is feasible through molecular tools, a new strategy consists in detecting genomic sequences responsible for the expression of quantitative traits (QTL or Quantitative Trait Loci). These sequences are identified by means of markers linked to them and a better knowledge of their action could offer the opportunity of a marker-assisted selection. Such approaches are particularly suitable for forest trees where most important traits (growth, wood quality, disease resistance ...) are assessed very late (Neale and Williams, 1991; Neale *et al.*, 1992; Williams and Neale, 1992).
The present study is based on a factorial mating design involving Japanese and European larches as parents, and hybrid larch families. Hybrid larch often associates parental interesting characteristics and some full-sib families are particularly outstanding. These hybrid families are field-tested and a reliable information about their phenotypic value can be obtained after some years. However, field-testing is quite expensive and time-consuming. Our final goal is to identify genitors likely to produce a high performing progeny, by evaluating markers linked to QTLs involved in economic traits. We have investigated the hybrid larch families by using molecular markers. The relationships between markers and hybrid vigour have been tested in two different approaches: the general effect of heterozygosity (assessed by genetic distance of parents), and the effect of marker alleles associated with QTLs.

M. R. Ahuja et al. (eds.), Somatic Cell Genetics and Molecular Genetics of Trees, 211–216.
© 1996 *Kluwer Academic Publishers. Printed in the Netherlands.*

2. Material and methods

2.1. PLANT MATERIAL

Twelve European larches (*Larix decidua* Mill.) as female parents and 12 Japanese larches (*Larix kaempferi* Lamb. Carr) as male parents were involved in a factorial mating design in order to produce interspecific hybrids. European larch originated from the Tyrol (1 tree) and from the Czech (7 trees) and Polish (5 trees) Sudeten mountains. For Japanese larch, 7 trees selected in Denmark were related (full-sibs from full-sibs), while the others came from different natural Japanese stands.

Fifty-nine crosses were carried out and produced 53 interspecific hybrid families tested in the field. Two-year-old seedlings were planted in a forest trial at Beaumont-du-Lac (Haute-Vienne, France) in April 1983. They were measured for various traits including (Pâques, 1992):

-growth characteristics: height (at 2 years in nursery, at 4, 5, 6, 8, 9, 10, 11 years in progeny test), B.H. girth (at 11 years), individual stem volume (at 11 years), diameter and number of branches of the 6th whorl (observed at 11 years),

-morphological characteristics: stem straightness score (at 6 and 11 years), tapper (ratio height on diameter) (at 11 years), branch angle of the two thickest branches of the 6th whorl (observed at 11 years);

-and wood quality: wood specific gravity (assessed by pilodyn at 11 years).

2.2. GENETIC MARKERS

Total DNA was extracted from parent tree needles according to Greenwood *et al.* (1989) with minor modifications. RAPD markers were obtained after amplification reactions containing 1 ng DNA, 83 µM of each dATP, dCTP, dGTP, dTTP, 1.4 µM of primer (Operon Technologies Inc., Alameda, USA), 1 x Taq Polymerase buffer, 0.1 u Taq Polymerase (Appligene), in a final volume of 12 µL. DNA amplification was achieved after 35 cycles (1 min at 94 °C, 1 min at 36 °C, 2 min at 72 °C).

Amplified fragments were revealed by ethidium bromide stain, only distinct bands with clear-cut segregation among tested individuals were recorded.

2.3. DATA ANALYSIS

Variation for each quantitative trait measured was significant among families (Pâques, 1992). Relationships between family performances for quantitative traits and genetic markers were tested according to the two following ways.

(1) RAPD markers were used for detection of Quantitative Trait Loci (QTL) by analysing the relationships between quantitative trait and marker frequency within family. The frequency of the RAPD marker within family was deduced from the presence and absence of this marker in parents: parents were presumed homozygous for the considered marker. Because of RAPD marker dominance, homozygous presence/presence and heterozygous presence/absence parental tree cannot be differentiated. In the lack of segregation distortion, the expected frequency of the marker in the family is thus 1.0 if both parents bear the marker band, 0.5 if only one parent bears it, and 0.0 if none of both parents exhibits the marker. Families were distributed into three different groups according to their frequency of the marker. Significant differences among these groups of families for a quantitative trait were tested by analysis of variance, they revealed a linkage disequilibrium between this marker and this quantitative trait locus. The genetic marker was then considered at a locus linked to a QTL. Heterozygosity of parent trees results in a decrease of expected frequency of marker within family but the three groups of families are maintained. Families could be distributed into five groups if the marker was codominant.

(2) The genetic distance between pairs of parents was assessed according to Jaccard (1908) similarity index Sj (number of RAPD markers shared by both parents on the number of different RAPD markers exhibited by both

parents). The genetic distance was calculated as Dj = 1 - Sj. Correlations between the genetic distances of the parents and the performances of their families were then tested for each quantitative trait.

3 Results

3.1. QUANTITATIVE TRAIT LOCI

Two-hundred-and-ten primers were assayed on the 24 parent trees. Ninety-eight marker bands generated by 52 primers were polymorphic in one species at least and represented the basis of this study. Families were often distributed into two different groups according to their marker frequency because most markers were monomorphic in one species. The analysis of variance revealed significant differences (P>0.999975) of performance between these groups for 21 combinations marker/quantitative trait. Nineteen RAPD bands are implicated and represent markers at loci likely to be linked to QTLs (Table1).

Two markers were associated with two quantitative traits at once: one with B.H. girth and wood specific gravity and the other with height at 8 years and wood specific gravity.

The markers are linked to QTL with a positive or negative effect on the character depending on the case. A positive effect means that the value for the quantitative trait increases with a higher frequency of the marker allele.

Different markers are involved depending on the life stage. One marker was detected associated with height at 5 and 6 years only: this marker may be linked to a QTL acting during the first years of growth solely, while 7 other markers were associated with height in older trees.

TABLE 1. Number of RAPD markers with significant positive (+) or negative (-) effects on quantitative traits

Quantitative trait	Age (years)	RAPD markers +	RAPD markers -
Height	5-6		1
	8-11	3	4
B.H. Girth	11		
Stem volume	11		
Tapper	11		
Branch angle	11	2	6
Branch diameter	11		
Number of branches	11		
Stem straightness	6		
	11		2
Wood specific gravity	11		3

One of the three markers linked to wood specific gravity was polymorphic in both species. This case is worthwhile as it allows the distribution of the families into three classes of marker allele frequency. An increase of the within family frequency of this marker O9A resulted in a decrease of the wood specific gravity (Figure 1). This marker was linked with an unfavourable allele of QTL.

Most markers were found to be polymorphic in one species only, and few markers associated with QTLs showed polymorphism in both species (Table 2). Polymorphism was inherited exclusively by the Japanese male parent for two characters: stem straightness and branch angle. Polymorphism was detected in both species for height and wood specific gravity, but a higher number of polymorphic markers were recorded in European larch for these two traits than in Japanese larch.

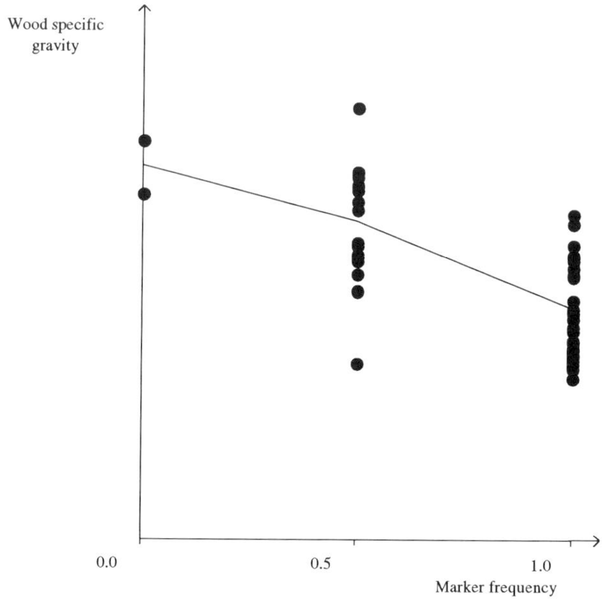

FIGURE 1. Relationships between wood specific gravity and frequency of RAPD marker O9A within families

TABLE 2. Distribution of intraspecific polymorphism (P) of RAPD markers related to quantitative traits

Quantitative trait	Age (years)	*Larix decidua*	*Larix kaempferi*
Height	5-6	P	
	8-11	P	P
Branch angle	11		P
Straightness	11		P
Wood specific gravity	11	P	P

3.2. QUANTITATIVE TRAIT AND GENETIC DISTANCE

Significant correlations between the genetic distances between parents and the family performances were observed for some quantitative traits (Table 3). They involved particularly growth traits (height, B.H. girth, individual volume), and only one morphological trait (branch angle).

TABLE 3. Significant correlations (P>0.999) between the quantitative trait of progenies and the genetic distance of their parents

Quantitative trait	Age (years)	Correlation
Height	6	0.55
	8	0.50
	9	0.48
	10	0.50
	11	0.45
Circumference	11	0.52
Stem volume	11	0.52
Branch angle	11	-0.45

The correlation between the genetic distance and height became significant and maximal for 6 year-old trees and decreased after this stage. Thus, the influence of heterozygosity level is evolving with the age and exerts favourable effects during few years.

The negative correlation between branch angle and genetic distance showed that a high level of heterozygosity in hybrid trees was associated with a reduced branch angle, which represents a disadvantage for breeding.

4. Conclusion - Discussion

This analysis of a larch factorial mating design provided insight into the inheritance of quantitative traits and their relationships with genetic markers. First, we have detected RAPD markers linked to QTLs of agronomic interest: height, stem straightness, branch angle, wood specific gravity. Genetic markers have been already used for detection and mapping of QTLs for several traits such wood specific gravity and vegetative propagation ability in forest tree species (Groover *et al.*, 1994; Han *et al.*, 1994; Grattapaglia *et al.*, 1995) by the analysis a F_2 family or of a pedigree, but not from a factorial mating design. The detection of association between RAPD marker and QTL revealed a linkage disequilibrium between the marker allele and the QTL allele in the studied population. As pointed out by Strauss *et al.* (1992), linkage disequilibrium should be very low in forest trees because of a predominantly outcrossing mating system. Linkage disequilibrium might persist among parents essentially if marker and QTL were tightly linked, or if original population size decreased dramatically in the past (bottleneck effect), or if parents were closely related (Japanese larches).

Moreover, changes of linkage phase between marker allele and QTL allele from parent to parent can be detected in a factorial mating design, as parental gametic contribution can be examined through several related half-sib families. We have not observed such changes of phase, which means that linkage disequilibrium was quite strong and stable.

Polymorphism of the marker related to height at 5 and 6 years was maternal, while polymorphism of markers related to stem straightness and branch angle was paternal. This observation was in agreement with the results presented by Pâques (1992), where for total height the female component (European larch) of total variation was predominant, and for stem straightness and branch angle, the male contribution (Japanese larch) was larger than the female one.

Secondly, we have shown the favourable effect of genetic distances between parents on hybrid growth vigour, while other characters, except branch angle, were not influenced by them. In general, similar analyses on plants including trees have not pointed out such relationships between parent distance or family heterozygosity level and quantitative traits (Aradhya and Phillips, 1995; Vaillancourt *et al.*, 1995; Verhaegen *et al.*, 1995), trees used in the present study had perhaps a different genetic structure. What was interesting to note in larch is that a greater amount of markers were found to be associated with these two traits, and this observation suggested that growth and branch angle were characters under multigenic control. Other traits like stem straightness or wood specific gravity would depend on fewer genes with major effects. Thus, if growth vigour of hybrid trees was the desired criteria, crosses should be carried out between parents with maximal genetic distances.

In conclusion, the results of this study confirm that the detection of QTL by means of a factorial mating design is feasible. This unusual method presents advantages compared with standard QTL detection in F_2 progenies, first because quantitative data are based on familial means which are more confident than usual individual values, and secondly because families' genotypes are deduced from parental genotypes and this reduces the sample size to be assessed by RAPD technique.

However, parents' genotypes need to be better specified and this represents the next step of the study. Indeed, as RAPD markers are dominant markers, homozygous parents (presence/presence) and heterozygous parents (presence/absence) were pooled in a same group of presence of the marker. This imprecision can be removed by the analysis of marker segregation in the megagametophyte or in progenies and will result in increasing the number of marker frequency groups.

The markers need to be mapped. This map is necessary for our knowledge of the exact numbers of QTLs detected, and to determine if several markers are related to the same QTL, or if several QTLs are linked to a same marker. A genetic map is also necessary to ensure an homogeneous distribution of markers on the genome, hence having a greater probability to detect additional QTLs.

These various prospects should broaden our knowledge of QTLs, of their expression and effects on quantitative traits. It will lead to practical applications in the field of marker-assisted selection.

5. References

Aradhya, K.M., and Phillips, V.D. (1995) Lack of association between allozyme heterozygosity and juvenile traits in *Eucalyptus*, *New Forests* **9**, 97-110.

Grattapaglia, D., Bertolucci, F.L., and Sederoff, R.R. (1995) Genetic mapping of QTLs controlling vegetative propagation in *Eucalyptus grandis* and *E. urophylla* using a pseudo-testcross strategy and RAPD markers, *Theor. Appl. Genet.* **90**, 933-937.

Greenwood, M.S., Hopper, C.A., and Hutchison, K.W. (1989) Maturation in larch. 1. Effect of age on shoot growth, foliar characteristics and DNA methylation, *Plant Physiol.* **90**, 406-412.

Groover, A., Devey, M., Fiddler, T., Lee, J., Megraw, R., Mitchel-Olds, T., Sherman, B., Vujcic, S., Williams, C., and Neale, D. (1994) Identification of quantitative trait loci influencing wood specific gravity in an outbred pedigree of loblolly pine, *Genetics* **138**, 1293-1300.

Han, K.H., Bradshaw, H.D., and Gordon, M.P. (1994) Adventitious root and shoot regeneration *in vitro* is under major gene control in an F_2 family of hybrid poplar (*Populus trichocarpa* x *P. deltoides*), *Forest Genetics* **1**, 139-146.

Jaccard, P. (1908) Nouvelles recherches sur la distribution florale, *Bull. Soc. Vaud. Sci. Nat.* **44**, 223-270.

Neale, D.B., and Williams, C.G. (1991) Restriction fragment length polymorphism mapping in conifers and applications to forest genetics and tree improvement, *Can. J. For. Res.* **21**, 545-554.

Neale, D.B., Devey, M.E., Jermstad, K.D., Ahuja, M.R., Alosi, M.C., and Marshall, K.A. (1992) Use of DNA markers in forest tree improvement research, *New Forests* **5**, 391-407.

Pâques, L.E. (1992) First evaluation of genetic parameters in a factorial mating design with hybrid larch (*Larix decidua* X *Larix kaempferi*), in H. Weisgerber (ed.), *Results and future trends in larch breeding on the basis of provenance research, IUFRO meeting of the IUFRO working party S2.02-07*, IUFRO, pp. 136-145.

Strauss, S.H., Lande, R., and Namkoong, G. (1992) Limitations of molecular-marker-aided selection in forest tree breeding, *Can. J. For. Res.* **22**, 1050-1061.

Vaillancourt, R.E., Potts, B.M., Watson, M., Volker, P.W., Hodge, G.R., Reid, J.B., and West, A.K. (1995) Detection and prediction of heterosis in *Eucalyptus globulus*, *Forest Genetics* **2**, 11-19.

Verhaegen, D., Kremer, A., and Vigneron, P. (1995) Relationships between heterosis and molecular polymorphism in interspecific crosses of *Eucalyptus urophylla* x *E. grandis*, in B.M. Potts., N.M.G. Borralho, R.J. Reid, R.N. Cromer, W.N. Tibbits and C.A. Raymond (eds.), *Eucalypt plantations: improving fibre yield and quality. Proc. CRCTHF-IUFRO Conf., Hobart, 19-24 Feb.*, CRC for temperate hardwood forestry, Hobart, pp. 434-437.

Williams, C.G., and Neale, D.B. (1992) Conifer wood quality and marker-aided selection: a case study, *Can. J. For. Res.* **22**, 1009-1017.

DEVELOPMENT OF MICROSATELLITE LOCI FROM *QUERCUS ROBUR* AND *QUERCUS PETRAEA*.

H. STEINKELLNER[1], S. FLUCH[1,2], E. TURETSCHEK[1], R. STREIFF[3], A. KREMER[3], K. BURG[2] and J. GLÖSSL[1]

[1]*Zentrum für Angewandte Genetik, Universität für Bodenkultur, A-1190 Vienna, Austria;* [2]*Forschungszentrum Seibersdorf, A-2444 Seibersdorf, Austria;* [3]*INRA, Station de Recherches Forestieres de Bordeaux-Cestas, Pierroton 33610 Cestas, France.*

Abstract:

The importance of genetic polymorphism detected by microsatellite sequences is now well established in mammalian and recently also in some plant genomes. In this study a size selected genomic library from *Quercus petraea* and *Q. robur*, the most common oak species in Europe, were screened for GA-microsatellite sequences and the resulting loci were analysed by PCR. Genetic polymorphisms of 5 GA-microsatellite DNA loci were examined among 34 individuals of *Quercus petraea* and *Q. robur* from different stands in Austria. The number of alleles per loci varied from 5 to 12. The same level of differentiation was detected for *Q. robur* and *Q. petraea*, none of the found alleles is restricted to one species. The range of allele frequency varies from 2% to 60%. Combining the genetic variation observed for the characterized loci provides a unique genotype for all the tested individuals. Using intraspecific controlled crosses Mendelian inheritance could be shown for three of the five loci. Our results suggest that microsatellites serve as a source of highly polymorphic genetic marker in *Querus*.

1. Introduction:

Simple sequence repeats (SSR) or microsatellites are tandemly arranged repeats of one or a few nucleotides. The high variability of the repeats makes microsatellite sequences to powerful genetic marker. Microsatellite sequences can be amplified by polymerase chain reaction (PCR) using primers from the conserved flanking regions. Polymorphic microsatellite loci are abundant throughout the genome of most eukaryotes [1] and in recent years analysis of these loci has become well established for mammalian genomes [2, 3]. However, the application of such analysis has been relatively rare in crop plants [e.g. 4, 5, 6] and has emerged only recently for natural plant populations [e.g. 7]. On the basis of genomic library screening, several authors roughly estimated a frequency of one di- or trinucleotide microsatellite every 50 to 480 kb, depending on the species and type of microsatellite investigated [8, 9]. Further analysis of microsatellite loci in different plant species showed that they are usually

M. R. Ahuja et al. (eds.), Somatic Cell Genetics and Molecular Genetics of Trees, 217–222.
© 1996 *Kluwer Academic Publishers. Printed in the Netherlands.*

inherited in a codominant Mendelian manner, show high allelic variation and are distributed randomly throughout the whole genome [10 - 15].

In the present paper, we describe the development of five GA-microsatellite markers in two oak species which have the largest distribution in Europe, *Quercus robur* and *Q. petraea*. The qualitative and quantitative features of these microsatellites are presented.

2. Materials and Methods

2.1. PLANT MATERIAL AND ISOLATION OF DNA

Fresh or frozen *Q. robur* and *Q. petraea* leaves were ground in liquid nitrogen and resuspended in 10 mM Tris/HCl, pH 8.0, containing 1 mM EDTA, 1 %(w/v) SDS and 50 μg/ml Proteinase K (20 ml buffer/1g of tissue) and gently shaken overnight at room temperature. The slurry was subsequently adjusted to 0.8 M potassium acetate and left on ice for 20 min. The precipitate was centrifuged at 10,000 rpm for 20 min and 4 °C in an SS34 Sorvall rotor, the supernatant brought to 10 %(w/v) PEG 6000 and 0.7 M NaCl, left on ice for 20 min and centrifuged as described above. The pellet was washed with 70 %(v/v) ethanol, air-dried and resuspended in 1 mM Tris/HCl, pH 8.0, 1mM EDTA (TE-buffer) for at least 1 hour at 50 °C. The solution was then extracted by buffer-saturated phenol, then by chloroform/isoamylalcohol (24:1) and finally precipitated with isopropanol (0.7-1.4 vol.). DNA fibres were then hooked on a small glass rod and soaked in 96% (v/v) ethanol. The pellet was briefly air dried and dissolved in TE-buffer as described above. Subsequently, a CsCl centrifugation step was carried out for four hours at 525,000 x *g* according to standard procedures. The yield was 20-200 μg of DNA per 1 g of leaf tissue with an absorbance ratio at 260/280 nm of around 1.9. Concentrations were determined fluorimetrically.

2.2. PREPARATION AND SCREENING OF THE LIBRARY

For the preparation of a genomic library, purified mixed genomic DNA from *Q. robur* and *Q. petraea* was cut with Sau3A, size selected (400-600bp) by agarose gel electrophoresis, ligated into pUC19 cut with BamHI and transformed into competent *E. coli* (DH5α) cells. Cells were plated on 12x12 cm square plates and filter replica were made using positive charged nylon filters (Amersham). Filters were prewashed for minimum 2 h in 50 mM Tris/HCl, pH 8.0, 1 M NaCl, 1 mM EDTA, 0,1 %(w/v) SDS at 60 °C, prehybridized and hybridized at 52 °C. As a probe a GA/TC polymer was used (kindly provided by C. Schlötterer, Munich, Germany). The probe was labelled with 30 μCi ^{32}P (3000Ci/mmol) by nick translation.

2.3. PCR, SEQUENCING AND ELECTROPHORESIS

A typical PCR reaction was carried out as follows: 30-50 ng template DNA, 2.5 mM dNTPs, 1 unit Taq polymerase (BiTaq, Biomedica), 5 μl 10x PCR-buffer (Biomedica) in a final volume of 50 μl. 28 cycles (1 min 94°C, 30 sec 58°C, 30 sec 72 °C) were performed and terminated by 10 min final extension. PCR products were denatured for 5 min at 92 °C, separated on a denaturing 15 %(w/v) polyacrylamide gel (Clean Gel®, Pharmacia Biotech) according to the manufacturer's instructions and visualized by silver staining.

Cycle sequencing was done according to the kit manufacturer's instructions (ABI-Cycle Sequencing Kit, Applied Biosystems, Inc.) and sequence analysis was performed on an ABI 377 automatic sequencer (Codon Genetic Systems, Vienna).

2.4. INHERITANCE ANALYSIS OF MICROSATELLITE MARKERS

The segregation of alleles at three loci AG 1/5, AG 3/64 and AG 16 was studied using three intraspecific crosses (*Q. robur*): 32P x A3 (22 progeny), 33P x A3 (45 progeny), 34P x A3 (30 progeny). The parents used for the crosses were adult trees from the campus of the Forestry Research Station at INRA Pierroton, France. Pollinations were made in 1989 for 33P x A3 and 1992 for 32P x A3 and 34P x A3. In each case the male parent was A3. For those three controlled crosses the inheritance of isoenzymes had been tested previously and showed a simple Mendelian segregation (personal communication A.Kremer, INRA France)). Genomic DNA was extracted from leaves or buds of three and five years old plants, following the method of Doyle and Doyle [16] as modified by Dumolin *et al.* [17].

3. Results and Discussion:

Southern blot analysis of Sau3A-digested genomic *Quercus* DNA using different repeating polymers as probes (consisting of (GA, GT, AAT, GGT, TAG or TCA)$_n$-repeats) showed high abundance of (GA)$_n$ repeats. A relatively weak signal was obtained when a (GT)$_n$ probe was used and no signal was detectable after screening with trinucleotide repeat probes.

Screening of a size-selected DNA library (insert size 400-800 bp) with GA-repeat sequence showed that there are about 10^5 microsatellite sites per haploid genome which is similar to other plant species that have been analyzed [8, 9]. GA repeats are more abundant than CA repeats which is in agreement with other reports [7, 15]. Finally, five GA-microsatellite loci were chosen for further characterization (Table 1). Primers were designed in a way that the size of amplified PCR products is between 100 and 200 bp. Four out of the five loci contain "perfect" microsatellite repeats. One locus, GA1/2, has a "perfect" microsatellite in the original subclone derived from the library but "compound" repeats were found when several alleles amplified from that locus were sequenced.

Although good resolution was obtained when PCR products were separated under denaturing conditions on a horizontal 15% polyacrylamide gel, it was usually not possible to distinguish alleles differing in one repeat unit, i.e. in two base pairs. Therefore, the number of alleles per locus given in Table 1 is most likely to be an underestimate and will increase by using high resolution gels, such as sequencing gels, and by investigating a higher number of trees.

TABLE 1. Summary of microsatellite primer sequences, number of alleles per locus and the size range of found alleles.

Locus	Repeat unit of cloned allele	No. of alleles found / No. of trees studied	Primers (F, forward; R, reverse)	Length of alleles (bp)
GA 1/2	$(AG)_{16}$	7/27	F: 5'-tcctccgctcactcaccatt R: 5'-aaacctccaccaaaacat tc	100-124
GA 1/5	$(GT)_5(GA)_9$	7/33	F: 5'-gcttgagagttgagatttgt R: 5'-gcaacaccctttaactacca	161-181
GA 16	$(AG)_{21}$	9/34	F: 5'-cttcactggcttttcctcct R: 5'-tgaagcccttgtcaacatgc	162-200
GA 3/64	$(AG)_{19}$	12/31	F: 5'-tagaaagcccaaaaccaaaacc R: 5'-ctttttggaagccgcttccgta	141-167
GA 3/62	$(AG)_4$	5/19	F: 5'-cttgagcatggaatcctatg R: 5'-tctagaggagctttcctttacac	95-105

A total of 34 individuals from both species *Q. robur* and *Q. petraea* sampled at 10 different stands in Austria (Table 2) were characterized with these 5 primer pairs. Altogether, 40 alleles were identified at the five loci investigated. None of the alleles were restricted to either *Q. robur* or *Q. petraea* and they seem to be evenly distributed across both species as far as it can be judged from the relatively small number of trees investigated so far. The number of alleles varied from 5-12 per locus. Even at locus GA 3/62, where $(GA)_4$ was identified as the core repeating unit, polymorphisms could be uncovered although it has been proposed that loci containing less than 10 repeats are unlikely to be polymorphic [1]. All five microsatellite loci appear to have similar allelic distribution patterns. As an example, Figure 1 shows the characterization of locus GA3/64. One to ten individuals from ten different stands (see Table 2) were analyzed. The locus shows a high degree of heterozygosity (0.74) in the trees tested.

TABLE 2. Austrian stands (from Lower Austria and Burgenland) from where samples of *Q robur* and *Q. petraea* were collected.

Predominantely *Q. robur*	Code	Predominantely *Q. petraea*	Code
Tulln	A/B/C	Ernstbrunn	EB
Niederweiden	NW	Stotzing	ST
Güssing	GU	Großschweinbarth	GS
Gralla	GR	Punitzer Wald	PW
Unterbildlein	UB	near Güssing	HW

FIGURE 1. Scan report from silver stained PCR fragments (locus GA 3/64) separated on a 15% Clean Gel. 29 *Q. robur* and *Q. petraea* samples were taken from 10 different stands. The numbers above bands indicate allele identity, e.g. the size of alleles 1 and 10 is 181 and 141 bp, respectively.

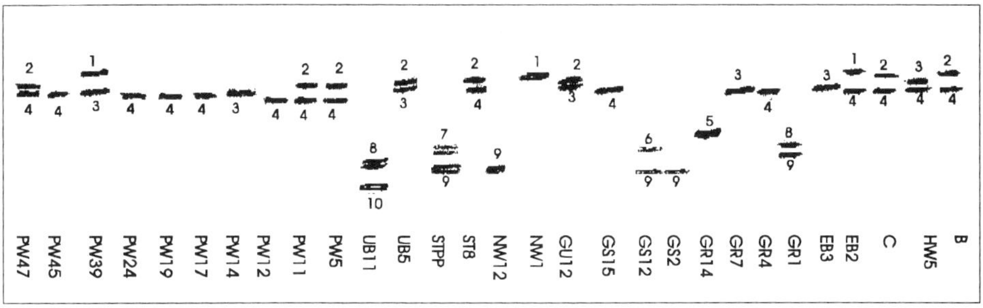

Segregation analysis clearly demonstrated codominant Mendelian inheritance for three loci studied so far. Figure 2 represents segregation analysis for locus GA 16 with 32 offspring (the data for 16 offspring are shown) of male parent A3 and female parent 34P. Oberved ratios of progeny genotypes were not significantly different from that expected under a simple codominant Mendelian inheritance hypothesis (χ^2 test).

FIGURE 2. Segregation analysis of microsatellite marker GA16. Alleles A and B derive from male parent A3, alleles C and D derive from female parent 34P. Allele distribution between the 16 offspring demonstrates Mendelian inheritance (CD x AB: AC,DA,BC,BD; 6:3:4:3).

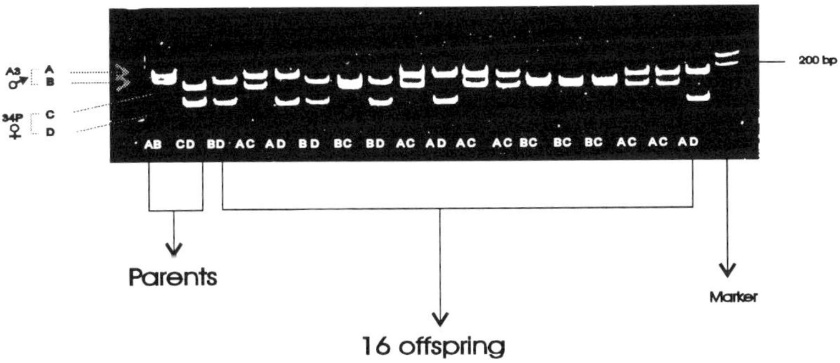

Our results indicate that a relatively small genomic library of a species, containing less than 4000 pfu, is sufficient to yield polymorphic $(GA)_n$ microsatellite loci. Although it has been reported that dinucleotide repeats with 10 or less repeats are not variable [1], 5 alleles could be identified at locus GA3/62, where the number of GA repeats varies between 4 and 9. Therefore, we suggest that sequences with smaller number of repeats should not necessarily be ignored when screening for polymorphic loci. It may be a good practice to carry out small-scale surveys for polymorphism before making judgements regarding the potential usefulness of particular repeat sequences.

Combining the genetic variation observed for the five identified (GA)-SSR loci produces a unique genotype for all of the individuals tested so far. All of the 34 individuals DNA-typed with the 5 loci could be discriminated genetically. We have demonstrated that microsatellites can serve as a powerful tool for genetic characterization of *Quercus* ssp. However, further investigation will be needed to develop and characterize enough markers for population genetic analysis, e.g. parental analysis or pollination studies.

Acknowledgment: This work was supported by grants from the Austrian Research Foundation (FWF), project numbers: P8477-MOB and P10800-MOB.

4. References

1. Weber, J.L.: Informativeness of human (dC-dA)n/(dG-dT)n polymorphism. *Genomics* 7 (1990), 524-530.
2. John,B. and Miklos, G.L.G.: *The Eucaryote Genome in Development and Evolutio,* Allen and Unwin, London, 1988.
3. Litt, M. and Lutty, J.A.A.: Hypervariable microsatellite revealed by in vitro amplification of dinucleotide repeat within the cardiac muscle actin gene. *Am. J. Hum. Genet.* **44** (1989), 397-401.
4. Yang, G.P., Saghai Maroof, M.A., Xu, C.G., Zhang ,Q., and Biyaswhev, R.M.: Comparative analysis of microsatellite DNA polymorphism in landraces and cultivars of rice. *Mol. Gen. Genet.* **245** (1994), 187-195.
5. Saghai Maroof, M.A., Biyaswhev, R.M., Yang, G.P., Zhang, Q., and Allard, R.W.: Extraordinarily polymorphic microsatellite DNA in barley: species diversity, chromosomal location, and population dynamics. *Pro.c Natl. Acad. Sci. USA* **91** (1994), 5466-5470.
6. Senior, M.L and Heun, M.: Mapping maize microsatellites and polymerase chain reaction conformation of the targeted repeats using a CT primer. *Genome* **36** (1993), 884-889.
7. Dow, B.D., Ashley, M.V., and Howe, H.F.: Characterization of highly variable $(GA/TC)_n$ microsatellites in the bur oak, *Quercus macrocarpa. Theor. Appl. Genet.* **91** (1995), 137-141.
8. Wu, K.S. and Tanksley, S.D.: Abuncance, polymorphism and genetic mapping of microsatellites in rice, *Molec. Gen. Genet.* **241** (1993), 225-235.
9. Morgante, M. and Olivieri, A.M.: PCR-amplified microsatellites as markers in plant genetics. *The Plant Journal* **3** (1993), 175-182.
10. Akkaya, M.S., Bhagwat, A.A., and Cregan, P.B.: Length polymorphisms of simple sequence repeat DNA in soybeans. *Genetics* **132** (1992), 1131-1139.
11. Thomas, M.A. and Scott, N.S.: Microsatellilte repeats in grapevine reveal DNA polymorphisms when analyzed as sequence-tagged sites (STSs). *Theo. Appl. Genet.* **86** (1993), 985-990.
12. Bell, C.J., and Ecker, J.R.: Assignment of 30 microsatellite loci to the linkage map of *Arabidopsis. Genomics* **19** (1994), 137-144.
13. Terauchi, R.A., and Konuma, K.: Microsatellite polymorphism in *Dioscorea tokoro*, a wild yam species. *Genome* **37** (1994), 794-802.
14. Smith, D.N. and Devey, M.E.: Occurence and inheritance of microsatellites in *Pinus radiata. Genome* **37** (1994), 977-983.
15. Terauchi, R.A.: Polymorphic microsatellite marker from the tropical tree *Dryobalanops lanceolata* (Dipterocarpaceae). *Jpn. J. Genet.* **69** (1994), 567-576.
16. Doyle, J.J. and Doyle, J.L.: Isolation of plant DNA from fresh tissue. *Focus* **12** (1990), 13-15.
17. Dumolin, S.B., Demesure, B., and. Petit, R.J.: Inheritance of chloroplast and mitochondrial genomes in pedunculate oak investigated with an efficient PCR-based method. *Theo. Appl. Genet.* (in press).

GENETIC DISSECTION OF PHENOTYPIC TRAITS IN MARITIME PINE USING RAPD AND PROTEIN MARKERS

C. PLOMION,[1,2] N. BAHRMAN[1], C-E. DUREL,[1]
D.M. O'MALLEY[2], A. KREMER[1]
[1]INRA, Station de Recherches Forestières, BP45, 33610 Cestas, France and [2]Forest Biotechnology Group, North Carolina State University, Box 8008 Raleigh, NC 27695 USA

Abstract Random Amplified Polymorphic DNA (RAPDs) were used for genetic mapping and QTL detection in an F2 selfed progeny of maritime pine. The seedlings were raised for two years in a greenhouse under accelerated growth conditions. Total height was measured at different maturation stages and corresponding QTLs were located in different genomic regions, providing an explanation for the absence of juvenile-mature correlation for height. Within each shoot cycle, QTL for traits related to the initiation of foliar primordia in the bud and the elongation of the stem were mapped in different chromosomes, suggesting that the activity of the apical and subapical meristems was controlled by separate genetic mechanisms.
Rapid mapping rely on the use of arbitrary primers and PCR-based technique to amplify polymorphic DNA fragments. These markers are anonymous and may be physiologically irrelevant if one want to understand the biological meaning of the detected QTLs. As part of our mapping programme, we are currently developping a map based on protein markers reveal by 2-D PAGE. Such translated regions of the genome might be very useful for the characterization of QTLs.

1. Introduction

Perennial organisms such as forest trees pass through a series of developmental phases during their life (Poethig, 1990). These phase changes are related to profound morphological, biochemical and physiological changes, mostly associated with maturation of the plants (reviewed by Haffner *et al.*, 1991). In addition to the obvious transition from vegetative to reproductive structures, progressive changes in vegetative characteristics occur during the lifetime (Hackett and Murray, 1993). In *Pinus* species, seedlings produce two kinds of stem units during the first two years of growth. In the first season, elongation of the stem occurs in the free growth phase, which ends when a terminal bud is produced during the second season. In the following seasons, the plant produces one or two morphogenetic cycles. The transition between free growth and fixed growth is an important change in *Pinus* species. Acceleration of this phase change and hastening of seedlings maturation under optimal environments has been used to identify early selection criteria of adult performance in the first season of growth (Lascoux *et al.*, 1993a). In particular, Lascoux *et al.* (1992; 1993b)

M. R. Ahuja et al. (eds.), Somatic Cell Genetics and Molecular Genetics of Trees, 223–231.
© 1996 *Kluwer Academic Publishers. Printed in the Netherlands.*

investigated the ontogenetical aging of maritime pine seedlings exposed to continuous light. Their results showed that first season seedlings raised under 18 weeks of continuous light moved rapidly from a free to a cyclic growth and exhibited morphological traits characteristic of mature trees. In addition, they showed that seedlings demonstrated a degree of maturation proportional to the number of morphogenetic cycles achieved during this period.

Although morphological changes have been well described (Romberger and Gregory, 1977), the physiological, biochemical and genetic basis of maturation processes remain elusive (Hackett, 1985; Poethig, 1990). Investigations were carried out to study the genome expression at distinct developmental stages. A differential expression of the same genes has usually been observed in juvenile and mature phases, few or no genes being found that were expressed exclusively at a single stage (*e.g.*, Greenwood *et al.*, 1989; Stabel *et al.*, 1990; Hutchison and Greenwood, 1991). In this paper, we use genetic mapping and quantitative trait dissection analysis (reviewed by Tanksley, 1993) to localize the genetic factors (Quantitative Trait Loci, QTLs) that control the variation of height growth during the maturation of young maritime pine seedlings. The first objective is to study the stability of QTLs for total height at different developmental stages: young stage (before applying an accelerating growth treatment) and more mature stages (after one and two periods of growth under continuous light). The second objective is to decompose the composite trait (*i.e.*, total height) into elementary components (number of stem units and mean stem unit length of each shoot cycle) and to determine whether the same or different QTLs control these traits under accelerated growth conditions.

2. Materials and Methods

2.1. PLANT MATERIAL AND GROWTH CONDITIONS

One hundred and ninety two selfed seeds were obtained from one hybrid tree (accession 'H12') between Corsican (accession 'C10') and Landes (accession 'L146') provenances of maritime pine. They were raised for two years in a greenhouse under controlled conditions. Fifteen weeks after the sowing date, a first continuous light treatment (natural light during the day and artificial light during the night with an average photon flux density of 150 μmol $m^{-2}s^{-1}$ at seedling level) was applied for 23 weeks during the first growth period (GP1). An second continuous light treatment was applied for 21 weeks during the second growth period (GP2).

2.2. LINKAGE ANALYSIS

A genetic map of 'H12' (Fig. 1) was constructed with RAPD (Random Amplified Polymorphic DNA, Williams et al. 1990) markers segregating in the expected 1:1 (presence:absence) Mendelian ratio in haploid megagametophytes of the F2 selfed progeny (Plomion *et al.*, 1995a; 1995b). This "haploid map" consisted on 436 RAPD markers from which 100%, 46% and 28% were assayed on 62, 94 and 126 megagametophytes, respectively. Linkage analysis were performed with the MAPMAKER software (Lander *et al.*, 1987).

2.3. QUANTITATIVE TRAITS MEASURED

Total height was measured before the first accelerated growth treatment was applied (H_0), after one (H_1) and two (H_2) periods of growth under continuous light. At the end of GP2, the plants were uprooted and height growth components measured for each morphogenetic cycle achieved during the second growth period. Components measured were: length of the fertile zone (LF_j) defined as the section of the shoot between the first secondary needle and the top of the cycle, number of stem units in the fertile zone (NSU_j, *i.e.* number of primary and secondary needles in the fertile zone) and mean stem unit length ($MSUL_j=LF_j/NSU_j$). In order to evalute pure growth independently from NSU_j, the adjusted variable LF_j^{ad} to NSU_j was derived from covariance analysis. The biological meaning of such transformed variable is not straightforward. Height growth of a shoot is the result of initiation and elongation of internodes. NSU is the result of initiation of foliar primordia in the bud. Adjusted data represent the growth potential of the shoot, initiation being removed.

2.4. MARKER-TRAIT ASSOCIATIONS

A subset of 120 evenly spaced RAPD markers (10 to 30 cM) assayed on 126 megagametophytes were used for quantitative trait dissection analysis with the MAPMAKER/QTL software (Lander and Botstein, 1990). A LOD score > 1.6 corresponding to a type I error rate of 0.0045 at the marker level was used for QTL detection. With this threshold, 0.5 false positive QTL on the average was expected to be found by chance for a given trait. Actually the markers are not independent, so this number is overestimated.

3. Results

3.1. QTLs FOR TOTAL HEIGHT

Associations were found between RAPD markers and all traits. One to three QTLs were mapped per trait (Fig. 1). For H_0 (total height before applying the first accelerated growth treatment), two QTLs were found in groups 2 and 3 and they accounted for 7% and 12% of the phenotypic variation of the trait, respectively. A multilocus model explained 17% of the variation. For H_1 (total height measured after one period of growth under continuous light) three QTLs were found in groups 5, 6 and 7. They accounted for 6.2%, 11.5% and 6.5% of the phenotypic variation of the trait, respectively and a multilocus model explained 21% of the variation. For H_2 (total height measured after two periods of growth under continuous light) one QTL that accounted for 10% of the phenotypic variation was detected on linkage group 9. Significant correlations were observed between H_0 and H_1 ($\rho = 0.30^{**}$) and between H_1 and H_2 ($\rho = 0.49^{***}$). Less correlated variables ($-0.01<\rho<0.21^*$) were derived from these traits: height increments ($\Delta H_{0,1}$ and $\Delta H_{1,2}$) and adjusted data H_1^{ad} (residual of H_1 adjusted to H_0, following the ANCOVA model: $H_{1i}=\mu_{1i}+\beta_1.H_{0i}+\varepsilon_i$) and H_2^{ad} (residuals of H_2 adjusted to H_1, following the ANCOVA model $H_{2i}=\mu_{2i}+\beta_2.H_{1i}+\varepsilon'_i$). The same QTLs were mapped for H_1, $\Delta H_{0,1}$ and H_1^{ad} and for H_2, $\Delta H_{1,2}$ and H_2^{ad}.

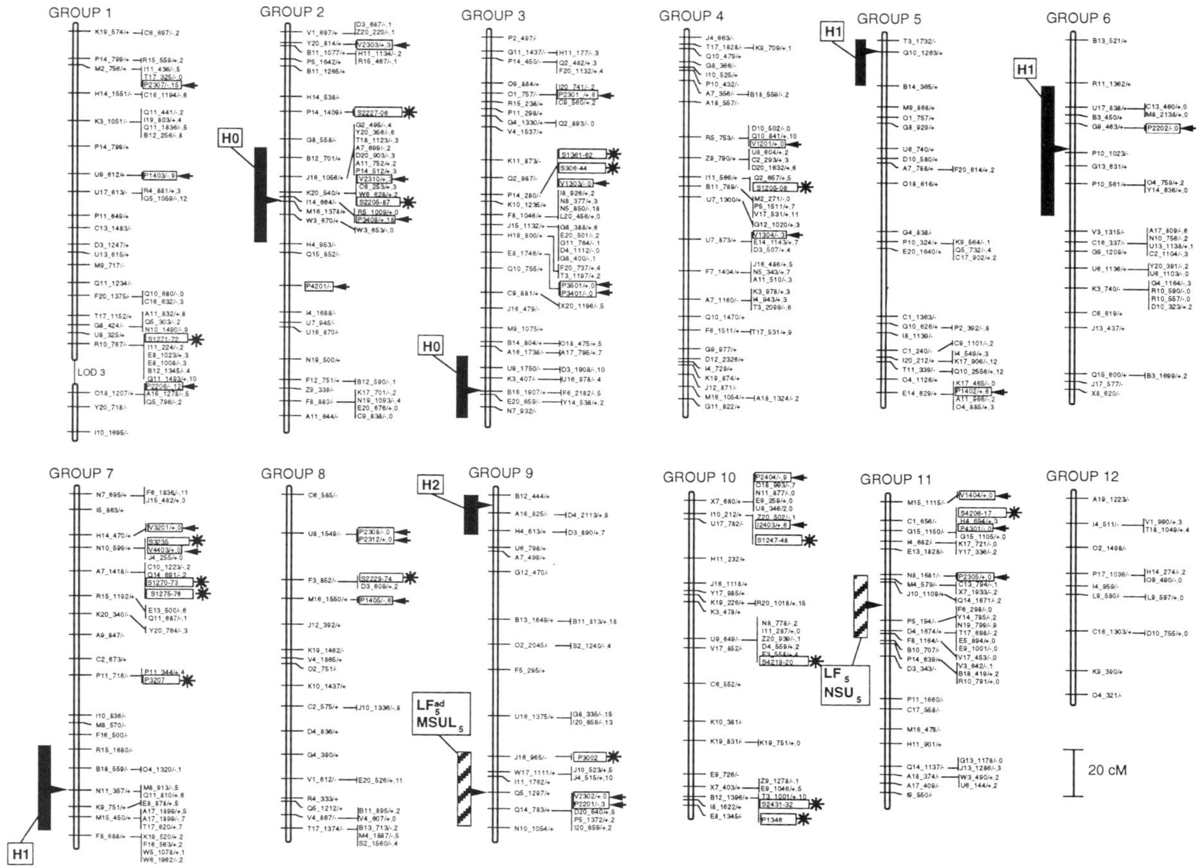

Figure 1 Genetic linkage map of maritime pine (12 linkage groups established with a LOD ≥ 5.0 and θ ≤ 0.30). Loci are named according to the nomenclature of Plomion et al. (1995b). The most likely position of a QTL is indicated by a black arrow. Boxes to the left of linkage groups represent the LOD 1.0 interval support of the putative QTLs. Filled boxes indicate QTLs for total height and hatched boxes indicate QTLs for height growth components of cycle 5. For the definition of the traits, see Materials and Methods. Protein markers are squared. Those assayed on megagametophytes are indicated by arrows whereas those assayed on needles are indicated by stars.

3.2. QTLs FOR HEIGHT GROWTH COMPONENTS

Two types of variables to be used for QTL analysis could be selected from the study of correlation coefficients between height components: components more involved in the initiation (NSU and LF) and components more involved in the elongation of the shoot (MSUL and LF^{ad}). For the four selected traits (NSU, LF, MSUL, LF^{ad}), permutation tests for association (Dietz, 1983) showed that correlation matrices did not significantly differ between the six cycles of GP2. Thus it was decided to present the QTL results for the most demonstrative cycle: cycle 5. The QTLs detected for the other cycles will be further discussed. One QTL effect was found for each growth component of cycle 5. QTLs for variables NSU_5 and LF_5 were mapped in group 11 (Fig. 1) and accounted for 20.4% and 19.6% of the phenotypic variation, respectively. QTLs for variables $MSUL_5$ and LF^{ad}_5 were mapped in group 9 and accounted for 6.0% and 11.8% of the phenotypic variation, respectively.

4. Discussion

4.1. HEIGHT GROWTH QTLs STABILITY OVER DEVELOPMENTAL STAGES

The first apical budset was observed about 50 days after applying the continuous light treatment. Total height of the plants was measured before this first phase change (H_0). This first measurement corresponded to a juvenile stage. The second measurement was completed after the first "light" treatment (H_1). The plants presented an average of 2 morphogenetic cycles interrupted by a nonresting terminal bud. Finally, the seedlings were measured after a second "light" treatment (H_2) and presented an average of 8 cycles. Thus, H_0, H_1 and H_2 corresponded to three distinct stages of development. Non-accelerated control seedlings presented the same characteristics at H_0, whereas they were on a free growth at H_1 and presented only two shoot cycles at H_2.

A differential QTL expression during the plant development was observed for total height. None of the QTL significant at one stage (either H_0, H_1 or H_2) was also significant at an other stage. This result supports the idea that different loci are likely to be involved in the genetic control of height growth at different ages. However, two kinds of QTLs could be observed: genomic regions showing a progressive expression during the plant growth and genomic regions strongly expressed only at one of the three stages of the development. Thus, rather than a complete change in QTL expression, maturation could also induce a progressive shift of the genetic control of height growth. In larch, Greenwood *et al.* (1989) and Hutchison *et al.* (1990) showed that changes in morphological and physiological phenotypes due to maturation were also associated with changes in gene expression in developing foliage. However, cross hybridization of mRNA suggested that either uniquely expressed transcripts must be present at a very low levels or that maturation did not arise from the induction of new maturation specific genes or repression of juvenile specific genes (Greenwood *et al.*, 1989; Hutchison *et al.*, 1990).

Finally, if the same metabolic pathways are assumed to be responsible for height growth in both juvenile and more mature developmental stages, our results suggest that different regulatory genes or differential expressions of the same set of regulatory genes are involved at different stages of maturation. This epigenetic variation (*i.e.* expression of phase-dependent genetic factors within an individual) provides an explanation for the low or absence of juvenile-mature correlation for growth performance in forest tree species (Kremer *et al.*, 1991; Greenwood and Volkaert, 1992).

4.2. QTLs FOR HEIGHT GROWTH COMPONENTS

For cycle 5, QTLs affecting initiation (NSU and LF) or elongation (MSUL and LF^{ad}) were mapped on different chromosomes (Fig. 1). For the five other cycles, QTL analysis was also performed for variables NSU, LF, MSUL and LF^{ad}. A total of 21 significant QTLs were identified. As expected, the clustering of QTLs for traits mostly involved in initiation or elongation were consistent with the high correlations among the traits. These ranged from 0.78*** to 0.90*** between MSUL and LF^{ad} and from 0.76*** to 0.92*** between NSU and LF. In addition, no clear correspondence of QTL location was observed for the same traits across shoot cycles. This result agree with the low correlations for each components across cycles. The mapping results concerning traits related to both phenomena agreed with those of Lascoux *et al.* (1993b; 1994). These authors found low phenotypic correlations and variable family mean correlations between NSU and MSUL of different cycles. Leaf initiation (NSUL) is a function of apical meristem activity, whereas internode length (MSUL) is function of subapical meristem activity (Lanner, 1976; Kremer, 1984). The different QTL locations of NSU, LF (initiation-related traits) on the one hand, and MSUL, LF^{ad} (elongation-related traits) on the other hand, for each shoot cycle suggests the occurrence of specific genetic factors controlling the activity of both meristems in accelerated growth conditions.

At this point, it is difficult to distinguish whether the differential expression of QTLs controlling (i) the variation of the total height or (ii) the variation of the same components across morphogenetic cycles, are related to maturation effects, environmental changes or a combination of both. Indeed, environmental conditions have changed during the two years of growth and could potentially create QTL by environment interactions (natural/artificial light ratio and temperature being progressively different for each cycle). However, for maritime pine seedlings raised under accelerated growth conditions in a phytotron, Lascoux *et al.* (1992; 1993a; 1993b) also observed the appearance of morphological traits characteristic of mature trees. Thus, it is probable that maturation effects were predominant in our experiment.

4.3. LINKAGE MAPPING AND QTL ANALYSIS WITH HAPLOID MEGAGAMETOPHYTES

The limitation imposed by the dominant mode of inheritance of RAPD markers for genetic mapping in a F2 progeny was resolved by using the megagametophyte (haploid tissue in seeds that is derived from the same megaspore that gives rise to the maternal gametes) of each F2 individual. Each megagametophyte was collected from each seedling two weeks after germination just before the seed coat would be cast off.

Dominant RAPD markers are not efficient for QTL analysis in a F2 progeny because additive and dominant effects cannot be tested separately. In our experiment, only the maternal contribution (megagametophyte) was determined for each F2 individual. Hence, only additive effect of a putative QTL could be estimated whereas the dominance deviation could not be determined.

4.4. LIMITS OF THE "MEGAGAMETOPHYTE-RAPD" APPROACH AND FUTURE DIRECTIONS

The main limit of the "megagametophyte approach" is that it requires the development of specific populations and is not applicable to analyze QTL for economically important traits in existing plantations. Indeed, the megagametophyte is a temporary tissue that can only be collected from seedlings during germination of the embryo. Conversely, conifer trees can be grafted and propagated by cuttings, and diploid progenies can constitute "perpetual" populations, analogous to recombinant inbred lines in crop plants. Screening RAPD primers for informative markers which segregate 1:1 in diploid tissue of angiosperms or gymnosperms hybrid progenies has already proved to be a very powerfull strategy for establishing single tree maps with diploid tissue (Grattapaglia and Sederoff, 1994). In addition, because high amount of DNA can be extracted from diploid tissue, genetic analysis using the sporophyte will make it possible to gather different types of molecular markers (*e.g.*, RFLPs, microsatellites, AFLPs) and to map them on a single linkage map.

QTL mapping models are purely based on statistical association between traits and molecular markers. QTLs have been identified that account for variation in many phenotypic traits, yet there has been little analysis of the relationship of known genes with those QTLs (*e.g.,* Touzet *et al.*, 1995). The accessibility of the biological meaning of a QTL can proceed from co-location between QTL and qualitative mutations (*e.g.*, Beavis *et al.*, 1991), between QTL and known-function genes (Goldman *et al.*, 1994), or between QTL of both agronomic and biochemical traits (Causse et al. 1995). In maritime pine, we already demonstrated (Plomion *et al.*, 1995b) that most RAPD markers amplified from highly repetitive chromosomal regions (*i.e.*, mostly non coding DNA). Therefore, RAPD markers appear to be limited to define the biological meaning of a putative QTL. Conversely, the development of maps based on expressed genes (*e.g.*, anonymous proteins revealed by 2-D PAGE or RFLPs generated from a cDNA library, performe marker-trait associations, sequence those protein or DNA markers that fall within a putative QTL, and search for homology in gene databanks, should allow testing the relationships between genetic variation of quantitative traits and known-function genes. We are currently investigated such a " candidate gene " approach using protein markers.

A total of 150 RAPD markers distributed all over the "haploid map" was amplified from diploid tissue (needle) of 130-192 F2 seedlings. From these 192 plants, 68 that were genotypes with RAPD markers in both haploid and diploid phases were also genotyped with protein markers. Total protein extraction from needles tissue and two-dimensional gel electrophoresis (isoelectric focusing and SDS-PAGE dimensions) were performed following Bahrman and Petit (1995). Significant linkages (LOD>2.5) were established between these RAPD markers segregating 3:1 and 13 codominant and 4 dominant proteins markers segregating 1:2:1 and 3:1, respectively. The approximate position of these 17 proteins in the " haploid map " is indicated in figure 1. An other set of 27 proteins assayed on megagametophytes was also localized in the same map (Plomion *et al.*, 1995b). Since these 44 proteins were found to be well distributed in the genome, we expect that the development of a map composed of expressed sequences will help us to better understand the genetic basis of quantitative variations.

5. References

Bahrman, N. and Petit, R. (1995) Genetic polymorphism in maritime pine (*Pinus pinaster* Ait.) assessed by two-dimensional gel electrophoresis of needle, bud and pollen proteins. *J. Mol. Evol.* **41**, 231-237.

Beavis, W.D., Grant, D., Albertsen, M., and Fincher, R. (1991) Quantitative trait loci for plant height in four maize populations and their associations with qualitative genetic loci. *Theor. Appl. Genet.* **83**, 141-145.

Causse, M., Rocher, J.P., Henry, A.M., Charcosset, A., Prioul, J.L., and de Vienne, D. (1995) Genetic dissection of the relationship between carbon metabolism and early growth in maize, with emphasis on key-enzyme loci. *Molecular Breeding* **3**, 259-272.

Dietz, E.J. (1983) Permutation tests for association between two distance matrices. *Syst. Zool.* **32**, 21-26.

Goldman, I.L., Rocheford, T.R., and Dudley, J.W. (1994) Quantitative trait loci influencing protein and starch concentration in the Illinois long term selection maize strains. *Theor. Appl. Genet.* **87**, 217-224.

Grattapaglia, D., and Sederoff, R. (1994) Genetic linkage maps of *Eucalyptus grandis* and *E. urophylla* using a pseudo-testcross mapping strategy and RAPD markers. *Genetics* **137**, 1121-1137.

Greenwood, M., Hopper, C.A., and Hutchison, K.W. (1989) Maturation in larch. I. Effect of age on shoot growth, foliar characteristics and DNA methylation. *Plant Physiol.* **90**, 406-412.

Greenwood, M.S., and Volkaert, H.A. (1992) Morphophysiological traits as markers for the early selection of conifer genetic families. *Can. J. For. Res.* **22**, 1001-1008.

Hackett, W.P. (1985) Juvenility, maturation and rejuvenation in woody plants. *Hort. Reviews* **7**, 109-155.

Hackett, W. P., and Murray, J. R. (1993) Micropropagation of woody plants, in: M.R. Ahuja (ed.), *Maturation and Rejuvenation in Woody Species*, Kluwer Academic Publishers, Dordrecht, Netherlands, pp. 93-105.

Haffner, V., Enjarlic, F., Lardet, L., and Carron, M.P. (1991) Maturation of woody plants: a review of metabolic and genomic aspects. *Ann. Sci. For.* **48**, 615-630.

Hutchison, K.W., Sherman, C.D., Weber, J., Smith, S.S., Singer, P.B., and Greenwood, M.S. (1990) Maturation in larch. II. Effect of age on photosynthesis and gene expression in developping foliage. *Plant Physiol.* **94**, 1308-1315.

Hutchison, K.W., and M.S. Greenwood, (1991) Molecular approaches to gene expression during conifer development and maturation. *For. Ecol. Manage.* **43**, 273-286.

Kremer, A. (1984) Distribution of relative growth rates and variation of cytohistological zonation in apical meristems of seedlings of two contrasting open - pollinated jack pine (*Pinus banksiana*) families. *Can. J. For. Res.* **14**, 297-310.

Kremer, A., Lascoux, M., and N'Guyen, A. (1991) Morphogenetic subdivision of height growth and early selection in maritime pine , in : Proc. 21st Southern Forest Tree Improvement Conference. Knoxville, Tennessee, pp. 203-221.

Lander, E.S., Green, P., Abrahamson, J., Barlow, A., Daly, M.J., Lincoln, S.E., and Newburg, L. (1987) Mapmaker: an interactive computer package for constructing primary genetic linkage maps of experimental and natural populations. *Genomics* **1**, 174-181.

Lander, E.S., and Botstein, D. (1989) Mapping Mendelian factors underlying quantitative traits using RFLP linkage maps. *Genetics* **121**, 185-199.

Lanner RM (1976) Pattern of shoot development in *Pinus* and their relationship to growth potential, in M.G.R. Cannell and F.T. Last (eds.), *Tree Physiology and Yield Improvement*, Academic press, pp. 223-289.

Lascoux, D.M., Dormling, I., and Kremer, A. (1992) Germination and phenology of 1-year-old maritime pine (*Pinus pinaster* Ait.) seedlings under continuous light. *Trees* **7**, 48-58.

Lascoux, D.M., Kremer, A., and Dormling, I. (1993a) Growth and phenology of 1-year-old maritime pine (*Pinus pinaster*) seedlings under continuous light: implication for early selection. *Can. J. For. Res.* **23**, 1325-1336.

Lascoux, D.M., Novitol Paino, E., Sierra de Grado, R., Kremer A., and Dormling, I. (1993b) Maturation of maritime pine (*Pinus pinaster* Ait.) seedling after exposure to a period of continuous light. *Tree Physiol.* **12**, 363-378.

Lascoux, M., Kang, H., and Lundkvist, K. (1994) Growth of 24 full-sib families of *Pinus sylvestris* L. at six relative nutrient addition rates. *Scand. J. For. Res.* **9**, 115-123.

Plomion, C., O'Malley, D.M., and Durel, C-E. (1995a) Genomic analysis in maritime pine (*Pinus pinaster*). Comparison of two RAPD maps using selfed and open-pollinated seeds of the same individual. *Theor. Appl. Genet.* **90**, 1028-1034.

Plomion, C., Bahrman, N., Durel, C-E., and O'Malley, D.M. (1995b) Genomic mapping in *Pinus pinaster* (maritime pine) using RAPD and protein markers. *Heredity* **74**, 661-668.

Poethig, R.S. (1990) Phase change and the regulation of shoot morphogenesis in plants. *Science* **250**, 923-930.

Romberger, J.A., and Gregory, R.A. (1977) The shoot apical ontogeny of the *Picea abies* seedling. III. Some age-related aspects of morphogenesis. *Am. J. Bot.* **64**, 622-630.

Stabel P, Eriksson T, Engström P (1990) Changes in proteins synthesis upon cytokinin - mediated adventitious bud induction and during seedling development in norway spruce, *Picea Abies. Plant* Physiol. **92**, 1174-1183.

Tanksley, S.D. (1993) Mapping polygenes. *Ann. Rev. Genet.* **27**, 205-233.

Touzet, P., Winkler R.G., and Helentjaris T. (1995) Combined genetic and physiological analysis of a locus contributing to quantitative variation. *Theor. Appl. Genet.* **91**, 200-205.

Williams, J.G.K., Kubelik, A.R., Livak, K.J., and Rafalski, J.A. (1990) DNA polymorphisms amplified by arbitrary primers are useful as genetic markers. *Nuc. Acids. Res.* **18**, 6531-6535.

POLYMORPHIC SIMPLE SEQUENCE REPEATS IN NUCLEAR AND CHLOROPLAST GENOMES: APPLICATIONS TO THE POPULATION GENETICS OF TREES

M. MORGANTE[1], A. PFEIFFER[1], A. COSTACURTA[1], A.M. OLIVIERI[1], W. POWELL[2], G.G. VENDRAMIN[3], J.A. RAFALSKI[2]

[1]Dipartimento di Produzione Vegetale e Tecnologie Agrarie, Universita' di Udine, Via delle Scienze 208, I-33100 Udine, Italy; [2]Biotechnology Research, Agricultural Products Department, E.I. du Pont de Nemours and Co (Inc.), P.O. Box 80402 Experimental Station, Wilmington, DE 19880-0402; [3]Istituto Miglioramento Genetico Piante Forestali, C.N.R., Via Atto Vannucci 13, I-50134 Firenze, Italy

1. Abstract

Molecular markers, to be used in the population genetics of forest trees, should ideally allow easy and fast genotyping, be codominant, reproducible over time and space, be fully transferable among labs and have a high information content. None of the available molecular marker systems fully meets all these requirements. We decided to focus on Simple Sequence Repeats (SSRs or microsatellites) because they are codominant, reproducible, highly informative and easy to exchange even though they have high development costs. We isolated AC/GT and AG/CT SSRs from the Norway spruce (*Picea abies* K.) nuclear genome. Based on hybridization data we estimate that there is one AC SSR every 500-600 kb and one AG SSR every 200-300 kb. Given the genome size of spruce (30-40 pg) this corresponds to an extremely large number of SSRs available for analysis. We isolated several hundreds positive clones from a small-insert genomic library and following sequence analysis we designed primers for 36 of them, 24 containing AG and 12 AC SSRs. Twenty-two percent of the primer pairs produced a single-locus hypervariable pattern, with the remaining ones giving either a single monomorphic product (22%) or very poor amplification (19%) or amplification of multiple bands (37%). Mendelian segregation was demonstrated for all the loci amplified with the primer pairs giving a simple variable pattern. We screened a panel of 18 spruce trees at these loci. The average number of alleles per locus was 14 and expected heterozygosity 0.80, with up to 23 alleles per locus and heterozygosities exceeding 0.94. This shows that nuclear SSRs can be very useful markers for the population genetics of conifers even though the overall efficiency of the marker identification process is quite low due to the high complexity of the spruce genome.

We recently extended the use of SSR markers to the chloroplast genome. We demonstrated that mononucleotide poly(A/T) stretches are frequent in the chloroplast genomes of plants and show high levels of between and within population variation. This makes them ideal tools for cytoplasmic population genetics and overcomes the difficulties in finding within species variation that are frequently encountered when analysing the cpDNA molecule by RFLPs or PCR-RFLPs. We identified a set of 20 cpSSRs for the analysis of pine species and will discuss the possible applications of such markers for studying gene flow, plant population dynamics and history and for paternity analysis.

2. Molecular markers in the genetics of forest trees

Isozyme markers have proven their utility as genetic markers for the analysis of genetic variability, mating system and gene flow in forest trees and have been applied to a variety of problems in population and ecological genetics. However only a limited number of isozyme loci are available and only coding regions of the genome can be assayed. Many different DNA marker systems have been developed and used in the last years. Some of the most commonly used systems in plants (see Rafalski *et al.* 1996 for a detailed technical description of each marker type) are RFLPs (Botstein *et al.* 1980), Random Amplified

M. R. Ahuja et al. (eds.), Somatic Cell Genetics and Molecular Genetics of Trees, 233–238.
© 1996 *Kluwer Academic Publishers. Printed in the Netherlands.*

Polymorphic DNAs (RAPDs, Williams *et al.* 1990), Simple Sequence Repeats (SSRs or microsatellites, Weber and May 1989), Cleaved Amplified Polymorphic Sequences (CAPS or PCR-RFLP, Konieczny and Ausubel 1993), Amplified Fragment Length Polymorphisms (AFLPs, Zabeau and Vos 1993). Each one of these marker systems offers a unique combination of advantages and disadvantages (reviewed in detail in Rafalski and Tingey 1993), and differ in the type of sequence polymorphisms detected (insertion/deletions vs. point mutations), the information content, the dominance relationships between alleles (dominant vs. codominant markers), the amount of DNA required, the need for DNA sequence information in the species under analysis, the development costs, the ease of use and the extent to which they can be automated. The choice of a marker system is to a significant extent dictated by the specific application and there is probably not a single class of markers that can satisfy all the needs encountered by plant geneticists and breeders.

Molecular markers do not yet have a widespread application in population and ecological genetics of forest trees because technology has not developed to be able to satisfy all the requirements for their successful use. Because very large numbers of individuals usually have to be screened, speed, ease of use, low costs and the feasibility of automating the molecular assay are primary criteria to select the appropriate marker system. Furthermore, markers should be codominant because estimation of the levels of heterozygosity at the individual, population and species level is one of the key issues in the population genetics of trees. This cannot simply be achieved on the basis of the frequency of the homozygotes for the recessive allele under the assumption that populations are under Hardy-Weinberg equilibrium because this is frequently not the case. For Gymnosperms, where the diploid genotype of the mother tree can be inferred by looking at several (6-8) endosperms (haploid megagametophytes) from the same tree, even dominant markers could be used. However, this would increase the total number of markers assays as well as of DNA extractions by a factor of 6-8. Extraction of good quality DNA from forest trees is often difficult and time-consuming because of the presence of polyphenols and polysaccharides. The ideal marker system should therefore not require high quality and high quantities of DNA. When molecular markers have to be applied to mating sytem, paternity or gene flow analysis, where the markers are used to differentiate individuals rather than to assess their degree of relatedness, they should be highly informative, i.e. show a high number of alleles with similar frequencies. A high information content reduces the number of loci that have to be screened. When relationships at the above population level or even more at the subspecies or species level have to be determined, i.e. when genetic relatedness has to be estimated and phylogeny has to be inferred, less informative markers may be preferred, so that the problem of reversion and the possibility of homoplasy due to the high mutation rates of hypervariable markers are avoided. Finally, repeatibility over time and space is a key requirement for molecular markers to be used for monitoring variability in forest trees. This is very important when one considers that such studies frequently require analysing the same population in different years to look at the population dynamics or to compare results obtained by different labs in different populations. The use of a common set of markers that can be easily exchanged would enormously facilitate such tasks. Isozymes have proven so far to be repeatable and allow comparisons between results from different groups both in a qualitative (presence or absence of certain alleles) and quantitative (levels of genetic diversity or differentiation, etc.) way.

Based on the needs that we have just defined, we decided to focus our attention on PCR-based markers that require species-specific sequence information, namely on SSRs or microsatellites. They fully share the advantages of Sequence Tagged Sites (STSs, Olson et al. 1989) and their hypervariability makes them extremely informative codominant markers (Beckmann and Soller 1990). They are thus almost ideal markers for population genetics studies. The effort required for developing the markers should be paid back through the greater utility of such markers and the possibility to efficiently automate the assay. Presence and hypervariability of SSRs has been demonstrated in nuclear genomes of plants (Morgante and Olivieri 1993) and more recently also in the chloroplast genome (Powell *et al.* 1995a, 1995b). Very little information is available on SSRs in conifers (Smith and Devey 1994). We will here present results of our work aimed at identifying and using SSRs for the nuclear and chloroplast genome in conifers, mainly in Norway spruce (*Picea abies* K.).

3. Nuclear SSRs in Norway spruce (*Picea abies* K.)

Due to their greater relative abundance in plant genomes (Morgante and Olivieri 1993) we decided to look at dinucleotide repeats and characterize them in terms of frequency, localization, inheritance and informativeness.

3.1 FREQUENCY OF $(AG)_n$ AND $(AC)_n$ REPEATS IN THE SPRUCE GENOME

Frequency of dinucleotide repeats was estimated by screening a small insert size (200-600 bp) spruce genomic library with radioactively labelled $(AC)_n$ and $(AG)_n$ polynucleotides. 151 positive clones for $(AG)_n$ and 72 for $(AC)_n$ were found among the approximately 65000 plaques screened. Based on an average insert size of 450 bp this means that there are on average one $(AG)_n$ SSR every 194 kb and one $(AC)_n$ SSR every 406 kb in the size fraction of the spruce DNA we used for the library construction. Considering a genome size of 3×10^{10} bp per haploid genome, this corresponds to a total number of 15.5×10^4 $(AG)_n$ regions and 7.4×10^4 $(AC)_n$ regions in the spruce genome.

3.2 CHARACTERIZATION OF SPRUCE SSRs

Thirty-two $(AG)_n$ positives and 14 $(AC)_n$ positives were sequenced among those isolated from the genomic library. The average length of the longest uninterrupted strech of $(AG)_n$ repeats in the $(AG)_n$ SSRs was 20 repeat units, ranging from 9 to 39. Nine of the SSRs were compound, 6 with $(AT)_n$ and 3 with $(AC)_n$ repeats. Primer pairs were generated for 24 of these loci. Primers could not be designed for the remaining ones either because the repeated region was too close to one end of the insert (4 clones) or because they contained sequences very rich in T+C with complex repetitive patterns that appeared to belong to a repeated family (4 clones). $(AC)_n$ repeats were on average 20 units long, ranging from 10 to 46. Two of the SSRs were compound with $(AT)_n$ repeats. Primers could be designed for 12 loci.

The 36 primer pairs that were generated were tested for amplification of the expected fragment and for polymorphism on a panel of 6 trees representing different populations. When fragments other than the expected ones were obtained, PCR conditions were optimised by testing a range of annealing temperatures and Mg^{2+} concentrations. A touchdown PCR profile was also tested but it didn't prove to be useful. After all the optimisation work, 8 (22%) gave a single-locus polymorphic pattern. The other primer pairs amplified either a single monomorphic fragment of the expected size (8, 22%) or fragments of many sizes (13, 37%) or gave poor or no amplification (7, 19%). All of the primer pairs successfully amplified the expected fragment using as a template the DNA of the plasmid from which their sequence was obtained. New primer pairs were designed for 7 SSRs that had previously produced several fragments and tested on the same panel of individuals. No significant improvement in the complexity of the banding pattern was seen, showing that the patterns seen were not due either to bad primer design or bad quality of the sequence data in the region chosen for the primers.

Fragments amplified with some of the primer pairs giving a monomorphic or a complex pattern were Southern blotted on membranes and probed with the corresponding repeat probe to check if they did contain the expected SSR region. All six AG monomorphic fragments gave a positive signal on blots while neither of the AC fragments contained the SSR region. When amplifications from the primers giving multiple bands were tested, in two cases a hybridization signal was seen only in some of the individuals tested. In the remaining five primers pairs tested, SSR-hybridising bands were always seen with sizes that only in some cases corresponded to the ones expected on the basis of the sequence. The positive bands sometimes did not correspond to the most prominent ones on the gel.

In light of our previous experience with SSRs in other plant species (soybean, corn) and of the data reported in the literature, the proportion of primers giving a single-locus polymorphic pattern is low. A low proportion of successful primer paris has been reported for another species with a large genome size, wheat (Roeder *et al.* 1995). Our hypothesis is therefore that the main cause for such a low success rate may be the very high complexity of the spruce genome (30-40 pg DNA per haploid genome) and that many of the SSRs may be embedded in repetitive rather than single copy DNA. If primers were taken from the repetitive sequences, then this would result in either poor amplification or amplification of multiple fragments. Results of the Southern blotting experiments are in agreement with this hypothesis. Repetitive DNA content of the positive clones for which primers were designed was determined by dot blotting insert DNA amplified by PCR with vector primers onto a nylon membrane and probing such a blot with labelled total spruce DNA.

The signal intensity in such an experiment is proportional to the abundance of the sequences contained in each clone in the genome. Weak signals therefore correspond to clones present in single or low-copy number. Primers giving a simple polymorphic pattern are usually obtained from clones giving weak signals on the blot, while clones giving monomorphic or complex pattterns usually correspond to strong signals. This confirms our hypothesis and gives us an easy way to prescreen clones and increase the efficiency of the SSR identification process by selecting clones giving weak signals.

3.3 INHERITANCE AND VARIABILITY OF POLYMORPHIC SSR LOCI

Segregation of polymorphic fragments was checked by analysing single megagametophytes from heterozygous trees for 7 of the 8 single locus polymorphic primers. Observed segregation ratios were always in accordance with a simple codominant Mendelian mode of inheritance. Segregation of polymorphic bands was also looked at for two of the primers pairs giving multiple bands and a band segregating in a simple dominant fashion was observed for both of them. Weak linkage was detected between two of the 7 loci tested for segregation on the same tree (SpAGC1 and SpAGD1, recombination frequency 0.34, $P<0.05$).

In order to assess the level of polymorphism of the identified SSR loci in Norway spruce 18 spruce trees deriving from three Italian and one Swiss populations were screened with the 7 loci for which segregation had previously been tested. This was performed by using fluorescently labelled primers and a Pharmacia A.L.F. automated sequencer (Mansfield *et al.* 1994) together with the Fragment Manager v. 1.1 software (Pharmacia). The advantages of this kind of analysis are the non radioactive detection, the automated sizing of fragments with a 0.2 bp accuracy and the high throughput achievable. Variability data are shown in Table 1. The mean values for the number of alleles per locus and for the expected heterozygosity are much higher than those previously found using isozyme markers on the same populations (Giannini *et al.* 1991; Morgante and Vendramin 1991) where the average number of alleles per locus was 1.78 and expected heterozygosity 0.16. SSR markers therefore appear to be almost five times more informative than isozyme markers, if we look at expected heterozygosity.

TABLE 1. Variability data for 7 SSR loci determined on 18 spruce trees (H_e: expected heterozygosity)

Locus	Repeat type and length	N. of alleles	Range in repeat units	H_e
SpAGD1	$(AG)_{25}$	22	14-54	0.940
SpAC1H8	$(GT)_{27}$	18	9-43	0.912
SpAGG3	$(GA)_{24}$	17	11-30	0.924
SpAGC2	$(TA)_{11}\cdots(GA)_{20}$	11	11-39	0.861
SpAG2	$(TC)_{16}$	10	12-21	0.893
SpAGC1	$(TC)_5\cdots TT(TC)_9$	8	3-22	0.427
SpAC1F7	$(AC)_{12}$	6	8-17	0.569
Mean		13		0.789

4. Chloroplast SSRs

Chloroplast and mitochondrial genomes exhibit different patterns of genetic differentiation compared to nuclear alleles due to their generally uniparental mode of transmission. Analysis of uniparentally inherited genomes can provide a deep insight into the evolutionary histories of populations. Due to the lack of recombination the analysis of molecular lineages is not perturbed and relationships between such lineages can be defined in terms of their phylogenetic divergence. Sequence analysis of the hypervariable mitochondrial control region has allowed detailed analysis of the dynamics, relationships and history of populations in animals and in particular in humans (Brown *et al.* 1979). A corresponding molecular tool has been lacking in plants due to the high rate of structural rearrangements in plant mitochondrial genomes and the high conservation and low mutation rates of chloroplast genomes. Restriction site analysis of chloroplast DNA (cpDNA) has been widely used for inter-specific studies and only in a few cases the magnitude of intra-specific variation has been sufficient to allow population based studies. The availability of a high resolution chloroplast specific polymorphic assay would facilitate the analysis of population differentiation and history and of gene flow. Furthermore, in Gymnosperms, where chloroplasts are primarily paternally transmitted, such an assay would allow pollen movement to be monitored.

We have examined the polymorphism of such sequences in the chloroplast genomes of plants, using a PCR-based assay. GenBank searches identified the presence of several $(dA)_n \cdot (dT)_n$ mononucleotide stretches in chloroplast genomes. A chloroplast simple sequence repeat (cpSSR) was identified in three pine species (*Pinus contorta*, *Pinus sylvestris* and *Pinus thunbergii*), 312 bp upstream of the *psbA* gene. DNA amplification of this repeated region from 11 pine species identified 9 length variants (Powell *et al* 1995a). The polymorphic amplified fragments were isolated and the DNA sequence was determined, confirming that the length polymorphism was caused by variation in the length of the repeated region. Analysis of 305 individuals from 7 populations of *Pinus leucodermis* Ant., an endangered conifer with a restricted and discontinuous distribution in Southern Europe, revealed the presence of 4 variants with intrapopulational diversities ranging from 0.000 to 0.629 and an average of 0.320. Restriction fragment length polymorphism (RFLP) analysis of chloroplast DNA (cpDNA) on the same populations previously failed to detect any variation (Boscherini *et al.* 1994). These data show that chloroplast SSRs exhibit high levels of between and within population variation, making them ideal tools for cytoplasmic population genetics. They allow to overcome difficulties in finding within species variation that are frequently encountered when analysing the cpDNA molecule by RFLPs or PCR-RFLPs. Similar findings were also obtained in an Angiosperm species such as soybean (*Glycine max*) (Powell *et al.* 1995b).

The high sequence conservation of the chloroplast genome makes it easier to design primers that work even in relatively distant species. Taking advantage of the recent availability of the complete chloroplast genome sequence of *Pinus thunbergii* we designed primers that amplify 20 different SSR regions distributed over the whole cpDNA molecule (Vendramin *et al.* 1996) . We amplified succesfully the corresponding regions from three mediterranean pine species, *P. leucodermis*, *P. halepensis* and *P. brutia* and are currently screening different populations for variation. The use of appropriate statistical tools for the analysis of the molecular haplotypes obtained by assaying several hypervariable cpSSRs should achieve similar results to those obtained by sequencing the mitochondrial control region in animals and be of particular importance for reconstructing the history of tree populations in the post-glacial era. This information could in turn be of primary importance for directing the efforts in maximising diversity in programs of breeding and conservation of genetic resources.

5. Conclusions

We have shown that nuclear and chloroplast SSRs fulfill many of the requirements for molecular markers to be applied to the population genetics of trees. Retrieval of useful nuclear SSRs seems to be more time consuming in spruce than in other species possibly as a consequence of the very large size of conifer genomes and the high amount of repeated DNA they possess. More efficient ways to retrieve the desired sequences are, however, being worked out as well as techniques to speed up and automate the assays and the scoring of the individual genotypes. Dissemination of these results and methodologies to the tree population genetics community will be a key issue in determining a widespread utilization of these molecular tools. The use of such tools holds the promise of providing us with a much deeper insight into many of the population and ecological genetics processes of trees that have a profound impact on the dynamics and survival of forests.

6. Acknowledgments

Financial support from the European Commission (BIOTECH project BI02-CT93-0373) and from the North Atlantic Treaty Organization (CRG 940005) are gratefully acknowledged.

7. References

Beckmann J.S. and Soller M. (1990) Toward a unified approach to genetic mapping of eukaryotes based on sequence tagged microsatellite sites. *Bio/Technology* **8**:930-932.

Boscherini G., Morgante M., Rossi P. and Vendramin G.G. (1994) Allozyme and chloroplast DNA variation in italian and greek populations of Pinus leucodermis Ant. *Heredity* **73**:284-290.

Botstein D., White R.L., Skolnick M.H. and Davis R.W. (1980) Construction of a genetic map in man using restriction fragment length polymorphism. *Am. J. Hum. Genet.* **32**:314-331.

Brown W.M., George M. and Wilson A.C. (1979) Rapid evolution of animal mitochondrial DNA. *Proc. Natl. Acad. Sci. USA* **76**:1967-1971.

Giannini R., Morgante M. and Vendramin G.G. (1991) Allozyme variation in Italian populations of Picea abies K. *Silvae Genetica* **40**:160-165.

Konieczny A. and Ausubel F.M. (1993) A procedure for quick mapping of *Arabidopsis* mutants using ecotype-specific markers. *Plant J.* **4**:403-410.

Mansfield D.C., Brown A.F., Green D.K., Carothers A.D., Morris S.W., Evans H.J. and Wright A.F. (1994) Automation of genetic linkage analysis using fluorescent microsatellite markers. *Genomics* **24**:225-233.

Morgante M. and Olivieri A.M. (1993) PCR-amplified microsatellites as markers in plant genetics. *Plant J.* **3**:175-182.

Olson M., Hood L., Cantor C. and Botstein D. (1989) A common language for physical mapping of the human genome. *Science* **245**:1434-1435.

Powell W., Morgante M., Mcdevitt R., Vendramin G.G. and Rafalski J.A. (1995a) Polymorphic simple sequence repeat regions in chloroplast genomes: applications to the population genetics of pines. *Proc. Natl. Acad. Sci. USA* **92**:7759-7763.

Powell W., Morgante M., Andre C., Mcnicol G., Doyle J.J. and Rafalski J.A. (1995b) Hypervariable chloroplast simple sequence repeats provide a general source of polymorphic DNA markers for the chloroplast genome. *Current Biol.* **5**:1023-1029.

Rafalski J.A. and Tingey S.V. (1993) Genetic diagnostics in plant breeding: RAPDs, microsatellites and machines. *Trends Genet.* **9**:275-280.

Rafalski J.A., Morgante M., Powell W., Vogel J.M. and Tingey S.V. (1996) Generating and using DNA markers in plants. In: Birren B and Lai E (Eds.), *Analysis of Non-mammalian Genomes - A Practical Guide*. Academic Press, 75-134.

Roeder M.S., Plaschke J., Koenig S.U., Boerner A., Sorrells M.E., Tanskley S.D. and Ganal M.W. (1995). Abundance, variability and chromosomal location of microsatellites in wheat. *Mol. Gen. Genet.* **246**:327-333.

Smith D.N. and Devey M.E. (1994) Occurrence and inheritance of microsatellites in *Pinus radiata*. Genome **37**:977-983.

Vendramin G.G., Lelli L. and Morgante M. (1996) A set of primers for the amplification of 20 chloroplast microsatellites in *Pinaceae*. *Molecular Ecology*, in press.

Weber J. and May P.E. (1989) Abundant class of human DNA polymorphisms which can be typed using the polymerase chain reaction. *Am. J. Hum. Genet.* **44**: 388-396.

Williams J.G.K., Kubelik A.R., Livak K.J., Rafalski J.A. and Tingey S.V. (1990) DNA polymorphisms amplified by arbitrary primers are useful as genetic markers. *Nucleic Acids Res.* **18**: 6531-6535.

Zabeau M. and Voss P. (1993) Selective restriction fragment amplification: a general method for DNA fingerprinting. European Patent Application 92402629.7 (Publ. Number 0 534 858 A1).

RAPD-SCAR-SSCP: A METHOD TO DETECT MOLECULAR DIFFERENTIATION IN CLOSELY RELATED OAK SPECIES

C. BODENES, S. PRADERE, A. KREMER
INRA , Laboratoire de Génétique et d'Amélioration des Arbres Forestiers, BP 45 33610 Cestas, France

Abstract

Quercus petraea and *Quercus robur* are two closely related interfertile oak species, that showed extremely low genetic differentiation in previous studies, although morphological and ecophysiological differences were reported. We develop a technique to detect genomic regions, evenly rarely dispersed in the genome, involved in molecular interspecific differences. The method is a three step procedure: (1°) random amplification with 10 mers to detect discriminant fragments (2) conversion of informative RAPD fragmenst into SCAR fragments and (3) detection of polymorphisms within SCAR fragments by using SSCP. The method (RAPD-SCAR-SSCP) is applied to the *Q. petraea* - *Q. robur* example with particular emphasis to the construction of SCARs and allelic interpretation of SSCP fragments. Genomic regions differentiating the two species are extremely rare (less than 1% to the total number of fragments screened) and highly polymorphic (13 to 19 alleles for fragmenst varying in size from 450 to 790 bp).

1. Introduction

Sessile (*Q. petraea* (Matt.) Liebl.) and Pedunculate (*Q. robur* L.) oaks are two widespread European species. They are sympatric and generally occupy different but proximal ecological niches. Although leaf and fruit interpecific differences were clearly recognised (Dupouey and Badeau, 1993), the two species exhibit extremely low genetic differentiation (Zanetto *et al.*, 1994; Müller-Starck *et al..*, 1993). Gene diversity studies based either on allozymes (Zanetto *et al.*, 1994), or ribosomal DNA and chloroplast DNA (Petit *et al.*, 1993) show that both species share the same alleles, including the rare alleles. The extremely low differentiation may result from their interfertility. Extensive undirectional hybridization has been shown in a mixed oak stands (Bacilieri *et al.*, 1996) and confirmed by interspecific crosses (Steinhoff, 1993).

The obvious discrepancy of the level of differentiation at the phenotypic and genetic level raises the question of the molecular basis of species differentiation : how are interspecific polymorphic regions distributed in the genome of these two white oaks ? Are they numerous or rare ? Are they widespread or linked in a few groups? Answers to these questions required an intensive prospection of the genome in order to inventory genomic regions that show species differentiation. We applied DNA amplification with randomly amplified fragments (RAPD, Williams *et al.*, 1990) to identify the discriminant regions. As shown by previous experiments, 10 basepair primers allowed to amplify more than 12 fragments among which 10 on average could

M. R. Ahuja et al. (eds.), Somatic Cell Genetics and Molecular Genetics of Trees, 239–248.
© 1996 *Kluwer Academic Publishers. Printed in the Netherlands.*

be analysed (Moreau *et al.*, 1994). In the present study, 250 primers were screened for the identificatrion of discriminant regions. Because most of amplified fragments exhibit dominance, further technical developments were needed to transform dominant into codominant markers. This involved (1) sequencing of the RAPD discriminant fragments, (2) design of specific pairs of primers for target amplification of informative regions (SCARs, Sequence Characterized Amplified Regions, Paran and Michelmore, 1993), and (3) search of polymorphisms within the informative fragments. For the last step, SSCP (Single Strand Conformation Polymorphism, Orita *et al.*., 1989) was adapted to identify polymorphisms in amplified fragments.

The present contribution details the RAPD-SCAR-SSCP approach that was developed for the analysis of molecular differentiation in closely related species as oaks (Figure 1). Since the investigations are still underway, this contribution presents the results for only those loci for which SSCP polymorphism have been investigated in a range wide population survey, but outlines the principles of the approach.

2 Material and methods

2.1. SCREENING OF PRIMERS

Eleven populations of *Q. petraea* and *Q. robur* were sampled throughout the range of the species, from Spain to Poland. These samples were grouped in eleven pairs (one *petraea* and one *robur*) corresponding to a given region (Gallicia, Pyrenees foothills, Brittany, Lorraine, Niedersachsen, Bavaria, Slovenia, Serbia, Southern Poland, Northeastern Poland). Bulk seed collections of these populations were sown in the nursery and 2 trees from each population among the seedlings were used for DNA extractions and RAPD analysis.

DNA was extracted from the buds of one year old seedlings according to Moreau *et al.* (1994) , a method which was adapted from Saghai-Maroof *et al.*(1984). Two hundred and fifty decamers (Operon technologies Inc., Alameda Calif., USA) were utilized for amplification of DNA sequences of the oaks. Amplification protocols were described in Moreau *et al.*, (1994). Fisher's Exact test was used to identify discriminant fragments, e.g. fragments exhibiting significant frequency differences between the two species.

2.2. SINGLE STRAND CONFORMATION POLYMORPHISM (SSCP) ON SCARs

Single Strand Conformation Polymorphism is a technique based on the principle that the single-stranded DNA molecules take on specific conformation, based on secondary structure under non-denaturing conditions.

Single strand conformation polymorphism was analysed for the SCAR fragments. Two µl of PCR products were added to a solution containing 95% formamide, 10 mM NaOH, 0.05% of xylene cyanol, 0.05% of bromophenol blue. The samples were heated to 94°C for 4 minutes, then immediately cooled on ice and loaded onto a 0.75 mm × 16 cm × 18 cm non-denaturating polyacrylamide gel (0.5X MDE; Bioprobe systems, 0.6X TBE). Electrophoresis was run for 14 h at 15 V/cm or 10 V/cm (depending on the size of the fragment) at constant temperature of 15°C in

0.6X TBE running buffer. The gel was then silver stained according to Bassam *et al.* (1991).

2.3. INHERITANCE OF SSCP MARKERS

Inheritance of SSCP markers was tested for three fragments (P14-450, B12-500, U7-790) in three controlled crosses of *Q. robur*. The fragments were denominated according to the Operon decamer of the RAPD primer and their length in base pairs: P14-450 is the SCAR fragment that was detected on an amplification with decamer P14, and which is 450 bp long. Family sizes varied between 26 and 46 offsprings. Validity of crosses was earlier checked with isozymes (Zanetto *et al.*, 1995).

2.4. VALIDATION OF THE INFORMATIVE FRAGMENTS: INTERSPECIFIC SSCPs IN A POPULATION SURVEY

Based on the inheritance study, SSCP alleles were identified in different populations and allelic frequency and diversity statistics calculated. The diversity analysis was done on 8 pairs of populations of *Q. petraea* and *Q. robur*, among the eleven cited in paragraph 2.1, each represented by a sample of 50 individuals. The population survey was done for 3 SSCP loci, the same as for the inheritance study.

3. Results

3.1. SCREENING OF INFORMATIVE FRAGMENTS

A total of 250 decamers were screened to detect RAPD fragments that showed significant frequency differences between the two species. Among the 2800 amplified fragments, 36 (produced by 34 primer pairs) exhibited significant differences of frequency between the two species (Figure 1). These informative primers amplified a total of 412 fragments. Only 7 fragments among the 412 were monomorphic among the 43 trees used for the screening. These results indicate that the two species are extremely variable and that genomic regions involved in the species differentiation represent an extremely reduced part of the genome (about 1 %).

Nucleotide diversity within species and nucleotide divergence among the two species were estimated from these 412 fragments using the method adapted from Nei and Miller (1990). The nucleotide diversity was 0.025 in *Q. petraea* and 0.023 in *Q. robur,* and the nucleotide divergence between the two species was estimated to be 0.005 (Bodénès *et al*, 1996b)

3.2. CONSTRUCTION OF SCARs FOR INFORMATIVE FRAGMENTS

The 36 informative RAPD fragments varied in length from 300 to 1600 base pairs. They were partially sequenced at their extremetics and new primers (20 base pairs in length) were designed based on the DNA sequences. The new primer sites (corresponding to the SCAR fragments) may or may not partially overlap with the RAPD primer sites due to the amplification requirements. In general SCAR primers were a few tens of base pairs apart from the RAPD primer sites.

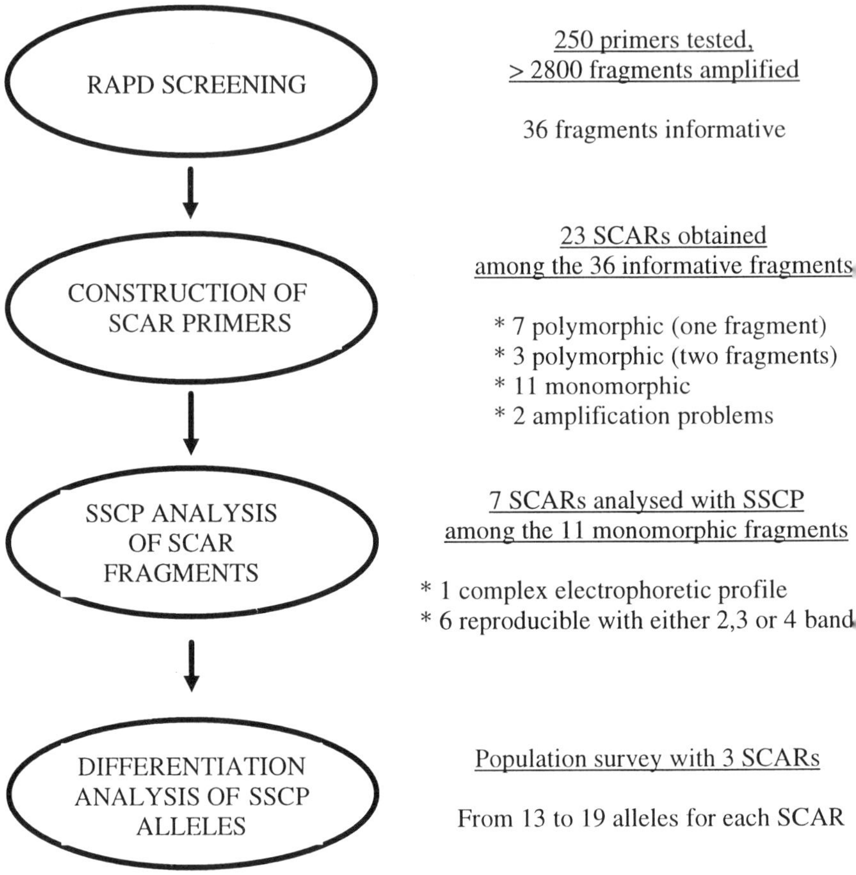

Figure 1 : General outline of the methodology used

SCAR fragments were obtained for 23 fragments among the 36 that were originally selected in the RAPD screening phase. For the 13 remaining fragments, various difficulties were encountered in the cloning or during the sequencing (mixture of different fragments of several sizes) of the RAPD fragments and were further discarded.

Cloned SCAR fragments were used as probes and hybridized with the RAPD fragments, and with the genomic DNA. Hybridization results with RAPD fragments clearly indicated that there was an homology in sequence between informative RAPD fragments in the two species. Hybridization with genomic DNA showed that informative fragments were only poorly repeated in the genome (less than 10 repetitions).

Target amplification of the 43 original samples with the SCAR primers resulted in three different patterns (Figure 1):

In the first pattern, a presence-absence type of polymorphism of one single band was observed which corresponds to a typical PCR dominant marker. Seven SCARs fragments presented this type of polymorphism.

In the second pattern, different samples exhibited either one or two bands, each one showing itself a presence-absence polymorphism. Different genetic controls may be advocated to interpret these results: either the two bands are the expression of two different alleles of a single locus (codominant fragment length polymorphism) or they correspond to two different locus, each one showing a presence-absence polymorphism as in pattern 1. Three SCARs fragments showed length polymorphism between two bands.

Finally in the third pattern, one single fragment was amplified for all samples, resulting in a complete monomorphic electrophoretic profile. PCR fragments that were monomorphic were further analysed with the SSCP technique in order to identify polymorphism within the amplified fragments. Eleven SCARs among the 23 corresponded to this case.

3.3. INHERITANCE OF SSCP FRAGMENTS

The results of the segregation data were presented elsewhere (Bodénès *et al.*, 1996a). We will only report here on the genetic interpretation of SSCP electrophoretic profiles as inferred from the segregation data. The conclusion from these analyses were that a homozygous tree presented 2 bands and a heterozygous tree either 3 or 4 bands. Because DNA is denaturated in the SSCP protocol and because the two strands have different base compositions, two fragments each one corresponding to a strand will be stained for a homozygous tree. A base substitution on the DNA molecule will however not result in the same mobility shift for the two strands. As a result two homozygous trees at different alleles may share a common band. A heterozygous tree may therefore exhibit either 3 fragments or 4. The 3 banded phenotype is the offspring of a cross of two homozygous tree sharing a band, whereas the 4 banded phenotype results from a cross of two trees having no common band (Figure 2).

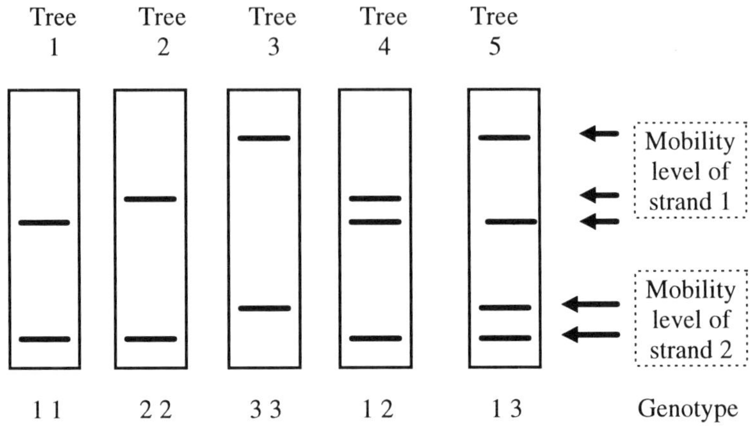

Strand 1 is more mobile than strand 2. Trees 1 and 2 are homozygous and share one band which corresponds to the less mobile strand, whereas tree 4 exhibits two different bands when compared to tree 1 or 2. As a result, heterozygous trees may present either 3 (tree 4) or 4 (Tree 5) bands (Bodénès *et al.*, 1996a.

Figure 2 : Genetic analysis of SSCP electrophoretic profiles

3.4. DIFFERENTIATION ANALYSIS IN A POPULATION SURVEY

Alleles were scored in the population survey according to the genetic interpretation of SSCP profiles (Figure 2). From 13 to 19 alleles were identified among the three SCAR fragments and there was apparently no relationship between the fragment length and the number of alleles (Table 1). As a result, high levels of within population diversity were found in natural oak populations for the SSCP markers. Interspecific differentiation values (G_{ST}, Nei 1987) ranged from 0.028 to 0.078. Allelic frequencies were significantly different between the two species in each of the 8 geographic regions.

Table 1 : Diversity statistics of SSCP fragments

SCAR Fragment	Number of alleles a)	Species differentiation (G_{ST} b)	Mean population diversity in *Q. petraea* c)	Mean population diversity in *Q. robur* c)
B12-500	19	0.078	0.547	0.865
P14-450	13	0.047	0.770	0.685
U7-790	15	0.028	0.716	0.810

a) Total number of alleles identified in 16 populations.
b) Species differentiation values averaged over all geographic regions (G_S, see Nei, 1987)
c) Population diversity averaged over all the 8 populations of *Q. petraea* (or *Q. robur*)

4. Discussion Conclusion

4.1. DETECTION OF MOLECULAR DIFFERENTIATION IN CLOSELY RELATED SPECIES WITH THE RAPD-SCAR-SSCP APPROACH.

We combined three techniques in a single approach to detect molecular differences between the genomes of *Q. petraea* and *Q. robur*, the ultimate objective being twofold; (1) describe the molecular organisation of species differentiation in the genome, and (2) obtain markers for large scale interspecific diversity analysis. In other words, the study aimed at a systematic search of discriminant markers that would allow them to be localized on a genetic map (search of *loci* involved in differentiation) and to be used in population genetics (assessment of *allele* frequency differences).

Amplification with randomly designed primers (RAPD technique) permitted investigation of the entire genome. Based on data of the physical genome size of *Quercus petraea* and *robur* (Ohri and Ahuja, 1990; Favre and Brown, 1995) and assuming that 1 cM corresponds to 400 to 1000 kb in oaks, one may estimate the genetic genome size to vary between 900 and 2250 cM. Since more than 2800 fragments were scored during the screening phase and assuming that these were regularly distributed in the genome, the search of informative fragments was performed at a density varying between 1.24 to 1.67 fragments per cM. Although the intensity of sampling within the genome depends on the level of linkage disequilibrium that exist in the complex of the two species (which is largely unknown), the total number of RAPD fragments that were scored in this study represent a significant part of the genome.

The conversion of informative RAPD fragments into SCARs has resulted in a different focus on molecular differences between the two species. While RAPD polymorphism reveal mutations on the primer sites, the development of SCARs led to the search of polymorphism within the amplified fragment (the original RAPD fragment). This change of focus does not affect the first objective of our study which was to identify *loci* involved in differentiation but may modify the assignment of *alleles* being different between the two species (second objective of the study). The experimental results based on three loci show however that the allele frequency differences were much higher than previously obtained with isozymes (Zanetto *et al.*, 1994), and therefore validate empirically our approach.

The conversion of informative RAPD fragments into SCARs did not always result in codominant markers. In half of the cases (Figure 1), when amplification was done with the SCAR primers, dominance was still present. The primer sites were too variable among the different samples. Hence a preliminary analysis would require the identified regions being highly conserved that could be used to design primers. This has not been undertaken in this study, the first choice of primer was the only one tested.

4.2. SSCP AS A TECHNIQUE TO IDENTIFY POLYMORPHISMS IN PCR PRODUCTS

While SSCP is extensively used to detect all single point mutations on small size fragments (<200 bp) in medical research (Michaud *et al.*, 1992 for an example), we extended the application of the method to the search of polymorphism in larger fragments corresponding to the average size of PCR products (800 bp). It is likely that all point mutations on 800bp long fragments will not result in different conformations, but our objective was not to identify all polymorphisms. The results obtained in the segegation analysis and in the population survey showed that SSCP is a reliable technique to convert monomorphic SCAR fragments into codominant markers. The technique identifies more alleles than the PCR-RFLP method, which is more routinely used for the search of polymorphism in PCR products. From 13 to 19 alleles were identified in the population survey, a number that would require numerous digestion with restriction enzymes in the PCR-RFLP method. Furthermore since many alleles are revealed, the method is better adapted to our second objective which is to assess allele frequency differences between the two species. Ultimately all recorded alleles will be sequenced in order to construct a phylogenetic gene tree that would estimate the molecular divergence between the two species.

However major drawbacks of the technique concern the complexity of electrophoretic profiles in certain cases. Different conformations may be obtained under different experimental conditions as temperature, ionic strength, DNA concentration, glycerol concentration, denaturated agents, concentration of primers in the PCR (Maekawa *et al.*, 1993). SSCP profiles of one SCAR (I14-250) exhibited complex patterns that could not be simplified by varying the experimental conditions.

4.3. MOLECULAR DIFFERENTIATION BETWEEN *Q. PETRAEA* AND *Q. ROBUR*

The RAPD screening procedure clearly confirmed earlier reports indicating that these species are poorly differentiated (Moreau *et al.*, 1994; Zanetto *et al.*, 1994). Although interspecific allele frequency differences assessed by SSCP (Table 1) were low, they exceeded those previously reported by other nuclear markers as isozymes (Zanetto *et al.*, 1994). The large coexistence of the two species in mixed stands and the extensive hybridization (Bacilieri *et al*, 1996) contributes to the maintenance of important interspecific gene flow reducing divergence between the two species. The distribution of the 36 discriminant RAPD fragments (and their corresponding SCARs) in the genome remains still to be analysed by cosegregation analysis in controlled crosses. Future investigations on the molecular basis of differentiation will depend on the number and the distribution of linkage groups of discriminant markers in the genome, if linkage groups are identified. Of particular interested will the relationship between these groups and the QTLs of leaf or fruit morphological traits that exhibit phenotypic differentiation between the two species.

5. References

Bacilieri, R., Ducousso, A., Petit, R., and Kremer, A., (1996) Mating system and asymmetric hybridization in a mixed stand of European oaks. *Evolution* **50** : 900-908

Bassam B.J., Caeteno-Anolles G., Gresshoff P.M., (1991) Fast and sensitive silver staining of DNA in polyacrylamide gel. *Anal. Biochem.* **196**: 80-83.

Bodénès, C., Laigret, F., and Kremer, A., (1996a). Inheritance and molecular variation of SSCP fragments in Pedunculate oak (*Q. robur* L.) *Theor. Appl. Genet.* (In press).

Bodénès, C., Joandet S., Laigret, F., and Kremer A., (1996b). Detection of genomic regions differentiating two closely related oak species *Quercus petraea* (Matt.) Liebl. and *Quercus robur* L. *Heredity* (in press)

Dupouey, J.L., and Badeau, V., (1993) Morphological variability of oaks (*Quercus robur* L, *Quercus petraea* (Matt.) Liebl., *Quercus pubescens* Willd) in northeastern France: preliminary results. *Ann. Sci. For.* **50** (Suppl 1): 35s-40s.

Favre, J.M., and Brown, S. , (1995) A flow cytometric evaluation of the nuclear DNA content and GC% in *Quercus petraea, Q. pubescens* and *Q. robur* (Fagaceae). *Ann. Sci. For.* (In press).

Maekawa, M., Sudo, K., and Kanno, T., (1993) Search for improved electrophoretic conditions for PCR-single-strand conformation polymorphism analysis: is an SDS buffer condition useful? *PCR Methods Appl.* **3**: 130-132.

Michaud , J., Brody, L.C., Steel G., Fontaine, G., Martyin, L.S., Valle, D., and Mitchell, G., (1992). Strand separating conformational polymorphism analysis: efficacy of detection of point mutations in the human ornithine d-aminotransferase gene. *Genomics* **13**: 389-394.

Moreau, F., Kleinschmit, J., and Kremer, A., (1994) Molecular differentiation between *Q. petraea* and *Q. robur* assessed by random amplified DNA fragments. *Forest Genetics* **1**: 51-64.

Müller-Starck, G., Herzog, S., and Hattemer, H.H., (1993) Intra- and interpopulational genetic variation in juvenile populations of *Quercus robur* L. and *Q. petraea* Liebl. (Matt.). *Ann. Sci. For.* **50** (Suppl 1): 25s-34s.

Nei, M., (1987) Molecular evolutionary genetics. Columbia University Press, NY.

Nei , M., Miller ,J.C., (1990). A simple method for estimating average number of nucleotide substitutions within and between populations from restriction data. *Genetics* :**125** : 873-879

Ohri, D., and Ahuja, M.R., (1990) Giemsa C-banded karyotype in *Quercus* L. (oak). *Silvae Genetica* **39**: 216-219.

Orita ,M., Suziki, Y., Sekiya, T., Hayashi, K., (1989) Rapid and sensitive detection of point mutations and DNA polymorphisms using the polymerase chain reaction. *Genomics* **5**: 874 - 879.

Paran, I., and Michelmore, R.W., (1993) Development of reliable PCR-based markers linked to downy mildew resistance genes in lettuce. *Theor. Appl. Genet.* **85**: 985-993.

Petit, R.J., Wagner, D.B., and Kremer, A., (1993) Ribosomal DNA and chloroplast DNA polymorphism in European oaks. *Ann. Sci. For.* **50** (Suppl. 1): 41s-47s.

Saghai-Maroof, M.A., Soliman, K.M., Jorgensen, R.A., and Allard, R.W., (1984) Ribosomal DNA spacer-length polymorphisms in barley: Mendelian inheritance, chromosomal location, and population dynamics. *Proc. Natl. Acad. Sci.* USA **871**: 8014-8018.

Steinhoff S., (1993) Results of species hybridization with *Quercus robur* L. and *Quercus petraea* (Matt.) Liebl. *Ann Sci For***50** (Suppl. 1): 137s-143s.

Williams, J.G.K., Kubelik, A.R., Livak, K.J., Rafalski J.A., and Tingey, S.V., (1990) DNA polymorphisms amplified by arbitrary primers are useful as genetic markers, *Nucleic Acid Res.* **18**: 6531 - 6535.

Zanetto, A., Roussel, G., and Kremer, A., (1994) Geographic variation of interspecific differentiation between *Quercus robur* L. and *Quercus petraea* (Matt.) Liebl. *Forest Genetics* **1**: 111-123.

Zanetto, A., Kremer, A., Müller-Starck, G., and Hattemer, H.H. (1995) Inheritance of isozymes in Pedunculate oak (*Quercus robur* L.). *Journal of Heredity* (In press).

MOLECULAR AND GENETIC APPROACHES TO RUST RESISTANCE (*Melampsora sp.*) IN POPLAR (*Populus sp.*)

GOUÉ-MOURIER M.C., FAIVRE-RAMPANT P., LE GUERROUÉ B., LEFÈVRE F. and M. VILLAR

INRA,
Station d'Amélioration des Arbres Forestiers
F-45160 Ardon, France

1. Abstract

Various *Melampsora* rust races are causing defoliations on susceptible cultivars in commercial poplar plantations. Genetic and molecular analyses of parents and families of a factorial mating design have been investigated to determine components of resistance to rust. One genomic area was revealed which was involved in qualitative resistance. But, this qualitative resistance is not sufficient to assure durability and other genes involved in quantitative resistance are masked. Then, our objective is to reveal and combine quantitative resistance genes to obtain durable resistant cultivars. A genetic map is in progress in order to reveal molecular markers linked to quantitative rust resistance genes.

2. Introduction

In Europe, poplar cultivation consists in a clonal forestry with euramerican (*P. deltoides x P. nigra*) and interamerican (*P. trichocarpa x P. deltoides*) hybrids. Clonal forestry may be affected by several diseases and the main objective of poplar breeders is to select for resistance or tolerance to these pathogens. Rusts caused by *Melampsora sp.* threaten European poplar culture (Pinon *et al.*, 1987). These foliar fungi attack and defoliate plants at the nursery and adult stages, leading to reduction of growth and even death of the trees. Several pathological races have been identified (Pinon *et al.*, 1987; Pinon and Peulon, 1989; Pinon and Lefèvre, 1994; Steenackers *et al.*, 1994) based on clone-race compatibility.

Two reactions can be differentiated in poplar/rusts interaction. The first is named race-specific incompatibility or total resistance and consists in poplar clones not being infected by certain rust races. The second is named partial resistance where different poplar clones are susceptible but not with the same infection level (Lefèvre *et al.*, 1994).

The occurence of new races is more frequent than renewal of commercial cultivars (Lefèvre *et al.*, 1994). With the apparition of race E4 in 1994 (Steenackers *et al.*, 1994), there are no more cultivars registered in Europe resistant to rust. Breeding strategy of poplars are now focused on the selection of genotypes possessing durable resistance rather than total resistance. In this way, we assume that clones combining several resistance genes would have a more durable resistance.

We present in this paper our approaches, combining both genetic and molecular analyses on progenies of families linked in an intra and interspecific factorial mating design.

M. R. Ahuja et al. (eds.), Somatic Cell Genetics and Molecular Genetics of Trees, 249–254.
© 1996 *Kluwer Academic Publishers. Printed in the Netherlands.*

3. Genetics of rust resistance

Genetics of rust resistance was studied using a factorial mating design involving ten *Populus deltoides* parents and eight *Populus trichocarpa* parents (Figure 1). Susceptibility to *Melampsora* was assessed both under natural conditions in nursery trials and after leaf disk inoculation with pathogen races (E1, E2, E3) in a controlled environment (Lefèvre *et al.*, 1994).

		POPULUS TRICHOCARPA				*POPULUS DELTOIDES*				
	♀ \ ♂	36 -100 (or)	19-73 (or)	19-77 (or)	101 -74 (w)	73021 15 (iln)	H200 10 (oh)	L123 011 (iln)	ALR (iln)	MSS 050 27 (mss)
POPULUS TRICHOCARPA	36-77 (or)	11	12	13	14	15	16	17	18	19
	36 -134 (or)	21	22	23	24	25	26	27	28	29
	212 -161 (bc)	31	32	33	34	35	36	37	38	39
	FPL (w)	41	42	43	44	45	46	47	48	49
POPULUS DELTOIDES	73028 62 (iln)	51	52	53	54	55	56	57	58	59
	L150 089 (iln)	61	62	63	64	65	66	67	68	69
	L123 041 (iln)	71	72	73	74	75	76	77	78	79
	L155 079 (iln)	81	82	83	84	85	86	87	88	89
	TNS 06042 (tns)	91	92	93	94	95	96	97	98	99

Figure 1. Parents and families of the factorial mating design. 8 *P. trichocarpa* clones were crosses with 10 *P. deltoides* clones. Both intra and interspecific crosses are realized. Families of an average of 40 individuals are coded from 11 to 99. The origins of the clones is mentionned in brackets (or : oregon, w : washington, iln : illinois, mss : mississippi, bc: british columbia, tns : tennessee).

Most of the *P. deltoides* parents were resistant to rust races whereas all *P. trichocarpa* parents were susceptible to the three races of rust. Three different distributions were observed in nursery trials within families (Figure 2). Intraspecific P. *trichocarpa* hybrids showed a Gaussian distribution suggesting that many genes are involved in resistance mechanisms. Intraspecific P. *deltoides* hybrids showed a distribution in "L" corresponding to resistant individuals and low susceptible individuals. Interspecific P. *trichocarpa* x P. *deltoides* hybrids showed generally a bimodal distribution with some total resistant individuals and very susceptible ones, more susceptible than the P. *trichocarpa* parent (Lefèvre *et al.*, 1994).

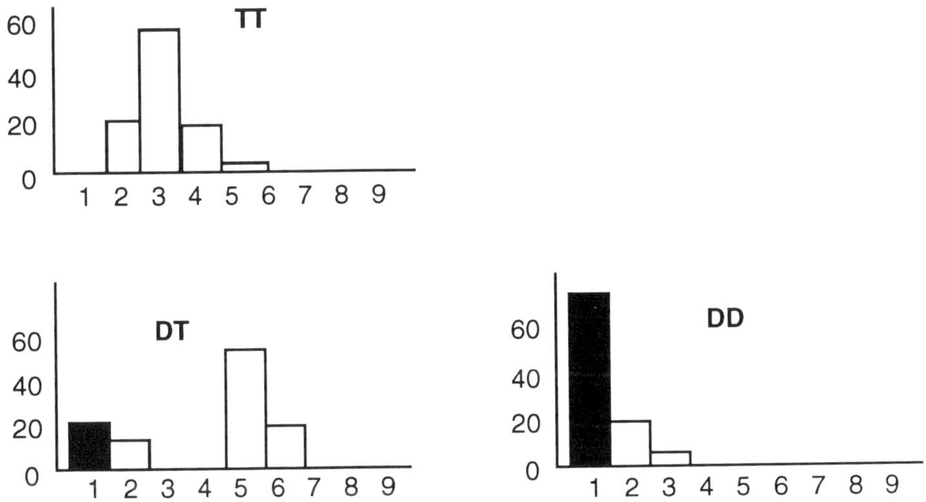

Figure 2 : Frequency distribution (%) of the clonal means within one interspecific progeny DT (83) and the two intraspecific-related families TT (23), DD (87). Non infected clones appear in the dashed bar; they were not found in the *P. trichocarpa* progeny (Lefèvre *et al.* 1994).

These different distributions suggest that *Populus deltoides* presents qualitative and quantitative resistance, *Populus trichocarpa* presents quantitative resistance but interspecific *Populus trichocarpa* x *Populus deltoides* have qualitative resistance controlled by a few genes. Other distributions for interspecific families were observed depending on the *Populus deltoides* parent, suggesting that a quantitative resistance can also exist. Genetic analyses of five different families from the factorial mating design indicate that few genes are involved in race-specific incompatibility.

4. Molecular approaches to rust resistance

Molecular markers are a useful tool to complete genetic analysis (Nance *et al.*, 1992). Identification of markers linked to rust resistance genes will be possible through two different strategies. In a first step, targetted mapping strategies are currently beeing used to find markers linked to qualitative resistance genes for all races and race by race. In a second step, a detailed genetic map of an interspecific family will permit to identify QTLs linked to quantitative resistance genes.

For qualitative resistance analysis, several families of the factorial mating design were used in a "Bulked Segregant Analysis" with RAPD markers (Michelmore *et al.*, 1991). We have revealed a marker linked to qualitative resistance to the three races E1, E2 and E3 (Villar *et al.*, 1996 and unpublished data), demonstrating that a unique chromosomic region could be involved in qualitative resistance to these three rust races. To study race by race qualitative resistance and to clarify gene for gene relationships (Flor, 1951), we are currently analysing segregation of progenies (150 individuals) of a cross between two discriminating clones. The two clones are: 'Ogy' incompatible to E1 and E3 races and compatible to E2 race and clone 'Luisa_Avanzo' incompatible to E1 and E2 races and compatible to E3 race. For quantitative resistance assumed to be under polygenic control, detection of linkage between molecular markers and QTLs is the more appropriate method (Mather and Jinks, 1971). We have constructed linkage maps with RAPD markers of an interspecific family *Populus deltoides* x *Populus trichocarpa* of 75 individuals following a pseudo test-cross strategy (this family will be enlarged to 200 individuals). 255 Operon primers were screened, 119 were selected to amplify a total of 274 markers. 40 markers were significantly different from 1:1 ratio by chi2 test. These markers were grouped at LOD = 3 and $\theta = 0{,}4$ resulting in the maternal *Populus deltoides* map having a total of 113 markers into 25 linkage groups and the paternal *Populus trichocarpa* parent map with 121 markers in 28 linkage groups. RFLPs and isozymes markers will be added to densify these two maps. These maps will be compared to the USA F2 generation linkage map, with common RFLP probes (Bradshaw *et al.* 1994).

5. Durable resistance

5.1. COEVOLUTION BETWEEN HOST AND PATHOGEN

Coevolution between host and pathogen exists. Plants are submitted to parasite pressure and this pressure involves an equilibrium between plant resistance genes and parasite virulences. This balance is dynamic because pathogens act with a reciprocal influence of host (Ouchi, 1983). Selection by breeders adds another pressure that modifies equilibrium between plant resistance and pathogen virulences: when a cultivar is selected for total resistance to a race, this resistant plant favours the appearance of new races, being able to overcome resistance. These new races appear fortuitously by mutation or by recombination. Then, durable resistance appears to be all the more difficult to obtain.

5.2. VERTICAL AND HORIZONTAL RESISTANCE

Van der Planck (1968) proposed distinction between vertical and horizontal resistance. The vertical resistance would correspond in our model to incompatibility. This resistance could be defined by cultivars with a high level of resistance for some races but totally susceptible to other races. This vertical resistance is a total qualitative resistance controlled by few genes but that do not confer durable resistance. Horizontal resistance would correspond in our rust/poplar model to compatibility. This resistance consists in non-specific resistance genes, but plants are susceptible to all races with different levels of susceptibility. It is a partial resistance controlled by many genes and giving durable resistance.

Our genetic and molecular studies showed that a single chromosomic area controls the resistance to all races. Such a genomic segment associated with several host pathogen interactions was already reported by several authors (Nodary *et al.*, 1993). These genomic regions could be allelic series in a specific locus or several tightly linked loci. In our case, this genomic area has a major effect in qualitative resistance and masks other genes involved in quantitative resistance. Therefore, to have access to quantitative resistance genes we have to overcome this mask (such as the the work realized in *Capsicum sp.* by Lefèbvre in 1993, where QTLs with low effect were detected by analyses of variance, fixing the major effects).

Moreover, genetic and molecular studies of different families of our factorial mating design and analysis of progeny from crossing between 'Ogy' and 'Luisa_Avanzo' should provide a better understanding of quantitative and qualitative resistance genes.

6. Consequences in breeding programs

Most of the commercial poplar cultivars are interspecific clones. Our studies demonstrated that these clones may express qualitative resistance and might not be the best strategy for breeders to select for durable resistance (Lefèvre *et al.*, 1995). A strategy would be to combine several genes involved in partial resistance. Molecular markers tightly linked to QTLs of components of partial resistance could be used for an early selection of the best genotype within a pre-selected family.

7. References

Bradshaw, H.D., Villar, M., Watson, B.D., Otto, K. and Stettler, R.F. (1994) Molecular genetics of growth and development in *Populus*. III. A genetic map composed of RFLP, STS and RAPD markers. *Theor. Appl. Genet.* **89**, 167-178.

Flor, H.H. (1951) Genes for resistance to rust in Victoria flax, *Agron. J.* **43**, 527-531.

Lefèbvre, V. (1993) Marquage moléculaire et analyse de résistances polygéniques. Intéraction piment (*Capsicum annuum* L.)-*Phytophtora capsici* Leon, PhD, Université de Paris Sud, Centre d' Orsay. 182pp.

Lefèvre, F., Pichot, C. and Pinon, J. (1994) Intra and interspecific inheritance of some components of the resistance to leaf rust (*Melampsora larici-populina* Kleb.) in poplars, *Theor. Appl. Genet.* **88**, 501-507.

Lefèvre F., Chartier S., Faivre-Rampant P. , Goué M.C., Laurans, F., Pilate G., Pinon, J., Valadon, A. Villar M. 1995. Composantes de la résistance aux rouilles chez les peupliers: utilisation en sélection, C. R. Acad. Agric. Fr. **81**: 111-121

Mather, K. and Jinks, J.L. (1971) Biometrical Genetics, Chapman and Hall, 2nd edition, London, England, pp. 382.

Michelmore, R., Paran, I. and Kesseli, R. (1991) Identification of markers linked to disease-resistance genes by bulked segregant analysis: a rapid method to detect markers in specific genomic regions by using segregating populations, *Proc. Natl. Acad. Sci.* **88**, 9828-9832.

Nance, W.L, Tuskan, G.A., Nelson, C.D., and Doudrick R.L. (1992) Potential applications of molecular markers for genetic analysis of host-pathogen systems in forest trees, *Can. J. For. Res.* , **22**, 1036-1043.

Nodari, R., Tsai, S., Guzman, P., Gilbertson, R. and Gepts, P. (1993) Toward an integrated linkage map of common bean. III. Mapping genetic factors controlling host-bacteria interactions, *Genetics* **134**, 341-350.

Ouchi, S. (1983) Induction of resistance or susceptibility, *Ann. Rev. Phytopathol.*, **21**, 289-315.

Pinon, J., Van Dam, D.C., Genetet, I. and De Kam, M. (1987) Two pathogenic races of *Melampsora larici-populina* in north western Europe, *Eur. J. For. Pathology*, **17**, 47-53.

Pinon, J. and Peulon, V. (1989) Mise en évidence d'une troisième race physiologique de *Melampsora larici-populina* Kleb. en Europe, *Crytogam. Mycol.* **10**, 95-106.

Pinon, J. and Lefèvre, F. (1994) A new virulence found among isolates of *Melampsora larici-populina, Réunion du groupe de travail sur les maladies, International Poplar Commission, FAO Izmit (Turquie)*, 3-7 Octobre 1994.

Steenackers, M., Steenackers, V. and Delporte, T. (1994) A new physiological race of *Melampsora larici-populina* in Belgium. *Réunion du groupe de travail sur les maladies, Commission internationale du peuplier, FAO, Izmit (Turquie)*, 3-7 Octobre 1994, pp. 8.

Van der Planck, J. (1968) Disease resistance in plants, *Academic Press, Inc., New York, USA*, pp. 206.

Villar, M., Lefèvre, F., Bradshaw Jr. , H.D. and Teissier du Cros, E. (1996) Molecular genetics of rust resistance in poplars (*Melampsora larici-populina* Kleb/*Populus sp.*) by bulked segregant analysis in a 2 x 2 factorial mating design. Genetics 143: (in press)

Section IV.

Stress-related gene expression

INDUCED DEFENSE RESPONSES IN NORWAY SPRUCE

A.LÖNNEBORG, P.SHARMA
Norwegian Forest Research Institute
Høgskoleveien 12, 1432 Ås, Norway

1. Abstract

In an effort to characterize induced defense responses in Norway spruce (*Picea abies*), we have developed an infection system that is reproducible, do not require advanced growth systems or large space, plants for experiments is accesible all the year, and the experimental time is short. The fungi used, *Pythium dimorphum*, is a pathogen which mainly penetrate the roots in the extracellular space but also penetrate through the plasmamembrane into the cytoplasm both in cortex and stele within 12 h. With this infection system we have shown that several low pH extractable proteins, including chitinases, chitosanases and ß-1,3-glucanases, not detected before infection, appeared within 1-4 days after infection. However, the appearance of these proteins was not mainly due to an increased synthesis of the proteins but due to release from a form, not accesible without the use of detergents or caffeine during the extraction. The release with these compounds are strong indications that these proteins are non-covalently bound in complex with polyphenols in non-infected roots but are released from these complex after infection.

In addition to the characterization at the anatomical and protein levels we have also studied genes activated after infection. To do so we constructed a subtracted cDNA library enriched for cDNA clones representing genes activated after infection with *P.dimorphum*. From this library we have isolated several clones that are currently being characterized. One of these cDNA clones, SPI1, encodes a protein with high similarity to γ-thionins from other plants. This clone showed a complex expression pattern with a possible negative regulation after pathogen infection.

2. Introduction

Induced plant defence systems are commonly known from several different plant species. Among the systems are probably the induction of pathogenesis related (PR) proteins one of the best characterized. The common features of the PR proteins are that in addition to their accumulation after pathogen infection they are also extractable in a buffer with a low pH, and resistant to degradation by proteases. Within hours after infection of a pathogenic virus or fungus the transcripts for these proteins starts accumulating and within 1-5 days the proteins can be detected. From tobacco 35 different PR proteins, grouped in 11 families, have been identified (van Loon et al., 1994). Also from several other plants many different PR proteins have been detected. A few of the proteins have been identified as ß-1,3-glucanases or chitinases (Kauffman et al., 1987; Legrand et al., 1987; Kombrink et al., 1988). These hydrolytic enzymes have been implicated in plant

M. R. Ahuja et al. (eds.), Somatic Cell Genetics and Molecular Genetics of Trees, 257–264.
© 1996 *Kluwer Academic Publishers. Printed in the Netherlands.*

defence, since many pathogens have ß-1,3-glucanases and/or chitinases as cell wall components. Together, ß-1,3-glucanases and chitinases have been shown in vitro to exert a synergistic effect in retarding the growth of fungal hyphae (Mauch et al., 1988). Some other of the PR proteins have also been shown to have a fungistatic effect (Woloshuk et al., 1991; Hejgaard et al., 1992; Vigers et al., 1992) although their *in vivo* identity not yet is known. In addition to the genes encoding PR proteins, several other genes have also been shown to be induced after pathogen infection (Collinge and Slusarenko, 1987). However, until recently the identification of PR proteins or genes activated by pathogen infection had only been done in herbaceous angiosperms but nothing in any gymnosperm or woody angiosperm. It was not known if also gymnosperms had PR proteins or if also they had genes that were activated after infection. Also very little was known about the responses in roots since most of the previous studies were made on leaves. To find out if also the gymnosperm Norway spruce (*Picea abies*) have an induced pathogen response we have developed a system to study this plant. In this paper the infection system and the results emanating from work with this system are presented.

3. Results and Discussion

3.1. EXPERIMENTAL SYSTEM

Seeds of Norway spruce were surface sterilized and germinated on solid maltose medium. About ten days after imbibition non-infected seedlings were transferred to liquid maltose medium in black microcentrifuge tubes with or without a high inoculum of homogenized hyphae of *P.dimorphum* (Sharma et al., 1993). The infection was followed for up to ten days (Fig.1). This infection system has several advantages. It is reproducible from experiment to experiment, it can be performed in a conventional laboratory without any advanced growth system or large space, the plants needed for the experiment can be germinated when needed and are available within 2 weeks at any time during the year, and the infection is fast, limiting the length of experiments. In addition is *P.dimorphum* a fast growing fungi, and the isolate used is from plants showing the symptoms of root dieback, an important disease in nordic forest nurseries.

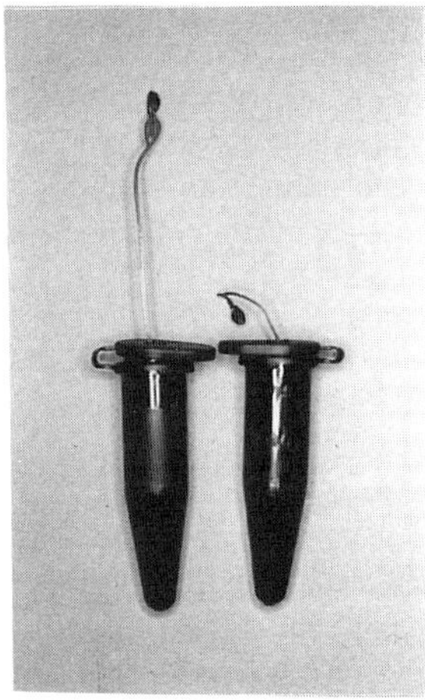

Figure 1. Spruce seedlings in black microcentrifuge tubes without or with hyphae of *P.dimorphum*.

3.2. INFECTION PROCESS

The root epidermis of the young seedlings was not a major barrier for the infecting hyphae (Børja et al., 1995). This cell layer had open clefts in which the hyphae could grow and reach the next cell layer of the cortex. The hyphae penetrated further into the roots using three alternative pathways. Most hyphae infected through the middle lammellae between adjacent cells (Fig.2). However, the hyphae also grew in the space between the cell wall and the plasmamembrane or formed appressoria and penetrated the plasmamembrane and grew into the cytoplasm (Fig.2,3). The same hyphae could probably use all three pathways in different areas of the root. The endodermis did not appear to be a strong barrier for the infecting hyphae since both cortex and stele were infected within 12 h (Børja et al., 1995).

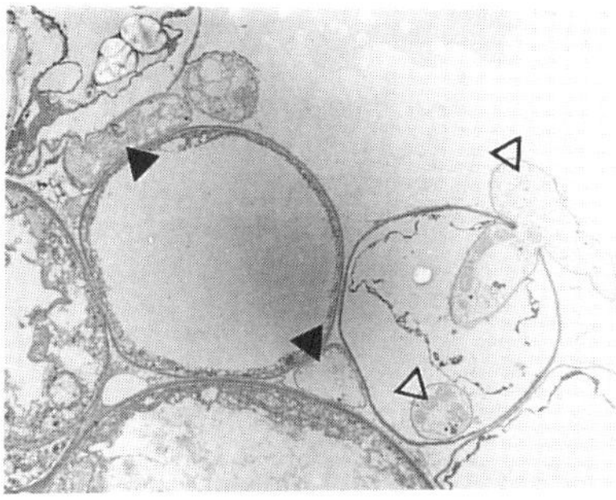

Figure 2. Transmission electron micrograph showing *P.dimorphum* hyphae entering the root between the cells (▸), between cell wall and plasmamembrane (▹) or penetrating the plasmamembrane entering the cytoplasm (▹).

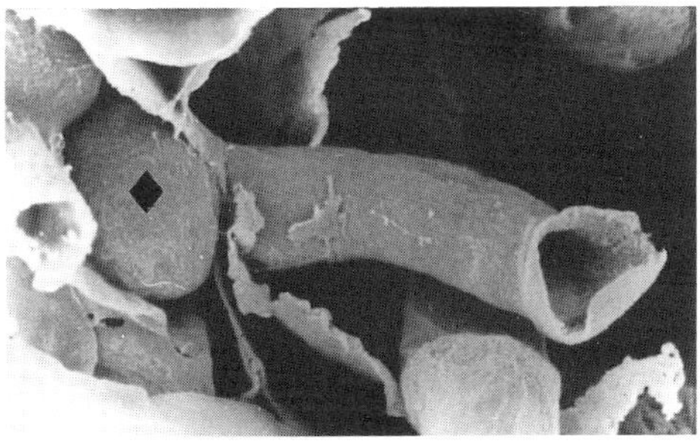

Figure 3. Scanning electron micrograph of hyphae of *P.dimorphum* forming an appresoria (♦) and entering a cortex cell of Norway spruce through the cell wall and plasmamembrane.

3.3. LOW pH EXTRACTABLE PROTEINS

From herbaceous angiosperms it is known that several low pH extractable proteins are detected specifically after infection. Since they appear after pathogen infection they have also been called pathogenesis related (PR) proteins (van Loon, 1985; Linthorst, 1991). Before infection only two very acidic low pH extractable proteins could be detected in low amounts in roots of Norway spruce seedlings. However, after infection these two proteins increased in amount and more than 30 different proteins with both acidic, neutral and basic pI were detected. All these proteins increased in amount till about the fourth day after infection when they reached a steady state level (Fig.4). All the proteins were maintained at this steady state level to the end of the experiment, ten days after the start of infection (Sharma et al., 1993). The function of most of the PR proteins are not known. However, some of them have been found to have chitinase or ß-1,3-glucanase activity (Kauffman et al., 1987; Legrand et al., 1987; Kombrink et al., 1988). Before infection and in control roots only two very acidic chitinase activities could be detected in Norway spruce. After infection these two increased in activity and 8 novel activities appeared. Four of these activities were acidic, two were neutral while two of them had a basic pI (Fig.4). Like the proteins the chitinase activities reached a steady state level about 4 days after infection (Sharma et al., 1993). No ß-1,3-glucanase activities could be detected before infection or in the uninfected roots. However, two days after infection one acidic activity appeared and the next day an additional acidic activity could be detected (Fig.4). Also the ß-1,3-glucanases reached a steady state level about 4 days after infection (Sharma et al., 1993).

Changing the substrate in the overlay gel from glycolchitin to glycolchitosanase, the roots could also be tested for presence of chitosanase activities. This enzyme has previously been associated with chemical stress (Grenier & Asselin, 1990) but not strictly as a PR protein. Since many fungi also contains chitosan in addition to chitin and ß-1,3-glucans it was possible that also this enzyme were among the PR proteins. Before infection no chitosanase activity was detected but 2 days after infection three different activities appeared on the gel. These activities increased to the end of the experiment (Sharma et al., 1993).

3.4. THE APPEARANCE OF ß-1,3-GLUCANASES AFTER INFECTION IS NOT MAINLY DUE TO GENE ACTIVATION.

Since the increased accumulation of PR proteins after infection in herbaceous angiosperms, like tobacco, arabidopsis and barley, is mainly due to transcriptional activation, the roots of Norway spruce were also tested for activation at this level. A northern blot with RNA from uninfected and infected roots were hybridized with a homologous probe for a ß-1,3-glucanase transcript. The results showed that this transcript was present after infection but also that it was present in the noninfected roots att nearly the same level (Sharma, P., Børja, I., and Lönneborg, A., 1995). This indicated that the appearance of this ß-1,3-glucanase is not mainly due to transcriptional activation but somewhere later in the pathway.

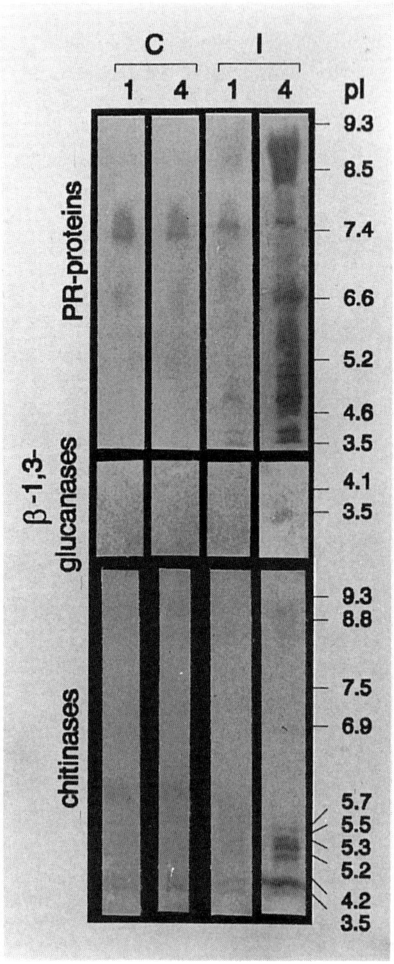

Figure 4. PR proteins extracted from uninfected (**C**) and infected (**I**) roots of Norway spruce, 1 and 4 days after infection, and ß-1,3-glucanase and chitinase activities among these proteins. **pI**, pI markers.

3.5. THE LOW pH EXTRACTABLE PROTEINS, INCLUDING CHITINASES AND ß-1,3-GLUCANASES, ARE BOUND TO A COMPOUND(S) WITH ALL FEATURES IN COMMON WITH POLYPHENOLS.

Norway spruce is rich in polyphenols and the polyphenols are located both in the vacuole (Børja et al., 1995) and the extracellular space (Strack et al.,1988). PR proteins are also located in these compartments, both in the vacuole and/or in the extracellular space. Since polyphenols are well known to non-covalently bind proteins (Swain, 1978; Spencer et al., 1988), a possible explanation for the control of appearance of PR proteins could be that they are bound to polyphenols in the non-infected roots and released from these complex after pathogen infection.

Including detergents like Triton X-100 or CHAPS or the more specific nitrogen heterocycle caffeine during the extraction, all the low pH extractable proteins and all the chitinase and ß-1,3-glucanase activities could be detected in uninfected roots at at least the same level as in infected roots. Furthermore, all the proteins were extracted from roots in low pH buffer by incubating the plants in Triton X-100 in vivo. Also, from embryonic cells low in polyphenols, all low pH extractable proteins, including all chitinase activities, could be recovered without detergents or caffeine included in the extraction while similar tissue high in polyphenols needed detergents for full recovery of these proteins (Sharma, P., Børja, I., and Lönneborg, A., 1995). All these results are in agreement with, and are strong indications that these proteins are non-covalently bound in vivo to a compound(s) with all features in common with polyphenols. It was also shown in vitro that the low pH extractable proteins, including chitinases, were capable of binding to polyphenols and could be released from these complex when Triton X-100 or caffeine were included, and the chitinases were still active.

The existence of low pH extractable chitinases in uninfected plants is not unique to Norway spruce but was also found in other gymnosperms like Scotch pine (*Pinus sylvestris*) and lodgepole pine (*P.contorta*) and also in the polyphenolrich angiosperm oak (*Quercus ruber*). However, they were not found in the herbaceous angiosperms barley (*Hordeum vulgare*) or tobacco (*Nicotiana tabaccum*) (Sharma, P., Børja, I., and Lönneborg, A., 1995).

3.6. CONSTRUCTION OF A SUBTRACTED cDNA LIBRARY ENRICHED IN cDNA CLONES REPRESENTING GENES ACTIVATED AFTER INFECTION.

In analogy with other infection systems, like TMV infected tobacco and mildew infected barley, it was expected that also in Norway spruce infected with *P.dimorphum*, genes would be activated after infection. To isolate cDNA clones representing infection activated genes, a subtracted cDNA library was constructed (Sharma et al., 1993, Lönneborg et al., 1995). The strategy of construction is shown in Fig.5 and have several advantages compared with other construction strategies. The binding of mRNA and/or cDNA on magnetic beads made the purification of mRNA more effective and the synthesis of cDNA more efficient and controlable. The utilization of PCR made it possible to work with smaller amounts of tissue.

3.7. ISOLATION OF A cDNA CLONE ENCODING A PROTEIN WITH HIGH SIMILARITY TO γ-THIONINS.

From the subtracted cDNA library several clones with preferential expression in infected roots were isolated. One of these clones were used to isolate a full length clone named Spruce Pathogen Induced 1 (SPI1) (Sharma,P & Lönneborg,A., 1995). This clone encodes a protein with high similarity to γ-thionins from wheat (Collila et al., 1990), barley (Mendez et al., 1990), and petunia, and plant defensins from radish (Terras et al., 1992). The expression of SPI1 increased in roots when transferred to liquid medium either containing or not containing hyphae of *P.dimorphum*. After an initial increase 12 h after infection the transcript rapidly decreased and were no longer detectable 2 days after infection. However, in the uninfected roots the transcript accumulated to a peak 24 h after transfer to liquid medium after which it had transiently decreased the second day and then increased again (Sharma,P & Lönneborg,A., 1995). A possible explanation for the increase when transferred to liquid medium could be that the gene are regulated by hypoxic stress. This is a stress that occur also in this infection system. The decrease and dissappearance of the transcript in infected roots after 12 h may suggest that this transcript were negatively regulated by the pathogen and might reflect a strategy employed by this fungus to evade the effects of toxic plant gene products. Since we do not yet have a probe for a transcript that is specific for spruce and is constitutively expressed, an 18S rRNA cDNA probe, which also can hybridize to the rRNA from the pathogen, is used to calibrate the amount of total RNA loaded on RNA blots. Although, no hybridization to the SPI1 cDNA probe was observed 2 and 4 days after infection,

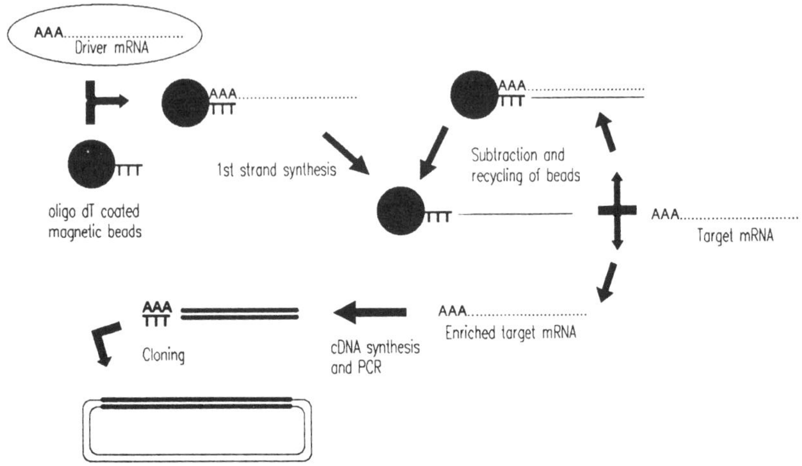

Figure 5. Construction of subtracted cDNA library using magnetic beads.

even when the amount of RNA loaded was doubled, this does not completely exclude the possibility that the relative amount of RNA from the plant in the total RNA pool drastically decreased after infection and that RNA from the pathogen became dominating. This could explain the reduction in the amount of SPI1 transcript following infection.

4. Conclusions.

The induced response upon pathogen infection has been described in a number of different plant species. However, all these plants have been angiosperms, either di- or monocotyledons. Gymnosperms have evolved separately from angiosperms for about 350 million years. This is a time long enough for the gymnosperms to evolve other strategies to defend themselves against infecting pathogens. Another limitation in previous work has been the lack of studies done on interactions between root pathogens and the plant root despite so many important pathogens infects the roots.

In our work to characterize the induced defense response in Norway spruce we have found that the appearance of low pH extractable proteins, including chitinases and ß-1,3-glucanases, are similar to what has been found in the angiosperms investigated. The appearance of these proteins after infection with the root pathogen *P.dimorphum* were, however, shown not to be due to a corresponding increase of the transcripts for these proteins. Instead it was found that the low pH extractable proteins are constitutively present in the healthy plant but are bound to a compound, with all features in common with polyphenols. After infection the proteins are released from these complex. A similar finding was found in other gymnosperms but also in oak, an angiosperm. The common feature of oak and the conifers are their high content of polyphenols further supporting the hypothesis of protein-polyphenol complexation.

Work has also been done to isolate and characterize genes induced after infection. The first charcterized cDNA clone encodes a protein with high similarity to γ-thionins or plant defensins. This group of proteins have been shown to be effective antimicrobial agents. Presence of this protein in Norway spruce indicates that it may have developed a defense with features in common with angiosperms.

5. Acknowledgement

We thank Marianne Jensen, Signe Drømtorp and Tore Gulliksen for providing excellent technical assistance. We also thank Carl Gunnar Fossdal and Rolf Gref for helpful discussions, Annika Sundås for donation of DNA probes and Harald Kvaalen for providing embryogenic callus of Norway spruce. This work was supported by grants from The Norwegian Research Council (P.S.) and The Norwegian Fund for Forest Development (A.L.).

6. References.

Børja, I., Sharma, P., Krekling, T., and Lönneborg, A. (1995) Cytopathological response in roots of *Picea abies* seedlings infected with *Pythium dimorphum*. *Phytopathol.* **85**, 495-501.

Colilla, F.J., Rocher, A., and Mendez, E. (1990) γ-purothionins: amino acid sequence of two polypeptides of a new family of thionins from wheat endosperm. *FEBS Lett.* **270**, 191-194.

Collinge, D.B., and Slusarenko, A.J. (1987) Plant gene expression in response to pathogens. *Plant Mol.Biol.* **9**, 389-410.

Grenier, J., and Asselin, A. (1990) Some pathogenesis-related proteins are chitosanases with lytic activity against fungal spores. *Mol.Plant-Microbe Interact.* **3**, 401-407.

Hejgaard, J., Jacobsen, S., Bjørn,S.E., and Kragh, K.M. (1992) Antifungal activity of chitin-binding PR-4 type proteins from barley grain and stressed leaf. *Fed.Euro.Biochem.Soc.Lett.* **307**, 389-392.

Kauffman, S., Legrand, M., Geoffroy, P., and Fritig, B. (1987) Biological function of pathogenesis-related proteins. four PR proteins of tobacco have ß-1,3-glucanase activity. *The EMBO J.* **6**, 3209-3212.

Kombrink, E., Schröder, M., and Hahlbrock, K (1988) Several pathogenesis-related proteins are 1,3-ß-glucanases and chitinases. *Proc.Natl.Acad.Sci.USA*, **85**, 782-786.

Legrand, M., Kauffman, S., Geoffroy, P and Fritig, B. (1987) Biological functions of pathogenesis-related proteins: 4 tobacco pathogenesis-related proteins are chitinases. *Proc.Natl.acad.Sci.USA*, **84**, 6750-6754.

Linthorst, H.J.M. (1991) Pathogenesis-related proteins in plants. *Crit.Rev.Plant Sci.* **10**, 123-150.

Lönneborg, A., Sharma, P., and Stougaard, P. (1995) Construction of subtractive cDNA library using magnetic beads and PCR. *PCR Meth.Appl.* **4**, S168-S176.

Mauch, F., Mauch-Mani, B., and Boller, T. (1988) Antifungal hydrolases in pea tissue. II. Inhibition of fungal growth by combinations of chitinase and ß-1,3-glucanase. *Plant Physiol.* **88**, 936-942.

Mendez, E., Moreno, A., Colilla, F., Palaez, F., Limas, G.G., Mendez, R., Sorano, F, Salina, M., de Haro, C. (1990) Primary structure and inhibition of protein synthesis in eucaryotic cell-free system of a novel thionin, γ-hordothionin, from barley endosperm. *Eur.J.Biochem.* **194**, 533-539.

Sharma, P., Børja, I., and Lönneborg, A. (1995) Low pH extractable proteins are present, bound to polyphenol-like compounds in Norway spruce and are released after pathogen infection. Manuscript in preparation.

Sharma, P., Børja, I., Stougaard, P., and Lönneborg, A. (1993) PR-proteins accumulating in spruce roots infected with a pathogenic *Pythium sp.* isolate include chitinases, chitosanases and ß-1,3-glucanases. *Physiol.Mol.Plant Pathol.* **43**, 57-67.

Sharma, P., and Lönneborg, A. (1995) Isolation and characterization of a cDNA encoding a γ-thionin-like protein. *Plant Mol.Biol.* submitted.

Sharma, P., Lönneborg, A., and Stougaard, P. (1993) PCR based construction of subtractive cDNA library using magnetic beads. *Biotechniques* **15**, 610-611.

Spencer, C.M., Cai, Y., Martin, R., Gaffrey, S.H., Goulding, P.N., Magnolato, D., Lilley, T.H., Haslam, E. (1988) Polyphenol complexation - some thoughts and observations. *Phytochem.* **27**, 2397-2409.

Strack, D., Heilemann, J., Klinkott, E.S., and Wray, V. (1988) Cell-wall bound phenolics from Norway spruce (*Picea abies*) needles. *Z.Naturforsch.* **43c**, 37-41.

Swain, T. (1978) Plant-animal coevolution: A synoptic view of the paleozic and mesozoic, in J.B.Harborne (ed.) *Biochemical Aspects of Plant and Animal Co-evolution*, Academic Press, London, pp.3-19.

Terras, F.R.G., Schoofs, H.M.E., de Bolle, M.F.C., van Leuven, F., Rees, S.B., Vanderleyden, J., Cammue, B.P.A., and Broekart, W.F. (1992) Analysis of two novel classes of plant antifungal proteins from radish (*Raphanus sativus* L.) seeds. *The Plant Cell* **7**, 573-588.

van Loon, L.C. (1985) Pathogenesis-related proteins. *Plant Mol.Biol.* **4**, 111-116.

van Loon, L.C., Pierpoint, W.S., Boller, T., and Conejero, V. (1994) Recommendations for naming plant pathogenesis-related proteins. *Plant Mol.Biol.Reporter* **12**, 245-264.

Vigers, A.J., Wiedemann, S., Roberts, W.K., Legrand, M., Selinrennikoff, C.P., and Fritig, B. (1992) Thaumatin-like pathogenesis-ralted proteins are antifungal. *Plant Sci.* **83**, 155-161.

Woloshuk, C.P., Meulenhoff, J.S., Sela-Buurlage, M., van der Eizen, P.J.M. and Cornelissen, B.J.C. (1991) Pathogen induced proteins with inhibitory activity toward *Phytophtora infestans*. *The Plant Cell* **3**, 619-628.

OZONE-INDUCED DEFENSE REACTIONS IN BIRCH (*Betula pendula* Roth)

J. KANGASJÄRVI, R. PELLINEN, J. TUOMAINEN,
R. JULKUNEN-TIITTO[1)] AND M. KIISKINEN

Department of Ecology and Environmental Science
University of Kuopio, Box 1627, FIN-70211 Kuopio, Finland
[1)]*Department of Biology,*
University of , FIN-80101 Joensuu, Finland.

1. Abstract

We have studied the responses of ozone sensitive and insensitive birch (*Betula pendula* Roth) clones to ozone and analyzed cell death, changes in enzyme activities, metabolite concentrations and gene expression. Cell death, measured by vital stain Evan's Blue permeability, increased in both clones similarly. The damaged individual cells were randomly distributed throughout the leaves but in middle aged leaves of the sensitive clone the cell death spread uncontrollable forming necrotic lesions. The activities of oxygen radical scavenging enzymes, superoxide dismutase (SOD), peroxidase (POX) and glutathione reductase (GR) increased concomitantly with the damage formation and the increase was higher in the sensitive clone. This suggests that the increase in the enzyme activities is more related to the cell damage rather than protection from ozone. We have isolated cDNA clones for birch phenylpropanoid and flavonoid pathways controlling genes [phenylalanine ammonia-lyase (PAL), chalcone synthase (CHS)] that are good indicators of the coordinated cellular defense responses in plants. PAL was induced rapidly but transiently in both clones during the ozone exposure indicating that the defense reactions were induced in both clones. CHS mRNA levels decreased as a response to ozone-exposure, but were increased after wounding. Ozone-exposure thus induced the programmed defense mechanisms in both birch clones but wide spreading cell death was induced only in one clone suggesting that there two processes are under separate regulation.

2. Introduction

Plants exposed to elevated atmospheric ozone undergo several biochemical changes belonging to general plant defense mechanisms normally associated with pathogen infection, even when no damage can be detected (Kangasjärvi *et al.*, 1994). Ozone exposure often causes changes in polyamine metabolism (Langebartels *et al.*, 1991), as well as a surge in the production of ethylene (Mehlhorn and Wellburn 1987;

M. R. Ahuja et al. (eds.), Somatic Cell Genetics and Molecular Genetics of Trees, 265–271.
© 1996 *Kluwer Academic Publishers. Printed in the Netherlands.*

Langebartels *et al.,* 1991) and increases in the activities and gene expression of phenylpropanoid and flavonoid pathway enzymes (Rosemann *et al.,* 1991; Galliano *et al.,* 1993; Eckey-Kaltenbach *et al.,* 1994; Sharma and Davis 1994; Tuomainen *et al.,* 1996). The activities of enzymes like superoxide dismutase (SOD), peroxidases (POX) and glutathione reductase (GR) that are good indicators of oxidative stress in the tissues change often, but not always, (Bowler *et al.,* 1992; Foyer *et al.,* 1994). Some of these changes correlate to plant's ozone-sensitivity, but its ultimate basis still remains unknown. Ethylene seems to play an important role in determining plant's ozone sensitivity (Mehlhorn and Wellburn 1987) and increase in ethylene synthesis is one of the very first reactions in ozone-exposed plants (Rodecap and Tingey 1983), thus studies of ethylene synthesis and its regulation in ozone-exposed plants may elucidate the basic reason for ozone-damage and the regulation of related processes.

Both ozone and UV-B irradiation induce the branches of phenylpropanoid metabolism, lignin synthesis and flavonoid accumulation in plants exposed to ozone or UV-B radiation. The role of flavonoids as protectants that absorb UV-B radiation in leaf epidermal cells is well documented in several studies, but the role of the lignification and flavonoid metabolism in ozone exposed plants, however, is less clear. Both pathways are induced during pathogen attack and it is possible that the oxygen radical formation in the apoplast from ozone just mimics (Kangasjärvi *et al.,* 1994) the role of the oxygen radicals that are formed in plants as the oxidative burst during the very first reactions of plant pathogen defense induction (Levine *et al.,* 1994).

Though plant-ozone interactions have been studied at biochemical and gene expression level in several studies in annual herbaceous plants, only a few studies have addressed these questions in forest trees. We have studied the defense reactions in a deciduous tree, birch (*Betula pendula* Roth) to ozone, wounding and UV-B stresses at biochemical and gene expression level. In previous studies (Tuomainen *et al.,* 1996) we have shown that ozone induces cell death and defense reactions differentially in two birch clones that differ in their sensitivity to ozone. Previously we cloned PCR fragments of birch *pal* and *rbcS* that were used as probes in studies with ozone-exposed plants (Tuomainen *et al.,* 1996). Here we report the cloning of PCR fragments for birch chalcone synthase (*chs*) and ACC oxidase (*aco*) and their use in gene expression analysis in ozone-exposed and wounded leaves.

3 Chalcone synthase and ACC oxidase in ozone-exposed birch

3.1. CHALCONE SYNTHASE (*chs*)

Birch DNA and RNA were isolated as described previously (Tuomainen *et al.,* 1996). The probe for birch *chs* was amplified with degenerate primers 5' AGGAGGATCCA-AGCAACC(TC)TG(TC)TG(AG)TACATCAT 3' and 5' AGCAGGATCCAA(AG)-GC(CT)AT(CT)AA(AG)GA(GA)TGGGG 3' corresponding to conserved amino acid sequences PPDLCGQQYM and AGSKAIKEWG. *chs* primers were kind gifts from Y.

Helariutta (Biotechnology Institute, University of Helsinki). The details of the cloning were as for *pal* (Tuomainen *et al.*, 1996). The PCR-amplified fragment of birch *chs* was sequenced. It was highly homologous to the corresponding *chs* part from other species. *chs* mRNA steady state levels were altered by ozone exposure (Figure 1); initial transcript-level was higher in the insensitive clone, KL-2-M, and ozone exposure decreased the mRNA levels during the eight hour ozone exposure. In the sensitive clone *chs* transcript levels increased initially, followed by a decline in the mRNA abundance.

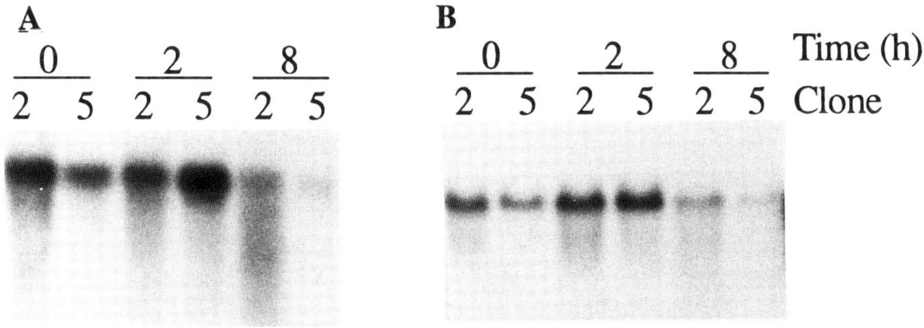

Figure 1. The effect of ozone exposure on the levels of chalcone synthase (**A**) and ACC oxidase (**B**) transcripts in ozone insensitive KL-2-M (clone 2) and ozone sensitive KL-5-M (clone 5) birch clones. Northern blots of total RNA (10 μg per lane) from birch leaves.

The induction of *pal* (Tuomainen *et al.*, 1996) and the changes in *chs* steady state mRNA levels in the ozone-exposed plants (Figure 1) were not strongly reflected in leaf flavonoid concentrations, that were measured as described in Julkunen-Tiitto & Meier (1992).The only prominent changes were seen in the concentrations of a flavan, catechin (Figure 2). In the ozone insensitive birch clone, the concentration of catechin increased first at 24 hours, declined somewhat at 48 hours and began to rise again towards the end of the experiment. In the ozone sensitive clone catechin concentration was not altered until after 48 hours, when it started to rise like in the insensitive clone. Catechin accumulation in ozone-exposed plants has been observed in other species too, but the significance of this accumulation is not clear.

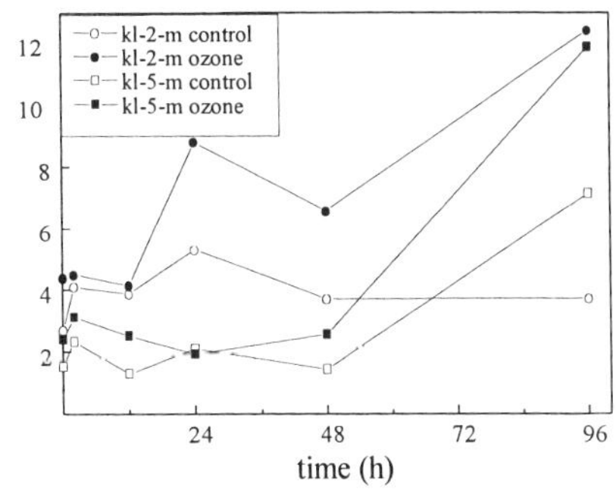

Figure 2. Effect of ozone exposure on the catechin concentration (mg/g dry weight) of the leaves in the ozone-exposed birch clones. Phenylpropanoids were extracted with methanol and separated with HPLC. Individual compounds were identified with spectrum and comparison to standards.

3.2. 1-AMINOCYCLOPROPANE-1-CARBOXYLATE OXIDASE (*aco*)

Birch *aco* cDNA fragments were cloned with RT-PCR. The first strand of cDNA was synthesized with AMV reverse transcriptase and used for amplification. *aco* was amplified with degenerate primers (a kind gift from Scot Peck, Kende lab, MSU, MI) 5'-CAC(AG)CTCTTGTA(TC)TT(AGCT)CC(AG)TT-3' and 5'-AG(TC)AA(TC)TA-(TC)CC(AGCT)CC(AGCT)TG(TC)CC-3' corresponding to conserved amino acid sequences SNYPPC and NGKYKSV. The PCR-amplified cDNA population was cloned and several independent clones were sequenced. The PCR-amplified cDNA population consisted of three different sequences, and in Southern analysis of birch genomic DNA, the cDNA fragments hybridized with three different fragments (data not shown). Thus birch genome seems to contain three genes for the ACC oxidase. At the nucleic acid level the three birch cDNA fragments were 61-77% similar. At the deduced amino acid level the amino acid identity was 76-86%. One of the birch clones is highly homologous to tomato ACO1 amino acid sequence (Figure 3). Similarly, the two other members of the birch *aco*-gene family showed high similarity to other members of the tomato *aco* gene family and ACC oxidase genes from various species (data not shown).

```
LeACO1  SNYPPCPKPD LIKGLRAHTD AGGIILLFQD DKVSGLQLLK DEQWIDV.PP MRHSIVVNLG DOLEVITNGK YKSV
        |||||||||: || ||||||| |||:|||||| |||||||||| |  |:|| || ||||||:||| |||||||||| ||||
BpACO1  SNYPPCPKPE LIQGLRAHTD AGGLILLFQD DKVSGLQLLK DGKWVDV.PP MRHSIVINLG DQLEVITNGK YKSV
BpACO3  *******E*D VAI**PP*** H*L*T**M*N E.L****VMH N****P*NH* PPN*FLV*M* *HM**L**** ****
BpACO2  *******E** **M*MPS*S* H*L*T**I*N G.I****VQH QR***N*N.S IPNAF*V*V* *HM*IMS*** ****
```

Figure 3. Comparison of three partial birch ACO amino acid sequences (BpACO1-3) deduced from the cDNA fragments to the corresponding part of the tomato ACO1 (LeACO1, Picton *et al.*, 1995) amino acid sequence. Amino acid identities between tomato ACO1 and birch ACO1 are indicated with a (|), conserved amino acid substitutions with a (:). Identical amino acids between the three birch ACO fragments are indicated with a star (*).

We have seen a rapid induction of the ACC oxidase gene in ozone-exposed tomato (Tuomainen, Kangasjärvi, Betz, Langebartels, Ernst and Sandermann, in preparation). The mRNA levels for the tomato *aco* increased immediately after the beginning of the ozone exposure and were already decreasing when the mRNA levels of a gene for ACC synthase were increasing. The birch genes for ACC oxidase had similar fast response (Figure 1, Table 1). The transcript levels increased immediately and were below the background-level at 8 hours indicating that changes in ethylene synthesis are among the very first reactions to ozone also in birch. Similar temporal response was observed in wounded leaves. We have screened a cDNA library constructed from ozone-exposed birch leaf-RNA with the *aco*-fragments. The isolated cDNA clones are being used to study in more detail birch *aco*-expression under stress and during development.

4. Timing of the biochemical and gene expression changes in ozone-exposed birch

The biochemical and gene expression changes in ozone-exposed plants can be separated into three phases: fast, intermediate and slow responses (Eckey-Kaltenbach *et al.*, 1994). The fast responses can be detected almost immediately at the beginning of the exposure. Changes taking place 6-12 hours after the beginning of the exposure can be classified as intermediate responses and the slow responses take place at 24 hours and later and are mostly related to the tissue damage formation. In our experiments with birch, tissue damage in the sensitive clone first appeared as watery lesions 24 hours after the beginning of the exposure (Table 1). Small brown/purple and necrotic flecks and sometimes larger necrotic sections were visible at 48 hours.

Fast increase in ethylene biosynthesis is one of the very first changes in ozone-exposed plants that has been observed in several experiments. Ethylene synthesis is controlled at the gene level, mostly by gene(s) for ACC synthase, but the ACC oxidase genes have also a role in the regulation. In tomato ACC synthase is encoded by at least nine genes (Zarembinski and Theologis 1994) and ACC oxidase by three genes (Picton *et al.*, 1995) that all are differentially regulated during plant development and as a response to external stimuli, such as wounding and pathogen attack. In trees, so far the only genes for ethylene biosynthesis enzymes have been isolated from apple (Dong *et al.*, 1992), where, however, the interest is to study their role in fruit ripening. The three *aco* PCR fragments and corresponding cDNA clones will allow detailed studies on ethylene synthesis and its genetic regulation in the vegetative parts of birch-trees.

TABLE 1. The fast (0-3 h), intermediate (6-12 h) and slow (> 24 h) biochemical and gene expression changes in birch exposed to ozone. Ozone sensitive (KL5) and -insensitive (KL2) birch clones were exposed to a single 8-hour ozone peak (150 ppb) in growth chambers. Several biochemical parameters and gene expression were analyzed from them (Tuomainen *et al.*, 1996). Abbreviations used: *aco*, ACC oxidase gene; *rbcS*, rubisco small subunit gene; *chs*, chalcone synthase gene; *pal*, phenylalanine ammoniumlyase gene; SOD, superoxide dismutase activity; POX, peroxidase activity; GR, glutathione reductase activity. Increase from the basal level is indicated with a +, decrease with a -.

	0-3 h		6-12 h		24 h		48 h	
Response	KL2	KL5	KL2	KL5	KL2	KL5	KL2	KL5
aco	++	++						
rbcS	-		-	-	-		-	
chs		+	-	--				
defense reactions, *pal*			+++	+++	+	+		
SOD					++	+++		
POX					+	+++	+	
GR				+	+	+++		++
water-soaked lesions						++		
necrosis								++

The end products of the phenylpropanoid pathway play several important roles in the growth and development of a plant, like for example as a part of the coordinated defense responses against pathogens. The first committed step of the pathway is catalyzed by phenylalanine ammonia-lyase, whose regulation is under strict genetic control. Since induction of *pal* in the defense responses is well characterized in severalstudies, it is widely used as an example of their activation (Hahlbrock and Scheel 1989). Both ozone (Table 1) and wounding of the tissue induced *pal* in birch. As a result of ozone, *pal* mRNA levels increased rapidly but transiently in both clones peaking at 8 hours (Tuomainen et al., 1996). A second, lower increase was visible at 24 hours. Ozone induced *pal* in both birch clones thus indicating activation of the programmed defense mechanisms, but only the sensitive clone showed wide spread cell death (Table 1). This suggests that the ozone-induced defense reactions in birch are not linked to cell death that was observed in the sensitive clone.

The basal levels of chalcone synthase transcripts were, in contrast to *pal*, clearly detectable in birch leaves. As a result of ozone-exposure, the mRNA levels decreased in both clones when *pal* mRNA levels increased (Figure 1, Table 1). This apparent down-regulation of the flavonoid branch would suggest that the carbon flow to the whole phenylpropanoid pathway as a result of PAL activation would not result in major increases in the flavonoid-substances. Indeed, when concentrations of several flavonoid substances were analyzed, changes were detected only in catechin concentrations (Figure 3). Increases in both flavonoid synthesis (Rosemann *et al.,* 1991) and lignification (Galliano *et al.,* 1993) have been observed in plants as a response to ozone-exposure. In further studies with birch it will be necessary to study flavonoid and lignin-branches together to elucidate whether ozone induces the lignin branch more preferably when the flavonoid branch (Figure 1, Table 1) seems to be down-regulated.

Changes in superoxide dismutase (SOD), peroxidases (POX) and glutathione reductase (GR) activities that have been observed in ozone-exposed plants, have been variable and dependent on several factors. Whether the SOD, POX and GR activities are protecting the cells from oxygen radicals produced from ozone, or whether they are related to damage created by ozone remains still open. According to Foyer et al., (1994), there is no evidence that the antioxidative defense systems would anticipate the potential damage caused by the AOS formed in various stress situations, but instead react only as a response to actual damage. In the ozone-sensitive birch clone the maximal activities of SOD, POX and GR (Table 1) coincided with the damage, first visible at 24 hours, i.e., 16 hours after the ozone-exposure was over. The magnitude of the enzyme activity increase was much higher in the sensitive clone. This suggests the presence and increased production of oxygen radicals in the tissues after the ozone exposure was over. Ozone can induce membrane-damaging lipid peroxidation and initiate chain reactions generating additional oxyradicals resulting ultimately in cell death. In birch the elevated SOD, POX and GR activities were probable related to oxygen radicals formed in this kind of processes rather than to protection from ozone.

5. References

Bowler, C., van Montagu, M., Inzé, D. (1992) Superoxide dismutase and stress tolerance, *Ann. Rev. Plant Physiol. Plant Mol. Biol.* **43**, 83-116

Dong, J.G., Olson, D., Silverstone, A., Yang, S.F. (1992) Sequence of a cDNA coding for a 1-aminocyclopropane-1-carboxylate oxidase homolog from apple fruit, *Plant Physiol.* **98**, 1530-1531

Eckey-Kaltenbach, H., Ernst, D., Heller, W., Sandermann, H., Jr. (1994) Biochemical plant responses to ozone. IV. Cross-induction of defensive pathways in parsley plants, *Plant Physiol.* **104**, 67-74

Foyer, C.H., Lelandais, M., Kunert, K.J. (1994) Photooxidative stress in plants, *Physiol. Plant.* **92**, 696-717

Galliano, H., Cabané, M., Eckerskorn, C., Lottspeich, F., Sandermann, H., Jr., Ernst, D. (1993) Molecular cloning, sequence analysis and elicitor-/ozone-induced accumulation of cinnamyl alcohol dehydrogenase from Norway spruce (*Picea abies* L.), *Plant Mol. Biol.* **23**, 145-156

Hahlbrock, K., Scheel, D. (1989) Physiology and molecular biology of phenylpropanoid metabolism, *Ann. Rev. Plant Physiol. Plant Mol. Biol.* **40**, 347-369

Julkunen-Tiitto, R., Meier, B. (1992) The enzymatic decomposition of salicin and its derivatives obtained from Salicaceae species, *J.Nat.Prod.* **55**, 1204-1212

Kangasjärvi, J., Talvinen, J., Utriainen, M., Karjalainen, R. (1994) Plant defense systems induced by ozone, *Plant, Cell Environ.* **17**, 783-794

Langebartels, C., Kerner, K., Leonardi, S., Schraudner, M., Trost, M., Heller, W., Sandermann, H., Jr. (1991) Biochemical plant responses to ozone I. Differential induction of polyamine and ethylene biosynthesis in tobacco, *Plant Physiol.* **95**, 882-889

Levine, A., Tenhaken, R., Dixon, R., Lamb, C. (1994) H_2O_2 from the oxidative burst orchestrates the plant hypersensitive disease resistance response, *Cell* **79**, 583-593

Mehlhorn, H., Wellburn, A.R. (1987) Stress ethylene formation determines plant sensitivity to ozone, *Nature* **327**, 417-418

Picton, S., Gray, J.E., Grierson, D. (1995) Ethylene genes and fruit ripening, in P.J. Davies, (ed.), *Plant hormones. Physiology, biochemistry and molecular biology*, 2nd edn., pp. 372-394 Kluwer Academic Publishers, Dordrecht / Boston / London

Rodecap, K.D., Tingey, D.T. (1983) The influence of light on ozone-induced 1-amionocyclopropane-1-carboxylic acid and ethylene production from intact plants, *Z. Pflanzenphys.* **110**, 419-427

Rosemann, D., Heller, W., Sandermann, H., Jr. (1991) Biochemical plant responses to ozone II. Induction of stilbene biosynthesis in scots pine (*Pinus sylvestris* L.) seedlings, *Plant Physiol.* **97**, 1280-1286

Sharma, Y., Davis, K. (1994) Ozone-induced expression of stress-related genes in *Arabidopsis thaliana*, *Plant Physiol.* **105**, 1089-1096

Tuomainen, J., Pellinen, R., Roy, S., Kiiskinen, M., Eloranta, T., Karjalainen, R., Kangasjärvi, J. (1996) Ozone affects birch (*Betula pendula* Roth) phenylpropanoid, polyamine and active oxygen detoxifying pathways at biochemical and gene expression level, *J. Plant Physiol.* In Press

Zarembinski, T.I., Theologis, A. (1994) Ethylene biosynthesis and action: A case of concervation, *Plant Mol. Biol.* **26**, 1579-1598

UV-B-INDUCED GENE EXPRESSION IN EUROPEAN BEECH AND SCOTS PINE

D. ERNST[1], I. LIEGL[1], J.-P. SCHNITZLER[1], T. P. JUNGBLUT[1], C. ZINSER[1], M. KÖFFERLEIN[2], E. SCHMELZER[3], W. HELLER[1], H. SANDERMANN JR.[1]
GSF - [1]Institut für Biochemische Pflanzenpathologie,
[2]Arbeitsgruppe Expositionskammern, D-85758 Oberschleißheim, Germany;
[3]MPI für Züchtungsforschung, Carl-von-Linné-Weg 10, D-50829 Köln, Germany
(Fax: +49-89-3187-3383; e-mail: ernst@gsf.de)

Abstract

The timing of changes in poly(A)$^+$ RNA pattern of UV-B-irradiated European beech (*Fagus sylvatica* L.) seedlings was analyzed by two-dimensional polyacrylamide gel electrophoresis (2D-PAGE) of *in vitro* translated poly(A)$^+$ RNA. Both, qualitative and quantitative differences in the translation products were observed. Expression of specific mRNAs, coding for UV-B-induced polypeptides between 14 kDa and 17 kDa, as well as repressed polypeptides ranging from 14 kDa to 46 kDa was demonstrated after 3 h of UV-B irradiation.

In UV-B-treated Scots pine (*Pinus sylvestris* L.) seedlings the mRNA of chalcone synthase (CHS), a key enzyme of flavonoid biosynthesis was found in epidermal, mesophyll, as well as vascular tissues and no quantitative changes, compared with controls were found.

Introduction

Reduction of stratospheric ozone is thought to increase the UV-B radiation reaching the earth's surface. Damaging effects of enhanced UV-B radiation equivalent to relatively high ozone depletion rates have been demonstrated (Tevini, 1993). The ability of UV-B radiation to damage DNA, RNA, proteins and membranes (Tevini and Teramura, 1989; Stapleton, 1992), or impair physiological processes, e.g. photosynthesis (Bornman, 1989; Strid *et al.*, 1990; Jordan *et al.*, 1994) is well established in UV-B sensitive plant species or cultivars. In addition to the protective mechanism of photoreactivation (Langer and Wellmann, 1990; Pang and Hays, 1991), the leaf epidermis is thought to play a major role in attenuating UV-B radiation, and flavonoids as well as hydroxycinnamic acid derivatives have been implicated to protect plants from the damaging effects of this radiation (DeLucia *et al.*, 1992; Day *et al.*, 1993; Li *et al.*, 1993; Tevini, 1993; Jungblut *et al.*, 1995). Measurement of UV-B penetration into the leaf, using a fibre-optic microprobe revealed that UV-B radiation is efficiently absorbed in the epidermal layer of conifer needles, while substantial amounts of UV-B radiation reach the mesophyll tissue in herbaceous plants (DeLucia *et al.*, 1992; Day *et al.* 1993). Molecular biological studies have so far been focused on flavonoid biosynthesis, showing an increased transcript accumulation (Hahlbrock and Scheel, 1989; Kubasek *et al.*, 1992), whereas transcripts for chloroplast proteins were reduced (Jordan *et al.*, 1994).

In an attempt to more fully describe the response of woody plants to UV-B radiation, this paper describes localization of CHS mRNA, and flavonoid pigments in tissue of Scots pine needles. *In vitro* translation analysis by 2D-PAGE revealed the influence of UV-B radiation in European beech leaves at the mRNA level.

M. R. Ahuja et al. (eds.), Somatic Cell Genetics and Molecular Genetics of Trees, 273–276.
© 1996 *Kluwer Academic Publishers. Printed in the Netherlands.*

Results and Discussion

2D-PAGE ANALYSIS OF *IN VITRO* TRANSLATION PRODUCTS FROM EUROPEAN BEECH

European beech seedlings were cultivated similar as described by Schraudner (1992), acclimated and UV-B-irradiated in a sun simulator (1300 μmol·m^{-2}·s^{-1}; 440 mW·m^{-2} UV-B_{BE})) (Seckmeyer and Payer, 1993). To get a better insight into the interaction of UV-B radiation and gene expression in this deciduous tree, changes of poly(A)$^{+}$ RNA isolated from UV-B-irradiated and control beech seedlings were analyzed by *in vitro* translation and 2D-PAGE. Poly(A)$^{+}$ RNA isolation and *in vitro* translation was according to Großkopf *et al.* (1994). In all cases poly(A)$^{+}$ RNA was isolated from two-month-old seedlings during 48 h after onset of UV-B irradiation. Neutral and acidic soluble polypeptides were separated by isolelectric focusing between pH 3 and 10.5. The M_r of the proteins studied was in the range of 12 kDa to 100 kDa. Most of the spots matched showed no strong difference in intensity after UV-B treatment of the plants. However, at least 10 different proteins were newly synthesized or accumulated, whereas about 5 proteins decreased during the UV-B irradiation period (Fig. 1). As indicated by arrows in Fig. 1, two polypeptides with a M_r of approximately 17 kDa increased from near zero levels, whereas two polypeptides with a M_r of about 16 kDa and 30 kDa, respectively, decreased dramatically. M_r, as well as isoelectric point of the two polypeptides increasing in amount fit very well to published data for 'pathogenesis-related' proteins of parsley (Somssich *et al.*, 1986), however, the characterization and identification of these proteins remains to be established by further analysis.

We show here at the mRNA level, that transcript repression is as transient as induction in beech leaves. Ten hours after onset of the experiments the *in vitro* translation pattern of UV-B treated and control seedlings were again very similar and no obvious differences were evident. Similar results have been obtained with UV-B-treated parsley leaves, showing an increase of CHS mRNA during the first hours and a resumption to the initial steady-state level after subsequent days (Jordan *et al.*, 1994). Therefore it is likely that beech seedlings rapidly adapted to the UV-B radiation applied.

TISSUE LOCALIZATION OF CHALCONE SYNTHASE mRNA AND OF ACYLATED FLAVONOIDS IN SCOTS PINE NEEDLES

Scots pine seedlings were cultivated according to Rosemann *et al.* (1991) and then acclimated for two weeks to high light conditions in the sun simulator, before the onset of UV-B lamps (218 mW·m^{-2} UV-B_{BE}) (Schnitzler *et al.*, 1994). CHS mRNA isolation, *in situ* hybridization and fluorescence microscopy of needle cross-sections have been described in detail (Schnitzler *et al.*, 1996). Hybridization with CHS antisense RNA was observed in the whole tissue. No preference for individual cell types, such as epidermal or mesophyll cells and cells of the vascular bundle was evident (Fig. 2). Similar results have been obtained in other plants where CHS transcripts were always detected in those tissues that contained flavonoid metabolites (Schmelzer *et al.*, 1988). UV-B irradiation resulted in no further increase in detectable CHS mRNA after 4 d, as can be seen in the uniform distribution of silver grains (compare Fig. 2A and B). This is similar to a transient induction of CHS mRNA in parsley, as well as in *Arabidopsis*, grown for several days under UV-B-supplemented white light (Schmelzer *et al.*, 1988; Kubasek *et al.*, 1992).

Cross-sections of cotelydons from UV-B-treated seedlings were stained with Naturstoffreagenz A and analyzed by fluorescence microscopy. A strong fluorescence was seen exclusively in the outer periclinal wall of the epidermal layer and could be assigned to kaempferol 3-glucoside. Furthermore, diacylated flavonol glycosides were the main soluble metabolites of the epidermal layer (Jungblut *et al.*, 1995; Schnitzler *et al.*, 1996). Similarly in petunia plants a strong fluorescence in the epidermal tissue was assigned to flavonol derivatives (Vogt *et al.*, 1994). The absorption properties at 300 nm for all epidermal screening pigments was calculated using Lambert-Beer's law. The calculated value of T_{300} ($2 \cdot 10^{-5}$ %) demonstrated a virtually complete elimination of UV-B radiation (Schnitzler *et al.*, 1996). These results suggest that Scots pine needles are well-protected against increased levels of UV-B radiation expected for the future (Blumthaler, 1993).

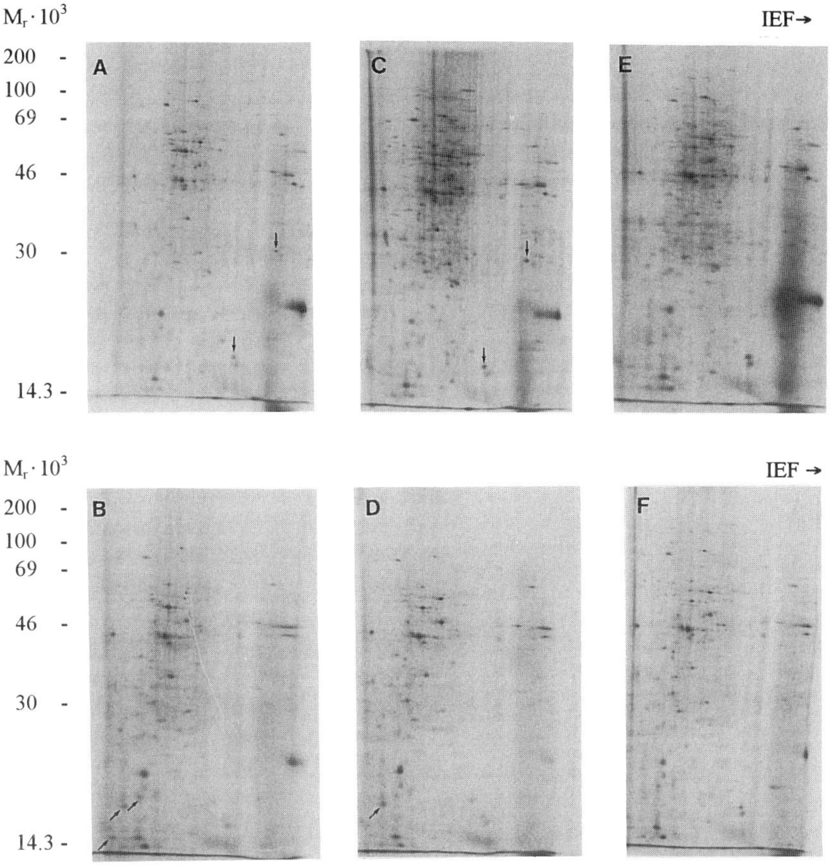

Figure 1. Fluorography of 2D-PAGE analysis of *in vitro* translated poly(A)$^+$ RNA isolated from leaves of 2-month-old European beech seedlings. Seedlings were irradiated with 440 mW·m^{-2} UV-B$_{BE}$ for 6 h and then further cultivated without UV-B radiation. After 3 h: control (A), UV-B (B), after 6 h: control (C), UV-B (D); after 10 h: control (E), UV-B (F). Isolated poly(A)$^+$ RNA was translated *in vitro* using a retiulocyte lysate extract, and approximately 500 000 dpm TCA-insoluble radioactivity was loaded on the first dimension. Upword arrows indicate polypeptides which increased or accumulated and downword arrows indicate polypeptides which were repressed by UV-B irraditation.

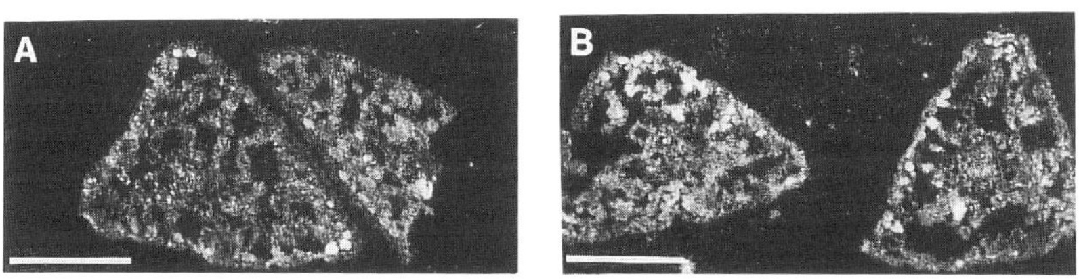

Figure 2. In situ localization of CHS mRNA in needles of UV-B-irradiated Scots pine seedlings. Five-week-old Scots pine seedlings were radiated with 218 mW·m^{-2} UV-B$_{BE}$ for additional 4 d (6 h/d) and cross-sections of UV-B-irradiated (A) or control (B) needles were hybridized with ^{35}S-labeled CHS antisense RNA. Bars represent 200 μm.

Acknowledgements. This work has been supported by EUROSILVA and Fonds der Chemischen Industrie.

References

Blumthaler, M. (1993) Solar UV measurements, in M. Tevini (ed.), *UV-B Radiation and Ozone Depletion*, Lewis Publishers, Boca Raton, pp. 71-94.

Bornman, J.F. (1989) Target sites of UV-B radiation in photosynthesis of higher plants. *J. Photochem. Photobiol.* **4,** 145-158.

Day, T.A., Martin, G., and Vogelmann, C. (1993) Penetration of UV-B radiation in foliage: evidence that the epidermis behaves as a non-uniform filter. *Plant Cell Environ.* **16,** 735-741.

DeLucia, E.H., Day, T.A., and Vogelmann, T.C. (1992) Ultraviolet-B and visible light penetration into needles of two species of subalpine conifers during foliar development. *Plant Cell Environ.* **15,** 921-929.

Großkopf, E., Wegener-Strake, A., Sandermann, H., and Ernst, D. (1994) Ozone-induced metabolic changes in Scots pine: mRNA isolation and analysis of in vitro translated proteins. *Can. J. For. Res.* **24,** 2030-2033.

Hahlbrock, K. and Scheel, D. (1989) Physiology and molecular biology of phenylpropanoid metabolism. *Annu. Rev. Plant Physiol. and Plant Mol. Biol.* **40,** 347-369.

Jordan, B.R., James, P.E., Strid, A., and Anthony, R.G. (1994) The effect of ultraviolet-B radiation on gene expression and pigment composition in etiolated and green pea leaf tissue: UV-B-induced changes are gene-specific and dependent upon the developmental stage. *Plant Cell Environ.* **17,** 45-54.

Jungblut, T.P., Schnitzler, J.P., Hertkorn, N., Metzger, J.W., Szymczak, W., and Sandermann, H. (1995) Struktur UV-B-induzierter Sonnenschutzpigmente der Kiefer (*Pinusy sylvestris* L.). *Angew. Chem., Int. Ed. English* **34,** 312-314.

Kubasek, W.L., Shirley, B.W., McKillop, A., Goodman, H.M., Briggs, W., and Ausubel, F.M. (1992) Regulation of flavonoid biosynthetic genes in germinating Arabidopsis seedlings. *Plant Cell* **4,** 1229-1236.

Langer, B. and Wellmann, E. (1990) Phytochrome induction of photoreactivating enzyme in *Phaseolus vulgaris* L. seedlings. *Photochem. Photobiol.* **52,** 861-863.

Li, J., Ou-Lee, T.-M., Raba, R., Amundson, R.G., and Last, R.L. (1993) Arabidopsis flavonoid mutants are hypersensitive to UV-B irradiation. *Plant Cell* **5,** 171-179.

Pang, Q. and Hays, J.B. (1991) UV-B-inducible and temperature-sensitive photoreactivation of cyclobutane pyrmidine dimers in *Arabidopsis thaliana*. *Plant Physiol.* **95,** 536-543.

Rosemann, D., Heller, W., and Sandermann, H. (1991) Biochemical plant responses to ozone. II. Induction of stilbene biosynthesis in Scots pine (*Pinus sylvestris* L.) seedlings. *Plant Physiol.* **97,** 1280-1286.

Schmelzer, E., Jahnen, W., and Hahlbrock, K. (1988) *In situ* localization of light-induced chalcone synthase mRNA, chalcone synthase, and flavonoid end products in epidermal cells of parsley leaves. *Proc. Natl. Acad. Sci USA,* **85,** 2989-2993.

Schnitzler, J.P., Jungblut, T.P., Heller, W., Langebartels, C., Köfferlein, M., and Sandermann, H. (1994) Effects of ozone and UV-B radiation on Scots pine. *Acta Hort.* **381,** 463-469.

Schnitzler, J.-P., Jungblut, T.P., Heller, W., Köfferlein, M., Hutzler, P., Heinzmann, U., Schmelzer, E., Ernst, D. Langebartels, C., and Sandermann, H. (1996) Tissue localisation of UV-B screening pigments and of chalcone synthase mRNA in needles of Scots pine (*Pinus sylvestris* L.) seedlings. *New Phytol.*, in press.

Schraudner, M. (1992) Frühe Reaktionen von Tabak und Buche auf Ozon: Induktion des Tyramin- und Polyaminstoffwechsels. PhD thesis, Universität München.

Seckmeyer, G. and Payer, H.D. (1993) A new sunlight simulator for ecological research on plants. *J. Photochem. Photobiol. B* **21,** 175-181.

Somssich, I.E., Schmelzer, E., Bollmann, J., and Hahlbrock, K. (1986) Rapid activation by fungal elicitor of genes encoding "pathogenesis-related" proteins in cultured parsley cells. *Proc. Natl. Acad. Sci USA* **83,** 2427-2430.

Stapleton, A.E. (1992) Ultraviolet radiation and plants: burning questions. *Plant Cell* **4,** 1353-1358.

Strid, A., Chow, W.S., and Anderson, J.M. (1990) Effects of supplementary ultraviolet-B radiation and photosynthesis in *Pisum sativum*. *Biochim. Biophys. Acta* **1020,** 260-268.

Tevini, M. (1993) Effects of enhanced UV-B radiation on terrestrial plants., in M. Tevini (ed.), *UV-B Radiation and Ozone Depletion*, Lewis Publishers, Boca Raton, pp. 125-153.

Tevini, M. and Teramura, A.H. (1989) UV-B effects on terrestrial plants. *Photochem. Photobiol.* **50,** 479-487.

Vogt, T. Pollak, P., Tarlyn, N., and Taylor, L.P. (1994) Pollination- or wound-induced kaempferol accumulation in petunia stigmas enhances seed production. *Plant Cell* **6,** 11-23.

STRESS-RELATED GENES IN WOODY PLANTS: TRANSCRIPTIONAL AND POST-TRANSCRIPTIONAL REGULATION

J. CAIRNEY, D.K. VILLALON, S. CHANG,
Institute of Paper Science and Technology,
500 10th Street, N.W
Atlanta, GA 30318, USA

M.A.D.L. DIAS, R.J. NEWTON
Department of Forest Science,
Texas A&M University,
College Station, TX 77843, USA

ABSTRACT

The induction of gene expression in response to environmental or hormonal cues, has been studied mostly at the of transcriptional level. The importance of post-transcriptional events in determining the final protein levels, however, is now becoming clear. As part of a program to investigate gene expression in woody plants growing under water deficit, we have isolated several cDNA clones of a Proteinase Inhibitor-like gene family from the woody desert shrub, *Atriplex canescens*. Northern analysis shows a broad inducible signal in response to water deficit. Reverse Transcription -PCR (RT-PCR) experiments show that levels of the individual transcripts are altered to different degrees by exposing plants to water deficit or Abscisic acid. Sequence analysis reveals several motifs in the 3'-untranslated region (3'-UTR) of the mRNA which may contribute to RNA stability or translatability. In a parellel program a cDNA for water deficit-repressible gene is shown to have unusual structure. The longest open reading frame begins over halfway through the mRNA and the 5'-untranslated region (5'-UTR) has extensive secondary structure and several upstream open reading frames (uORFs). These 5'-UTR features are involved in post-transcriptional regulation of gene expression in other eukaryotic systems. Elucildating the function of these sequences will assist in understanding the regulation of stress-inducible genes and in the design of plant transformation vectors to optimize the expression of foreign genes.

INTRODUCTION

The study of plant gene expression in response to environmental stress has resulted in the elucidation of a number of biochemical defense and repair mechanisms (Skriver and Mundy 1990, Keen 1992). The application of that knowledge, through the isolation of genes, their modification and subsequent transfer into plant cells, has brought spectacular successes (Chrispeels and Sadava 1994). As recognition of the efficacy of biotechnologies has spread, the desire among researchers and corporate enterprises to avail themselves of new opportunities has broadened the field. An increasing numbers of researchers now conduct molecular experiments with plants and with their arrival, the less highly publicized technical problems of gene transfer and expression have begun to surface (Matzke and Matzke 1995). The recognition of the complexities of gene regulation has led researchers to seek more fundamental understandings of gene expression. In this paper we discuss the regulation of two gene families whose members display variable responses to environmental and hormonal cues. The

M. R. Ahuja et al. (eds.), Somatic Cell Genetics and Molecular Genetics of Trees, 277–283.
© 1996 *Kluwer Academic Publishers. Printed in the Netherlands.*

sequence and structural features of the genes and mRNAs suggest tiers of regulation which would allow rapid modulation of protein synthesis in response to stress.

METHODS

Plant material and water deficit stress treatments. Saltbush (Atiplex canescens (Pursh Nutt.): Whole plants were desiccated after growth in pots for 4-5 weeks. After washing gently many times to remove soil, plants will be left to desiccate on Whatman 3MM paper at 22C and 60% humidity under dim light following published protocols (Yamaguchi-Shinozaki and Shinozaki, 1993 a and b). Plants treated with ABA were transferred from SEM directly to AM agar, then removed and grown hydroponically in half strength either MS media or MS media supplemented with 10(M ABA for 14 h.
Loblolly pine (Pinus taeda L.):Seedlings were full-sibling, resulting from the cross of S6PT2 and S6PT3 sources from east Texas. The seeds were sown in 5-liter cylindrical containers filled with a fritted clay medium adapted for pine seedlings (Meier et al. 1992). The seedlings were randomly separated into 7 groups, water was then withheld from the seedling groups in staggered fashion. Seedlings were harvested pre-dawn on the same day at which point the water potentials were taken. (Chang et al 1995).
Molecular methods and cDNA library construction. For Atriplex a cDNA library was made from the leaves of seedlings with water potential - 0.94 MPa (Adair et al 1992). The pine library was made from the entire root tissue of an 8-month-old seedling with a water potential of -1.1 MPa (Cairney et al 1993). In both cases the vector was the (gt10-derived vector, Lambda-ZAP (Stratagene, La Jolla, CA).
Reverse Transcription-Polymerase Chain Reaction. One microgram of total RNA was used for Reverse transcription using MMLV Reverse Transcriptase (Promega, Madison, WI). One tenth of the reaction products were then used for PCR. Conditions were, 940C, 15 seconds, 50-550C, 15 seconds, 720C 30 seconds, 35 cycles. Reaction products were transferred to membranes, probed with a labelled cDNA and the membrane scanned and signals quantitated.
DNA sequencing and homology comparison. DNA sequencing was performed by the dideoxy chain termination method (Sanger et al. 1977). Standard molecular methods were used in other cases (Sambrook et al. 1989).

RESULTS AND DISCUSSION

Halophytes which survive long periods with little moisture, possess both physiological and biochemical adaptations which permit them to prosper under conditions where most plant would perish. A study of halophyte physiology may shed light on novel molecular mechanisms of stress tolerance or may reveal a source of new genes. In parallel projects we have chosen to study gene expression under water deficit in the woody desert shrub, Saltbush (Atriplex canescens) and the commercially important conifer, Loblolly Pine (Pinus taeda L.). In both cases cDNA libraries were constructed from water stressed plants and differentially screened to isolate clones of genes whose steady-state mRNA levels fluctuate under water deficit. Details of clone isolation have been reported elsewhere (Adair et al 1992, Cairney et al 1993. Chang et al 1995).

An Atriplex clone, p23-3, isolated from a cDNA library by differential screening was shown to encode a protein with strong similarity to Proteinase Inhibitor-I (PI-I) from soybean and potato. The clone proved distinctive in hybridizing to a broad mRNA band in Northern analysis. This signal appeared be divided between two principal bands, of about 0.5 and 0.7kb, suggesting closely related transcripts. By rescreening the library four additional cDNA clones which hybridized to p23-3 were isolated. DNA sequencing showed them to be 95% identical at the nucleotide level. They appeared to be part of a multigene family. Most of these clones were around 0.4 kb however clone p8-3 was 0.65 kb and differed from the other clones, principally, in possessing a longer 3'UTR which contained a AU-rich region (Fig.1).

Since the translation products of these mRNA molecules were very similar, the presence of transcripts of varying size and sequence within a given organ begs questions about their role. Several explanations are possible: The PI-1-like proteins may differ in subtle but significant ways, in their substrate specificity and/or efficiency of action, thus fulfilling different roles within the cell. These transcripts may be produced in different cell types within an organ, their cognate genes responding to different cues each with a appropriate stability and translational efficiency. The altered cytoplasmic conditions of water deficit may favor the stability or translation of one transcript over another. A number of AU-rich sequences have been identified in the 3'UTRs of transcripts. These sequences have been assigned roles in mRNA instability (Green 1993), translational enhancement (Gallie 1993) regulated cytoplasmic polyadenylation and deadenylation (Bachvarova 1992)

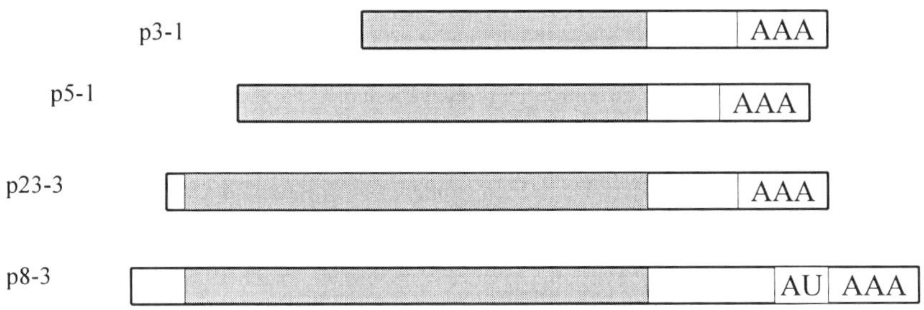

Figure 1. Diagram showing homology among PI-1-like cDNA clones. Darker shaded area represents the open reading frames (commencing ATG). The lighter shaded area marked AU is an AU-rich sequence. The AAA boxes at the 3' terminal of the clones represent the polyA tails. The clones are approximately 95% identical at the nucleotide level.

Genomic Southern blots gave a complex pattern indicating many copies of the PI-I genes. To shed light on the transcriptional regulation of these genes genomic clones were isolated. Two clones, pG12-95 and pG18-1 were sequenced. The exons and introns of both genomic clones were assigned on the basis of nucleotide sequence comparison with the cDNA clones. Neither genomic clones matches a cDNA exactly; their putative exonic sequences show 95% identity to the cDNAs. Both pG12-95 and pG18-1 contain one intron of 931 and 422 bp in length, respectively. The first exon of both clones are identical in length, coding for 17 amino acids. The principal difference between the clones, within the transcribed region, was the length of the introns. Preliminary analysis of the promoters reveals a number of motifs which have been identified in other plants as transcription factor binding sites. These include ABA-Responsive Elements (ABREs) (Skriver and Mundy 1990). A number of unusual repeat structure are conspicuous, their function is being investigated. There is no AU-rich region in the second exon of pG12-95 however pG18-1, while differing in sequence from the other clones, contains a 3'AU-rich sequence which is almost identical to that of cDNA clone p8-3.

Multigene families contain many pseudogenes and sequence analysis alone cannot distinguish these from transcriptionally active form. We have sought a swift, sensitive and specific assay of gene expression through the use of specific oligonucleotides and Reverse-Transcription-PCR (RT-PCR). Oligonucleotides specific to each transcript and suitable for PCR were synthesized. Each was designed to amplify a product of a characteristic size, facilitating product identification. Preliminary assay using the cDNA and genomic clones showed that these primers displayed fidelity, amplifying the correct fragment from a mixture of plasmid templates.

RT-PCR experiments were carried out with RNA from Atriplex seedlings dehydration-stressed for varying periods of time. Each transcript showed some level of induction however the magnitude of the response and its' triggering point were greatly different. The same individual patterns were observed during ABA induction. These results suggest that our technique is sufficiently sensitive to discriminate between individual members of a multigene family and at present these experiments are being repeated to establish levels of confidence with the assay. Such a technique will be of great use in determining which members are induced under different circumstances. Such assessment, in combination with other biochemical and physiological data, would allow workers to chose the most appropriate protein for their purposes from among a family of similar polypeptides. In gene regulation studies, pseudogenes or genes whose level of induction is low, specific to a tissue or environmental signal, could be identified.

As part of a parallel project, examining gene expression in response to water deficit in Loblolly Pine, a cDNA library was constructed from the roots of eight month old pine seedling which had been deprived of water for 11 days. The library was differentially screened using polyA+ RNA from control seedling (water potential, -0.4MPa) and stressed seedlings (-1.3MPa) and examples of inducible and repressible clones were selected (Cairney et al 1993, Chang et al 1995). Our experiments sought to simulate dehydration experienced by plants during periods of prolonged stress. Further by making our cDNA library from plants subject to water deficit for about two weeks, we would select mRNA from genes which were expressed at elevated levels some time after water deficit stress began. These may be "adaptation genes". Of six cDNA clones which we have analysed in detail, most seem implicated in cell wall re-inforcement (Chang et al 1995). Our library screening did not select any of the Lea family of genes, often isolated by such procedures (Baker et al 1988, Skriver and Mundy 1990). In many of these studies, water-deficit is imposed rapidly and, in addition, plants are harvested soon after the imposition of stress. Recently, Leone et al (1994) showed that in potato cell suspension culture, rapid vs gradual stress induced different sets of polypeptides. Proteins induced by rapid stress were also ABA-inducibe, those induced by gradually imposed stress were not. Either the "window" of expression examined by earlier studies, or the manner in which stress was imposed, could explain the absence of lea clones from our library screening.

One clone, pLP6, is strongly expressed in the roots and stems of well-watered plants but mRNA levels decline rapidly as plants dehydrate. The same pattern is seen in needles although the absolute level of gene expression is much lower. A single transcript of 1.5kb is dectected, the same size as the cDNA suggesting that almost all the information in the mRNA is present in the cDNA clone. We also showed that steady state RNA levels of pLP6 are greatly reduced 6h after wounding.

The nulceotide sequence of pLP6 consists of 1488 nulceotides, concluding with a polyA tail. The longest open reading frame commencing with an ATG could encode a polypeptide of 216 amino acids, which has a predicted molecular weight of 24.2 kD and PI of 5.05. However this open reading frame does not start until nucleotide 721, almost halfway through the mRNA (Fig. 2). To demonstrate that pLP6 is not a "double clone" fortuitously hybridizing to an mRNA of similar size we repeated hybridization using a 500 bp 5' probe generated using a convenient EcoRI site and found a similar hybridization pattern . To eliminate any possibility of an abundant mRNA masking the expression of a rare transcript, we performed RT-PCR using primers near the 5' terminal of the cDNA and from within the open reading frame. A fragment of predicted size was amplified and cloned. The sequence of approximately 150 nucleotides from either end of this molecule was determined, this is identical to the corresponding region of pLP6 (data not shown). These results confirm that pLP6 is derived from a single mRNA molecule.

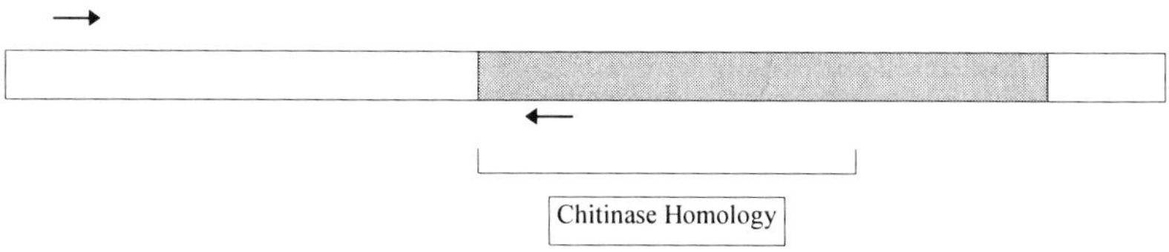

Figure 2. Schematic showing structure of pLP6 cDNA. The shaded region indicates the largest open reading frame. The region labelled "Chitinase Homolgy", shows 31-44% identity and 50-63% similarity to various Class 1 Chitinases from Bean (Broglie et a 1986), Tobacco (Shinshi et al 1987) and Poplar (Parsons et al 1989). The Arrows indicate the location of primers used in RT-PCR, to establish that pLP6 is derived from a single mRNA molecule.

The polypeptide encoded by pLP6 shows strong homology to a number of Class I Chitinases from bean, tobacco and poplar (Broglie et al 1986, Shinshi et al 1987, Parsons et al 1989) however the simililarity is only with the carboxy half of these proteins. The signal peptide, cysteine-rich chitin binding domain and Glycine/Proline rich "hinge" region, all located in the amino terminal portion of the Class I chitinases (Collinge et al 1993, Raikhel et a 1993) are absent from LP6. Neither the putative catalytic site nor the carboxy terminal sequence involved in translocation to the vacuole is present. In addition the pLP6 protein has a carboxy extension of 69 amino acids not present in any of the chitinases though no homolgies with any sequence in the gene bank could be detected.

At present we have no information on the function of the putative LP6 protein. The physiological function of chitinases has long proved elusive however recently, with the elucidation of the biochemical nature of Nod factors (D nari and Cullimore 1993) and the demonstration that a protein capable of rescuing a developmentally blocked carrot embryo mutant was a chitinase (De Jong et al 1992), a role for chitinases in development is being suggested. It is possible that the LP6 protein fulfills some role in develpment and to investigate this possibility expression in early embryos is being investigated.

The sequence of the 5'UTR of pLP6 revealed several inverted repeats which could form stem loop structures of moderate to high stabiliy (-11kcal/mol to -17kcal/mol). In addition eight upstream open reading frames (uORFs) were identified. Both these feature are found often in genes exhibiting post transcriptional regulation (Gallie 1993). The 600 nucleotide 5'UTR from Cauliflower Mosaic Virus posseses extensive secondary structure and several uORFs and recently a novel translation mechanism, a "ribosome shunt", was proposed to control expression of the downstream cistron (Fütterer et al 1993). At present we are cloning copies of the LP6 5'UTR into plant expression vectors to determine whether reporter gene expression can be influenced by this sequence.

CONCLUSIONS

Sequence anaysis and preliminary regualtory studies show stress genes are regulated at several levels. *In vivo* analysis of gene expression in transgenic plants and additional environmental and hormonal treatments will illuminate the relative contribution of each step and provide information for efficient biotechnological applications in forest species.

ACKNOWLEDGEMENTS

Acknowledgement. The authors thank Dr. Larry Harris-Haller of DNA Technologies Laboratories at Texas A&M University. This work was supported by the Temple Inland Foundation, Diboll, Texas, the Texas Agricultural Experiment Station and USDA-BARD grant US-2212-92.

LITERATURE CITED

Adair, L. S., D.L.Andrews, J. Cairney, E.A. Funkhouser, R.J. Newton, and Aldon E.A. (1992) Characterizing gene responses to drought stress in fourwing saltbush [Atriplex canescens (Pursh.) Nutt.]. J. Range Manage. 45, 454-459.

Bachvarova, R.F. (1992) A maternal tail of poly(A): the long and the short of it. Cell 69, 895-897.

Baker, J., Steele, C. & Dure III, L. 1988. Sequence and characterization of 6 Lea proteins and their genes from cotton. - Plant Mol. Biol. 11: 277-291.

Broglie, K.E., Gaynor J.J., Broglie, R.M. (1986). Ethylene-regulated gene expression: molecular cloning of the genes encoding an endochitinase from Phaseolus vulgaris. Proc Natl Acad. Sci USA 83, 6820-6824

Cairney, J., Chang, S. , Dias, M.A.D.L., Funkhouser, E.A. and Newton, R.J. (1993) cDNA Cloning and Analysis of Loblolly Pine Genes Induced by Drought Stress. Proceedings, 22nd Southern Forest Tree Improvement Conference 14-17 June, Atlanta, Georgia. pp357-369.

Chang S., Puryear J., Cairney J. (1993) A simple and efficient method for isolating RNA from pine trees. Plant Molecular Biology Reporter. 11, 114-117.

Chang, S., Puryear, J.D., Dias, M.A.D.L., Funkhouser, E.A., Newton, R.J. , and Cairney, J. .(1995) Gene expression under water deficit in loblolly pine (Pinus taeda L.): Isolation and characterization of cDNA clones. (Physiol. Plant. in press)

Chrispeels, M.J., and Sadava, D.E. (1994) Plant genetic engineering: new genes in old crops. in Plant, Genes and Agriculture. Jones and Bartlett Publishers.

Collinge, D.B., Kragh, K.M., Mikkelsen, J.D., Nielsen, K.K., Rasmussen, U., and Vad, K. (1993) Plant chitinases. Plant J. 3, 31-40.

De Jong, A.J., Cordewener, J., Schiavo, F.L., Terzi, M., Vandekerckhove, J., Van Kammen, A., and De Vries, S.C. (1992) A carrot somatic embryo mutant is rescued by chitinase. Plant Cell 4, 425-433.

Dénarié J., and Cullimore, J. (1993) Lipo-oligosaccharide nodulation factors: a minireview. New class of signaling molecules mediating recognition and morphogenesis. Cell 74, 951-954.

Fütterer, J, Kiss-László, Z., and Hohn, T. (1993) Nonlinear ribosome migration on cauliflower mosaic virus 35S RNA. Cell 73, 789-802.

Gallie, D.R. (1993) Posttranscriptional regulation of gene expression in plants. Annu. Rev. Plant Physiol. Plant Mol. Biol. 44: 77-104.

Green, P. (1993) Control of mRNA stability in higher plants. Plant Physiol. 102, 1065-1070.

Keen, N.T. (1992) The molecular biology of disease resistance. Plant Mol. Biol. 19, 109-122.

Leone, A., Costa, H., Tucci, M. and Grillo, S. (1994) Comparative analysis of short- and long-term changes in gene expression caused by low water potential in potato (Solanum tuberosum) cell-suspension cultures. Plant Physiol. 106, 703-712

Matzke, M.A. and Matzke, A.J.M. (1995) Homology-dependent gene silencing in transgenic plants: what does it really tell us? Trends Genet. 11, 1-3.

Meier, C.E., Newton, R.J., Puryear, J.D. and Sen, S. (1992) Physiological responses of loblolly pine (Pinus taeda L.) seedlings to drought stress: osmotic adjustment and tissue elasticity. J. Plant Physiol. 140, 754-760

Parsons, T.J., Bradshaw, H.D. and Gordon, M.P. (1989) Systemic accumulation of specific mRNAs in response to woudning in poplar trees. Proc Natl Acad.Sci USA 86, 7895-7899.

Raikhel, N.V., Lee H-I., and Broekaert, W.F. (1993) Structure and function of chitin-binding proteins. Annu. Rev. Plant Physiol. Plant Mol. Biol. 44, 591-615.
Sambrook, J., Fritsch, E.F., and Maniatis, T. (1989) Molecular Cloninng: A laboratory Manual, 2nd Ed. Cold Spring Harbor Laboratory Press, Cold Spring Harbor, NY.
Sanger, F., Nicklen, S., and Coulson, A.R. (1977) DNA sequencing with chain-terminating inhibitors. Proc.Natl. Acad. Sci. USA 74, 5463- .
Shinishi, H., Mohnen, D., and Meins Jr., F. (1987) Regulation of a plant pathogenesis-related enzyme: inhibition of chitinase and chitinase mRNA accumulation in cultured tobacco tissues by auxin and cytokinin. Proc. Natl. Acad. Sci. USA 84, 89-93.
Skriver, K. and Mundy, J. (1990). Gene expression in response to abscisic acid and osmotic stress Plant Cell 2, 503-512.
Stiekema, W.J., Heidekamp, F., Dirkse, W.G., Van Beckum, J., de Haan, P., Bosch, C, and Louwerse. (1988). Molecular cloning and analysis of four potato tuber mRNAs. Plant Mol. Biol. 11, 255-269.
Yamaguchi-Shinozaki, K., and Shinozaki, K. (1993a) Characterization of the expression of a desiccation-responsive rd29 gene of Arabidopsis thaliana and analysis of its promoter in transgenic plants. Mol. Gen. Genet. 236, 331-340.
Yamaguchi-Shinozaki, K., and Shinozaki, K. (1993b) The plant hormone abscisic acid mediates the drought-induced expression but not the seed-specific expression of rd22, a gene responsive to dehydration stress in Arabidopsis thaliana. Mol. Gen. Genet. 238, 17-35.

Index

FORESTRY SCIENCES

1. P. Baas (ed.): *New Perspectives in Wood Anatomy.* Published on the Occasion of the 50th Anniversary of the International Association of Wood Anatomists. 1982 ISBN 90-247-2526-7
2. C.F.L. Prins (ed.): *Production, Marketing and Use of Finger-Jointed Sawnwood.* Proceedings of an International Seminar Organized by the Timber Committee of the UNECE (Halmar, Norway, 1980). 1982 ISBN 90-247-2569-0
3. R.A.A. Oldeman (ed.): *Tropical Hardwood Utilization.* Practice and Prospects. 1982 ISBN 90-247-2581-X
4. P. den Ouden (in collaboration with B.K. Boom): *Manual of Cultivated Conifers.* Hardy in the Cold- and Warm-Temperate Zone. 3rd ed., 1982 ISBN Hb 90-247-2148-2; Pb 90-247-2644-1
5. J.M. Bonga and D.J. Durzan (eds.): *Tissue Culture in Forestry.* 1982 ISBN 90-247-2660-3
6. T. Satoo: *Forest Biomass.* Rev. ed. by H.A.I. Madgwick. 1982 ISBN 90-247-2710-3
7. Tran Van Nao (ed.): *Forest Fire Prevention and Control.* Proceedings of an International Seminar Organized by the Timber Committee of the UNECE (Warsaw, Poland, 1981). 1982 ISBN 90-247-3050-3
8. J.J. Douglas: *A Re-Appraisal of Forestry Development in Developing Countries.* 1983 ISBN 90-247-2830-4
9. J.C. Gordon and C.T. Wheeler (eds.): *Biological Nitrogen Fixation in Forest Ecosystems.* Foundations and Applications. 1983 ISBN 90-247-2849-5
10. M. Németh: *Virus, Mycoplasma and Rickettsia Diseases of Fruit Trees.* Rev. (English) ed., 1986 ISBN 90-247-2868-1
11. M.L. Duryea and T.D. Landis (eds.): *Forest Nursery Manual.* Production of Bareroot Seedlings. 1984; 2nd printing 1987 ISBN Hb 90-247-2913-0; Pb 90-247-2914-9
12. F.C. Hummel: *Forest Policy.* A Contribution to Resource Development. 1984 ISBN 90-247-2883-5
13. P.D. Manion (ed.): *Scleroderris Canker of Conifers.* Proceedings of an International Symposium on Scleroderris Canker of Conifers (Syracuse, USA, 1983). 1984 ISBN 90-247-2912-2
14. M.L. Duryea and G.N. Brown (eds.): *Seedling Physiology and Reforestation Success.* Proceedings of the Physiology Working Group, Technical Session, Society of American Foresters National Convention (Portland, Oregon, USA, 1983). 1984 ISBN 90-247-2949-1
15. K.A.G. Staaf and N.A. Wiksten (eds.): *Tree Harvesting Techniques.* 1984 ISBN 90-247-2994-7
16. J.D. Boyd: *Biophysical Control of Microfibril Orientation in Plant Cell Walls.* Aquatic and Terrestrial Plants Including Trees. 1985 ISBN 90-247-3101-1
17. W.P.K. Findlay (ed.): *Preservation of Timber in the Tropics.* 1985 ISBN 90-247-3112-7
18. I. Samset: *Winch and Cable Systems.* 1985 ISBN 90-247-3205-0

FORESTRY SCIENCES

19. R.A. Leary: *Interaction Theory in Forest Ecology and Management.* 1985 ISBN 90-247-3220-4
20. S.P. Gessel (ed.): *Forest Site and Productivity.* 1986 ISBN 90-247-3284-0
21. T.C. Hennessey, P.M. Dougherty, S.V. Kossuth and J.D. Johnson (eds.): *Stress Physiology and Forest Productivity.* Proceedings of the Physiology Working Group, Technical Session, Society of American Foresters National Convention (Fort Collins, Colorado, USA, 1985). 1986 ISBN 90-247-3359-6
22. K.R. Shepherd: *Plantation Silviculture.* 1986 ISBN 90-247-3379-0
23. S. Sohlberg and V.E. Sokolov (eds.): *Practical Application of Remote Sensing in Forestry.* Proceedings of a Seminar on the Practical Application of Remote Sensing in Forestry (Jönköping, Sweden, 1985). 1986 ISBN 90-247-3392-8
24. J.M. Bonga and D.J. Durzan (eds.): *Cell and Tissue Culure in Forestry.* Volume 1: General Principles and Biotechnology. 1987 ISBN 90-247-3430-4
25. J.M. Bonga and D.J. Durzan (eds.): *Cell and Tissue Culure in Forestry.* Volume 2: Specific Principles and Methods: Growth and Development. 1987 ISBN 90-247-3431-2
26. J.M. Bonga and D.J. Durzan (eds.): *Cell and Tissue Culure in Forestry.* Volume 3: Case Histories: Gymnosperms, Angiosperms and Palms. 1987 ISBN 90-247-3432-0
    Set ISBN (Volumes 24-26) 90-247-3433-9
27. E.G. Richards (ed.): *Forestry and the Forest Industries: Past and Future.* Major Developments in the Forest and Forest Industries Sector Since 1947 in Europe, the USSR and North America. In Commemoration of the 40th Anniversary of the Timber Committee of the UNECE. 1987 ISBN 90-247-3592-0
28. S.V. Kossuth and S.D. Ross (eds.): *Hormonal Control of Tree Growth.* Proceedings of the Physiology Working Group, Technical Session, Society of American Foresters National Convention (Birmingham, Alabama, USA, 1986). 1987 ISBN 90-247-3621-8
29. U. Sundberg and C.R. Silversides: *Operational Efficiency in Forestry.* Vol. 1: Analysis. 1988 ISBN 90-247-3683-8
30. M.R. Ahuja (ed.): *Somatic Cell Genetics of Woody Plants.* Proceedings of the IUFRO Working Party S2.04-07 Somatic Cell Genetics (Grosshansdorf, Germany, 1987). 1988 . ISBN 90-247-3728-1
31. P.K.R. Nair (ed.): *Agroforestry Systems in the Tropics.* 1989 ISBN 90-247-3790-7
32. C.R. Silversides and U. Sundberg: *Operational Efficiency in Forestry.* Vol. 2: Practice. 1989 ISBN 0-7923-0063-7
    Set ISBN (Volumes 29 and 32) 90-247-3684-6
33. T.L. White and G.R. Hodge (eds.): *Predicting Breeding Values with Applications in Forest Tree Improvement.* 1989 ISBN 0-7923-0460-8
34. H.J. Welch: *The Conifer Manual.* Volume 1. 1991 ISBN 0-7923-0616-3
35. P.K.R. Nair, H.L. Gholz, M.L. Duryea (eds.): *Agroforestry Education and Training.* Present and Future. 1990 ISBN 0-7923-0864-6
36. M.L. Duryea and P.M. Dougherty (eds.): *Forest Regeneration Manual.* 1991 ISBN 0-7923-0960-X

FORESTRY SCIENCES

37. J.J.A. Janssen: *Mechanical Properties of Bamboo.* 1991 ISBN 0-7923-1260-0
38. J.M. Bonga and P. Von Aderkas: *In Vitro Culture of Trees.* 1992 ISBN 0-7923-1540-5
39. L. Fins, S.T. Friedman and J.V. Brotschol (eds.): *Handbook of Quantitative Forest Genetics.* 1992 ISBN 0-7923-1568-5
40. M.J. Kelty, B.C. Larson and C.D. Oliver (eds.): *The Ecology and Silviculture of Mixed-Species Forests.* A Festschrift for David M. Smith. 1992 ISBN 0-7923-1643-6
41. M.R. Ahuja (ed.): *Micropropagation of Woody Plants.* 1992 ISBN 0-7923-1807-2
42. W.T. Adams, S.H. Strauss, D.L. Copes and A.R. Griffin (eds.): *Population Genetics of Forest Trees.* Proceedings of an International Symposium (Corvallis, Oregon, USA, 1990). 1992 ISBN 0-7923-1857-9
43. R.T. Prinsley (ed.): *The Role of Trees in Sustainable Agriculture.* 1993 ISBN 0-7923-2030-1
44. S.M. Jain, P.K. Gupta and R.J. Newton (eds.): *Somatic Embryogenesis in Woody Plants,* Vol. 3: Gymnosperms. 1995 ISBN 0-7923-2938-4
45. S.M. Jain, P.K. Gupta and R.J. Newton (eds.): *Somatic Embryogenesis in Woody Plants,* Vol. 1: History, Molecular and Biochemical Aspects, and Applications. 1995 ISBN 0-7923-3035-8
46. S.M. Jain, P.K Gupta and R.J. Newton (eds.): *Somatic Embryogenesis in Woody Plants,* Vol. 2: Angiosperms. 1995 ISBN 0-7923-3070-6
    Set ISBN (Volumes 44-46) 0-7923-2939-2
47. F.L. Sinclair (ed.): *Agroforestry: Science, Policy and Practice.* Selected Papers from the Agroforestry Sessions of the IUFRO 20th World Congress (Tampere, Finland, 6–12 August 1995). 1995 ISBN 0-7923-3696-8
48. J.H. Goldammer and V.V. Furyaev (eds.): *Fire in Ecosystems of Boreal Eurasia.* 1996 ISBN 0-7923-4137-6
49. M.R. Ahuja, W. Boerjan and D.B. Neale (eds.): *Somatic Cell Genetics and Molecular Genetics of Trees.* 1996 ISBN 0-7923-4179-1

KLUWER ACADEMIC PUBLISHERS – DORDRECHT / BOSTON / LONDON